T0301766

Theory and Design of Fermentation Processes

Theory and Design of Fermentation Processes

Davide Dionisi

CRC Press is an imprint of the
Taylor & Francis Group, an **informa** business

First edition published 2022
by CRC Press
6000 Broken Sound Parkway NW, Suite 300, Boca Raton, FL 33487-2742

and by CRC Press
2 Park Square, Milton Park, Abingdon, Oxon, OX14 4RN

ISBN: 978-1-032-10828-5 (hbk)
ISBN: 978-1-032-10832-2 (pbk)
ISBN: 978-1-003-21727-5 (ebk)

DOI: 10.1201/9781003217275

Typeset in Times
by KnowledgeWorks Global Ltd.

Dedication

To Luca and Valerio

Contents

Preface..........xi
About the Author..........xiii

Chapter 1 Introduction to Fermentation Processes..........1

1.1 Fermentation Processes and Microorganisms..........1
1.2 Classification of Microorganisms..........4
1.3 Structure and Components of Microbial Cells..........5
1.4 Stoichiometry and Metabolism of Cell Growth and Product Formation..........10
1.4.1 Metabolic Products..........14
1.4.2 Maintenance and Endogenous Metabolism..........15
1.5 Qualitative Time Profiles of Microbial Growth and Product Formation..........15
1.6 Processes and Reactors Used in Fermentation..........16
1.7 Starting Materials..........21
1.8 Culture Cultivation and Maintenance..........22
1.9 Extraction and Separation of Metabolites..........23
1.10 Sterilisation of Media and Equipment..........24
1.11 Geometry and Materials for Fermentation Vessels..........24
Questions and Problems..........25

Chapter 2 Kinetics and Stoichiometry..........27

2.1 Kinetics of Fermentation Processes..........27
2.2 Effect of Inhibitors, Temperature and pH on Growth Rate..........34
2.3 Experimental Measurement of Kinetic Parameters..........36
2.4 Yield Coefficients and Growth Stoichiometry..........39
2.5 Rate Equations for Secondary Metabolites, Maintenance and Endogenous Metabolism..........43
Questions and Problems..........50

Chapter 3 Mass Balances and Design for Batch, Continuous and Fed-Batch Reactors..........53

3.1 General Equation for Mass Balances..........53
3.2 Mass Balances for Batch Reactors..........54
3.3 Mass Balances for Continuous Reactors..........57
3.4 Mass Balances for Fed-Batch Reactors..........63

3.5 Modifications of Mass Balances for Secondary Metabolites, Maintenance and Endogenous Metabolism ... 67
3.6 Use of Continuous Reactors for the Estimation of Kinetic Parameters ... 72
3.7 Comparison of Batch, Continuous and Fed-Batch Reactors ... 75
Questions and Problems ... 77

Chapter 4 Oxygen Transfer ... 83
4.1 Theory ... 83
4.2 Measurement of Mass Transfer Coefficients ... 89
4.3 Correlations for Mass Transfer Coefficients ... 92
4.3.1 Correlations for Power Draw in Agitated Vessels ... 95
4.4 Oxygen Balances in Batch Reactors ... 103
4.5 Oxygen Balance in Continuous Reactors ... 110
4.6 Considerations for Scale-Up and Scale-Down ... 115
Questions and Problems ... 121

Chapter 5 Heat Generation and Heat Balances ... 129
5.1 Enthalpy of Reaction ... 129
5.2 Rate of Heat Generation of Fermentation Reactions ... 133
5.3 Heat Balances ... 139
5.3.1 Heat Balances for Continuous Fermentation ... 139
5.3.2 Heat Balances for Batch Fermentation ... 155
5.4 Effect of Variable Temperature on Reaction Rates ... 160
5.5 Heat Transfer Coefficients in Agitated Vessels ... 160
5.6 Considerations for Scale-Up and Scale-Down ... 164
Questions and Problems ... 168

Chapter 6 Design Summary and Examples of Industrial Fermentation Processes ... 175
6.1 Summary of Design Considerations ... 175
6.2 Bioethanol ... 176
6.3 Penicillin ... 181
6.4 Whisky ... 182
6.5 Citric Acid ... 184
6.6 L-Glutamic Acid ... 185

6.7 Cell-Based Virus Vaccines 186
6.8 Processes with Undefined Mixed Cultures 187
Questions and Problems 190

Appendix A: Numerical Solutions of Differential Equations in Excel 193

Appendix B: Numerical Solutions of Systems of Equations in Excel 197

Appendix C: Answers and Solutions 201

Bibliography 257

Index 261

Preface

This book is based on my lectures on fermentation engineering at the University of Aberdeen. These lectures are part of the EX4016 Biochemical Engineering course for Level 4 undergraduate students in Chemical Engineering. The book covers the kinetics and design of fermentation processes, defined in the broad sense as any industrial processes that use living microorganisms or cells, both under aerobic and anaerobic conditions. After a concise introduction on microorganisms and their metabolism (Chapter 1), rate equations and stoichiometry of fermentation processes are covered in detail (Chapter 2). Chapter 3 shows how to derive and use mass balances for the design of the most common types of fermentation processes. Chapter 4 covers oxygen transfer and oxygen mass balances, beginning with the fundamental theory to the use of oxygen transfer concepts to design and scale-up/scale-down aerobic fermentations. Chapter 5 does the same for heat transfer and includes heat generation, heat transfer and heat balances, showing how to use these concepts for the design and scale-up/sale-down of fermentation processes. The final chapter (Chapter 6) shows some industrially relevant process examples. The book includes over 100 solved examples, questions and problems. The examples are solved in their respective chapters, answers and solutions to questions and problems are provided in Appendix C. Appendices A and B also cover the solutions of differential equations and systems of equations in Excel.

A key feature of this book is the study of fermentation processes with the same tools used for the design of non-fermentation chemical reaction processes, i.e. kinetics and stoichiometry, mass and energy balances, mass and heat transfer. Therefore, this book can be considered an extension of chemical reaction engineering concepts to fermentation processes. The design of fermentation processes shares many similarities with the design of non-fermentation processes; however, there are also important differences due to peculiarities of fermentation reactions, e.g. their autocatalytic nature. The book covers these similarities and differences, showing the specific aspects of fermentation process design.

Like most models used in engineering, the value of the models and equations presented is not that they give us accurate numerical results. In fact, mathematical models of fermentation processes are inevitably great simplifications of very complex biochemical and physical processes and should not be expected to give accurate numerical predictions. Similarly, as for other areas of engineering, the value of models and equations applied to fermentation processes is that they help us in identifying the effect of operating variables on process performance and so they help us with process design and optimisation. With this is mind, models and equations are essential tools in the study of fermentation processes but should always be used together with experimental validation.

This book is not as comprehensive as other textbooks in biochemical engineering, as it doesn't cover enzyme kinetics and covers microorganism biochemistry and metabolism more concisely. However, compared with other textbooks, this book covers process design, oxygen and heat transfer and scale-up/scale-down topics in more detail and more systematically, with more examples and problems.

I believe that this book can be a useful resource as a textbook for undergraduate and graduate students in chemical engineering and can also be useful to process engineers working with fermentation reactions.

I am greatly indebted to my students of the course Biochemical Engineering at the University of Aberdeen for their useful comments over the years on my teaching materials, which are the basis for this book. I also acknowledge the help and assistance of my PhD students Ifeoluwa Bolaji, Chinedu Casmir Etteh, Rita Noelle Moussa, Adamu Rasheed, Igor Silva, Serena Simonetti, who over the years contributed to the teaching of my course and to the revision of this book. A special thanks to Serena Simonetti and Rita Noelle Moussa for the prompt and careful revision of the draft chapters. I would also like to thank my family in Italy and my wife Federica for her support during all the weekends that it took to complete the writing of this book.

Professor Davide Dionisi

About the Author

Professor Davide Dionisi is Personal Chair in Chemical Engineering at the University of Aberdeen, Scotland, UK. He obtained his MEng in Chemical Engineering and PhD in Industrial Chemical Processes from the School of Engineering, Sapienza University, Rome, Italy. Between 2001 and 2003, he worked as Post-Doctoral Research Assistant for the School of Mathematical, Physical and Natural Sciences at Sapienza University, before entering the same School as Lecturer in 2004. In 2007, he joined Syngenta's Process Studies Group in the UK, as a Process Engineer and then as a Principal Process Engineer. He joined the University of Aberdeen as a Lecturer in Chemical Engineering in 2012.

Prof Dionisi's research interests are focused on the application of chemical engineering principles to the development of improved, more sustainable and environmental-friendly processes, with particular focus on biological processes with open mixed cultures. He has published over 50 papers in international peer-reviewed journals which have received over 2,500 citations and his *h*-index is 25. He has also published two books, two book chapters and holds two patents and has written many internal industrial documents on various chemical engineering topics. He has supervised to successful completion five PhD students and has obtained research funding for over £1M.

At the University of Aberdeen, Prof Dionisi teaches biochemical engineering, wastewater treatment and energy from biomass in the undergraduate programme in Chemical Engineering and in the postgraduate programme in Renewable Energy Engineering, which he also co-ordinates. He has supervised over 50 undergraduate and postgraduate students in their individual projects.

1 Introduction to Fermentation Processes

This chapter introduces the main concepts on fermentation processes used in this book. Humans have been using fermentation processes for thousands of years. There's evidence of the use of fermented alcoholic beverages in Neolithic China, and the fermentation of milk to make yogurt was probably discovered even before then. Nowadays, industrial fermentation processes are carried out at very large scales, for example, bioethanol is produced from starchy or sugary feedstocks using yeasts at rates of millions of cubic metres per year. Commercial-scale fermentation processes need, therefore, to be designed and optimised taking into consideration their kinetics and stoichiometry, mass and energy balances and mass and heat transfer, which are the main focus of this book.

1.1 FERMENTATION PROCESSES AND MICROORGANISMS

In this book, we will use the term "fermentation" in the broadest sense to indicate any reactions that involve the growth of living microorganisms. In a stricter sense, the term fermentation is often used to indicate anaerobic reactions which produce a compound of interest in the liquid phase (e.g. alcoholic fermentation), but in this book we will refer to both aerobic and anaerobic processes.

Microorganisms (Figure 1.1) are unicellular (mostly) or multicellular organisms, in the order of micrometres in size, which grow (duplicate) on a carbon source and mineral nutrients.

For engineering purposes, in general terms the metabolism of any microorganisms can be schematised in a simple way (Figure 1.2). The microorganisms live in water and absorb from the external environment the carbon source (often called the "substrate") and the mineral elements that they need to grow (increase in number). The growth of microorganisms occurs by duplication. When they grow, microorganisms generate products that are released into the environment. Depending on the type of microorganisms and the environmental conditions, different products can be generated by the microorganisms, e.g. carbon dioxide (which is produced in most cases), organic acids, methane and many others. In this book, we will generally use "substrate" to express the carbon source and we will use the terms "microorganisms" and "biomass" with the same meaning. In addition to microorganisms, the theories and models presented in this book can be applied to cultures of animal cells, e.g. mammalian or stem cells, which are used in medical applications.

DOI: 10.1201/9781003217275-1

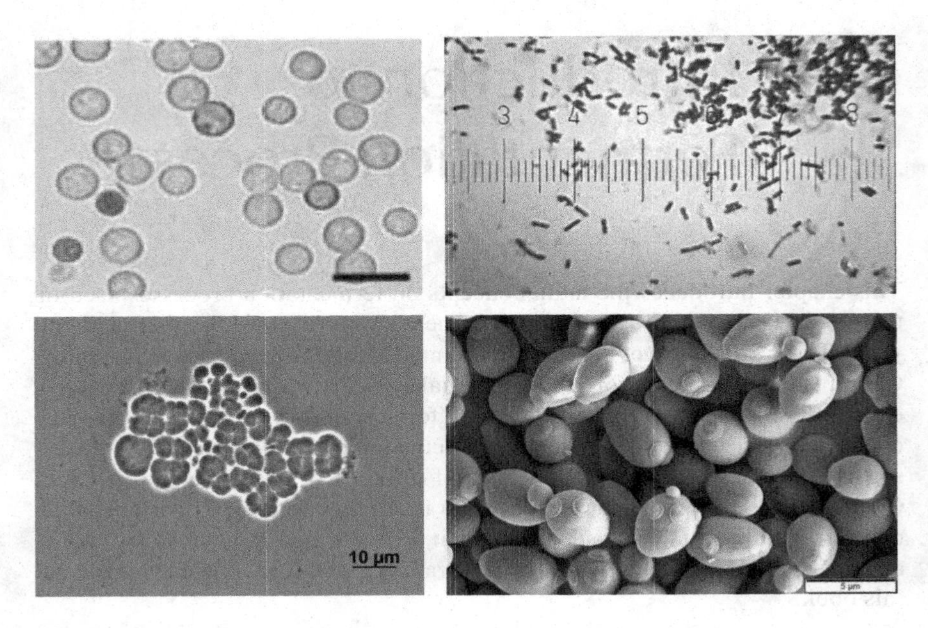

FIGURE 1.1 Examples of microorganisms. Top left: *Aspergillus niger* conidia (van Leeuwen et al., 2013, Creative Commons license). Bar is 10 µm. Top right: *Lactobacillus acidophilus* (Bob Blaylock, CC BY-SA, https://creativecommons.org/licenses/by-sa/3.0, https://commons.wikimedia.org/wiki/File:20101212_200110_LactobacillusAcidophilus.jpg). Numbered ticks are 11 µm apart. Bottom left: *Methanosarcina mazei* (https://bacdive.dsmz.de/strain/7096, Copyright Leibniz-Institut DSMZ-Deutsche Sammlung von Mikroorganismen und Zellkulturen GmbH). Bottom right: *Saccharomyces cerevisiae*, SEM image (Mogana Das Murtey and Patchamuthu Ramasamy, CC BY-SA 3.0, https://commons.wikimedia.org/w/index.php?curid=52254246).

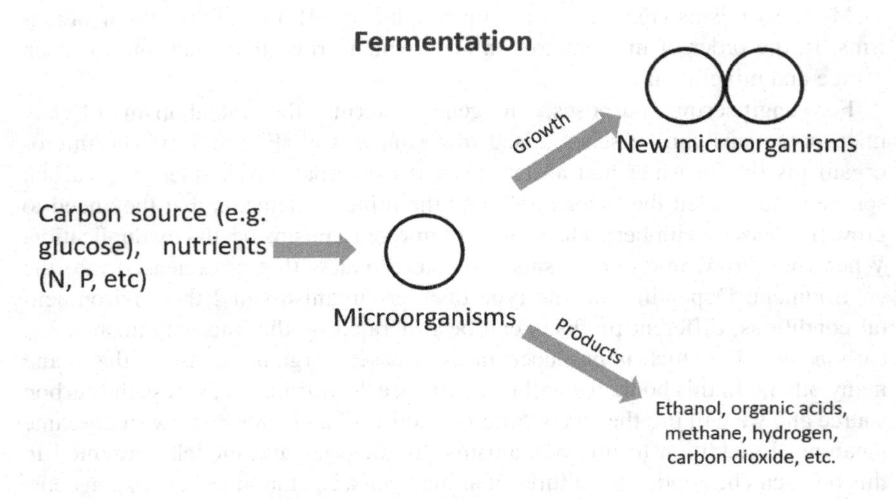

FIGURE 1.2 Conceptual scheme of microbial metabolism.

Fermentation processes are carried out for two main reasons:

- We are interested in the fermentation products, which are of industrial relevance, for example, ethanol, antibiotics, enzymes. In medical applications, viral vaccines and therapeutic proteins are among the products of the fermentation of mammalian cells.
- We want to remove polluting organic substances from the water phase. This is what happens in biological wastewater treatment processes.

When they are used to produce some products, fermentation processes are alternative to processes which use chemical synthesis. Some products can be produced by both fermentation and chemical processes and the process chosen to be used at industrial scale depends on the cost and availability of the feedstocks. For example, lactic acid (Figure 1.3) can be produced by fermentation

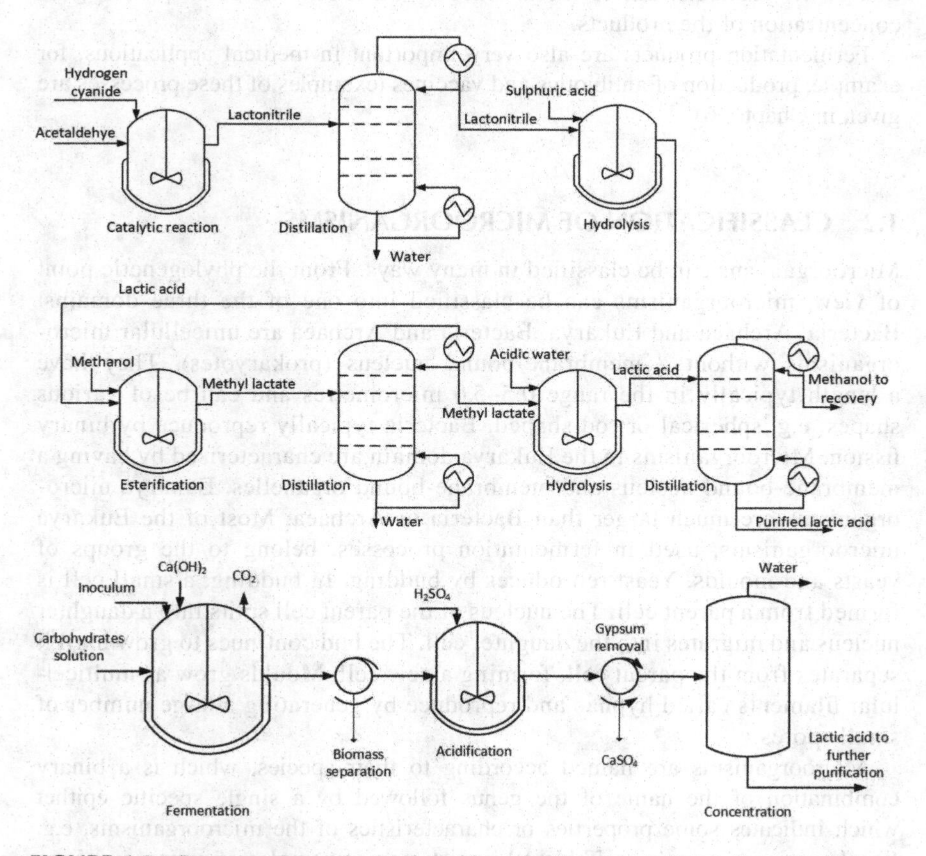

FIGURE 1.3 Process scheme for lactic acid production using chemical synthesis (top) and fermentation processes (bottom).

of carbohydrates by microorganisms of *Lactobacillus* or other species or by chemical synthesis from acetaldehyde, which is derived from coal or petroleum. Glutamic acid is another example of a substance which can be produced by either fermentation or chemical synthesis, the fermentation process being currently the preferred one in the industry. Generally, fermentation processes have the advantage over chemical synthesis processes, of requiring fewer reaction stages. When fermentation processes are available to produce a certain chemical, microorganisms may be able to convert the feedstock into the desired chemical in just one stage. On the other hand, chemical synthesis may require many reaction stages to obtain the same chemical. Another advantage of fermentation processes is that they are able to work with wet feedstocks (many organic materials are present in nature with a large fraction of moisture) and in some cases with mixed feedstocks as organic waste, while chemical processes usually require purified feedstocks. On the other hand, advantages of chemical processes are the faster reaction rates and often the easier separation of the products, due to the absence of water and the consequent higher concentration of the products.

Fermentation products are also very important in medical applications, for example, production of antibiotics and vaccines (examples of these processes are given in Chapter 6).

1.2 CLASSIFICATION OF MICROORGANISMS

Microorganisms can be classified in many ways. From the phylogenetic point of view, microorganisms can be classified into one of the three domains: Bacteria, Archaea and Eukarya. Bacteria and Archaea are unicellular microorganisms without a membrane-bound nucleus (prokaryotes). They have a length typically in the range 0.5–5.0 micrometres and can be of various shapes, e.g. spherical or rod-shaped. Bacteria typically reproduce by binary fission. Microorganisms in the Eukarya domain are characterised by having a membrane-bound nucleus and membrane-bound organelles. Eukarya microorganisms are much larger than Bacteria or Archaea. Most of the Eukarya microorganisms, used in fermentation processes, belong to the groups of yeasts and moulds. Yeast reproduces by budding. In budding, a small cell is formed from a parent cell. The nucleus of the parent cell splits into a daughter nucleus and migrates into the daughter cell. The bud continues to grow until it separates from the parent cell, forming a new cell. Moulds grow as multicellular filaments called hyphae and reproduce by generating a large number of small spores.

Microorganisms are named according to their species, which is a binary combination of the name of the genus followed by a single specific epithet which indicates some properties or characteristics of the microorganisms, e.g. *Saccharomyces cerevisiae*. Table 1.1 reports some examples of microorganisms used in industrial fermentations.

TABLE 1.1
Examples of Microorganisms and Fermentation Products

Microorganism	Classification	Carbon Source	Product	Industry
Penicillium spp.	Mould	Sucrose, lactose, glucose, and others	Penicillin	Pharmaceutical
Saccharomyces cerevisiae	Yeast	Glucose and others	Ethanol	Energy, chemical
Bacillus subtilis	Bacterium	Proteins and carbohydrates	Proteases	Food, detergents
Candida antarctica	Yeast	Lipids, glucose, lactose	Lipases	Detergents
Lactobacillus delbruekii	Bacterium	Lactose	Lactic acid	Food, plastics
Various	Various	Sugars, molasses	Amino acids	Food, pharmaceutical, chemical

1.3 STRUCTURE AND COMPONENTS OF MICROBIAL CELLS

The main distinction in the structure of microorganisms' cells is between prokaryotic and eukaryotic cells (Figure 1.4). Prokaryotic cells (Bacteria and Archaea) do not have a nucleus or internal compartmentalised organelles, which are instead present in eukaryotic cells (Eukarya).

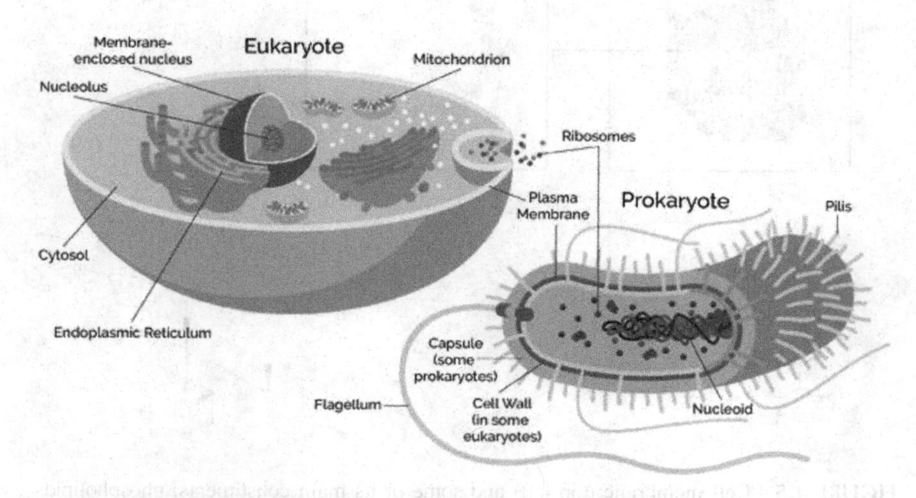

FIGURE 1.4 Differences between eukaryotic and prokaryotic cells. (Adapted from N. Gleitchmann, 2020, https://www.technologynetworks.com/cell-science/articles/prokaryotes-vs-eukaryotes-what-are-the-key-differences-336095.)

The main components of cells are listed below.

Cell membrane. The cell membrane (Figure 1.5) separates the cell from the external environment, allowing the flux of the molecules in and out of the cell. The cell membrane is selectively permeable, allowing only the in and out flow of substances, which are useful for cellular metabolism. Cell membranes are mainly composed of lipids and proteins. The most abundant lipids are phospholipids, but glycolipids and sterols are also present. Phospholipids are characterised by the presence of fatty acids and phosphoric groups.

Cytoplasm. Cytoplasm includes all the materials inside the cell. Cytoplasm is composed of water and all the organic and inorganic constituents

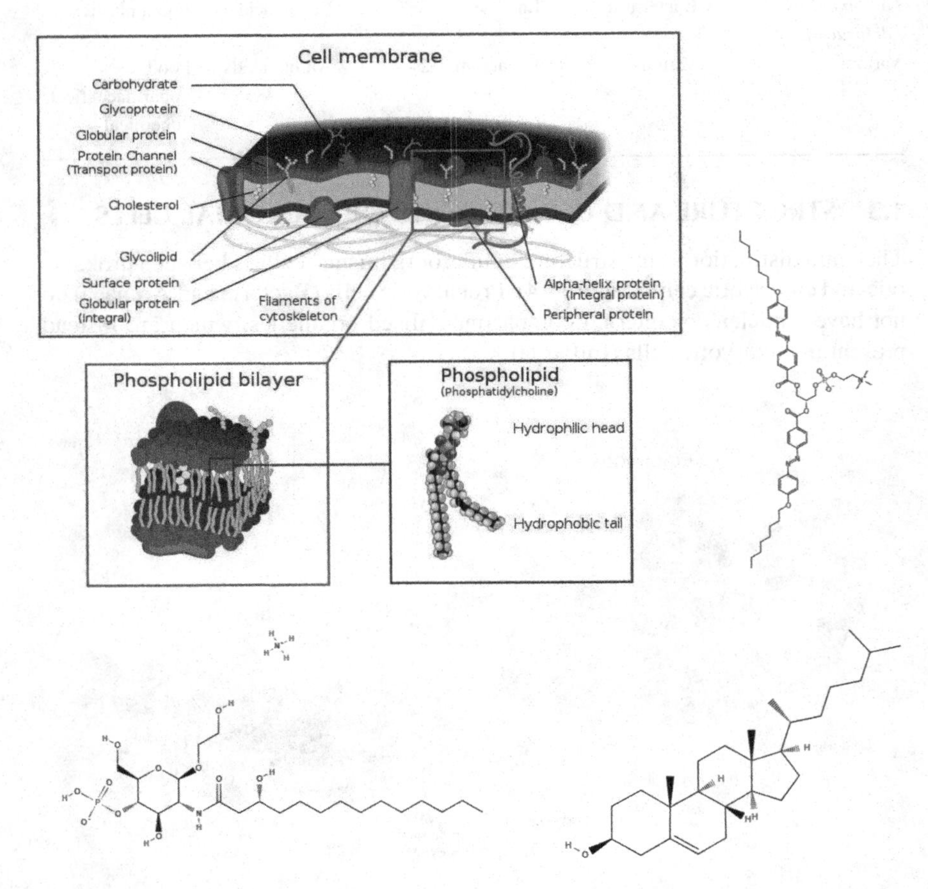

FIGURE 1.5 Cell membrane (top left) and some of its main constituents, phospholipids (top right, Cdpc-phospholipid), glycolipids (bottom left, Hedht-glycolipid), sterols (bottom right, cholesterol). (Images from https://commons.wikimedia.org/wiki/File:Cell_membrane_detailed_diagram_4.svg [cell membrane] and from Molview.org [chemical structures].)

(enzymes, DNA, RNA, organelles) of the cell. Approximately 80% of the cytoplasm is made of water.

Organelles. Organelles (Figure 1.5) are internal cell compartments separated by the rest of the cytoplasm by membranes. Organelles are present almost exclusively in eukaryotic cells. Examples of organelles are the nucleus, which contains the DNA; mitochondria, which control the energy generation and Golgi apparatus, which controls the processing and modifications of proteins.

Enzymes. Enzymes (Figure 1.6) are proteins which increase the rate of the metabolic reactions inside the cell. Enzymes are biological catalysts which decrease the activation energy of the reaction. Enzymatic reactions occur on the active sites of the enzymes, where the substrate(s) bind and is destabilised, allowing for new bonds to form. Enzymes are named after their substrate or after the chemical reaction they catalyse, with the word ending in ase.

DNA (deoxyribonucleic acid). DNA (Figure 1.7) contains the genetic information that regulates the growth and metabolism of all living organisms.

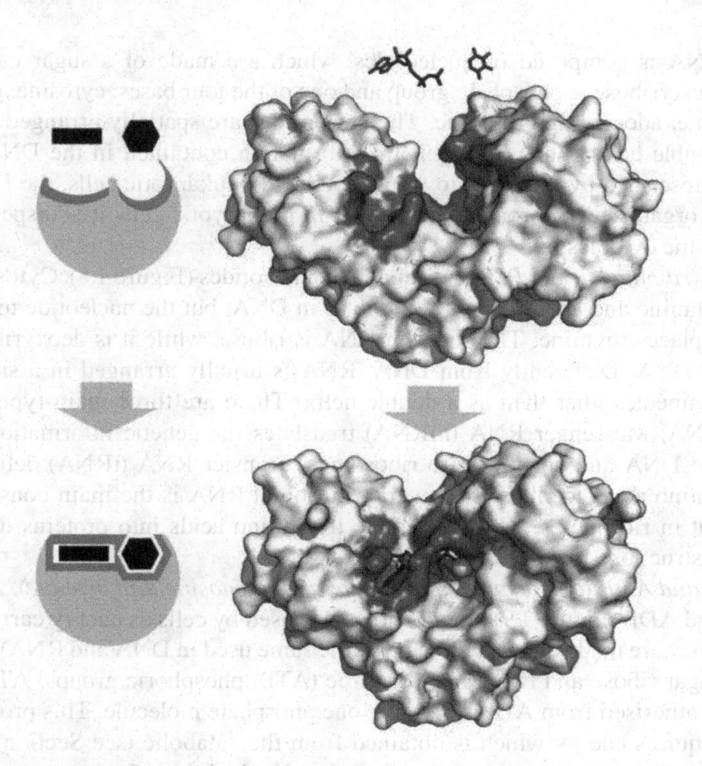

FIGURE 1.6 Scheme of an enzymatic reaction (hexokinase enzyme that binds adenosine triphosphate and glucose. (From https://commons.wikimedia.org/wiki/File:Hexokinase_induced_fit.svg.)

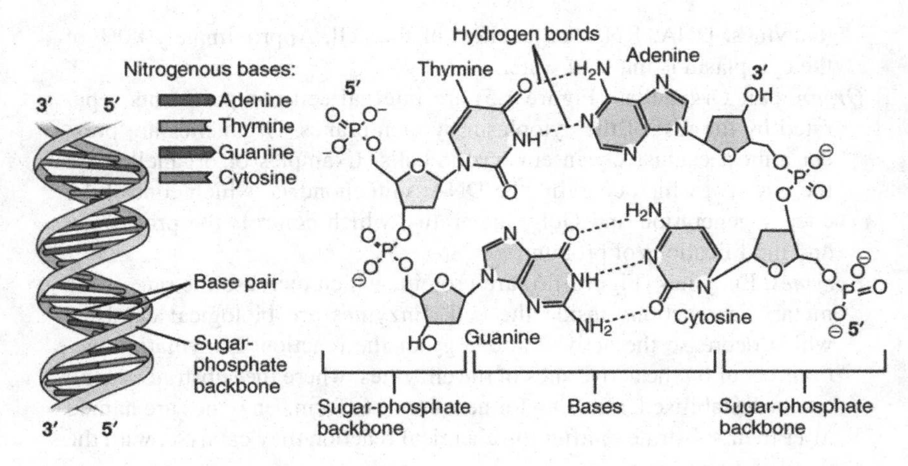

FIGURE 1.7 DNA double helix and the four nucleotides, joined by their phosphate groups and by double bonds. (From https://en.wikipedia.org/wiki/Nucleotide#/media/File:0322_DNA_Nucleotides.jpg.)

DNA is composed of nucleotides, which are made of a sugar called deoxyribose, a phosphate group and one of the four bases: cytosine, guanine, adenine and thymine. The nucleotides are spatially arranged in a double helix shape. The genetic information contained in the DNA is transcribed by the RNA to form proteins. In eukaryotic cells, the DNA is organised in chromosomes, while in prokaryotic cells it is dispersed in the cytoplasm.

RNA (*ribonucleic acid*). RNA is made of nucleotides (Figure 1.8). Cytosine, guanine and adenine are the same as in DNA, but the nucleotide uracil replaces thymine. The sugar in RNA is ribose, while it is deoxyribose in DNA. Differently from DNA, RNA is usually arranged in a single filament, rather than as a double helix. There are three main types of RNA: Messenger RNA (mRNA) translates the genetic information in the DNA and conveys it to ribosomes. Transfer RNA (tRNA) delivers amino acids to ribosomes, while ribosomal RNA is the main constituent in ribosomes, which assemble the amino acids into proteins using instructions from the mRNA.

ATP and ADP (*adenosine triphosphate and adenosine diphosphate*). ATP and ADP (Figure 1.9) are the molecules used by cells as energy carriers. They are made of an adenine base (the same used in DNA and RNA), the sugar ribose and two (ADP) or three (ATP) phosphoric groups. ATP is synthesised from ADP by adding one phosphate molecule. This process requires energy which is obtained from the catabolic (see Section 1.4) reactions. ATP is an energy-rich molecule which transfers energy to the reactions that produce cell materials (anabolic reactions, Section 1.4). When energy is required, ATP transfers one of its phosphate groups to

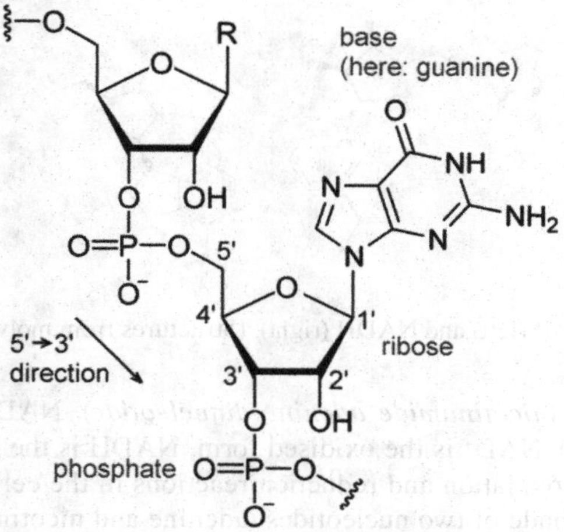

FIGURE 1.8 Scheme of a RNA nucleotide with the ribose sugar, a base (here: guanine) and a phosphate group which joins two consecutive nucleotides. (From https://commons.wikimedia.org/wiki/File:RNA_chemical_structure.GIF.)

the molecule that needs energy. When this molecule releases the phosphate group, energy is generated, and the biosynthetic reaction proceeds. The reactions that produce and consume ATP are schematised as follows:

$$ADP + P_i \rightarrow ATP \tag{1.1}$$

$$ATP \rightarrow ADP + P_i \tag{1.2}$$

In these reactions, P_i stands for the phosphate group PO_4^{3-}. For simplicity, Equations (1.1) and (1.2) and other metabolic reactions in this chapter are shown without the water molecules and the protons, which also take part in the reactions.

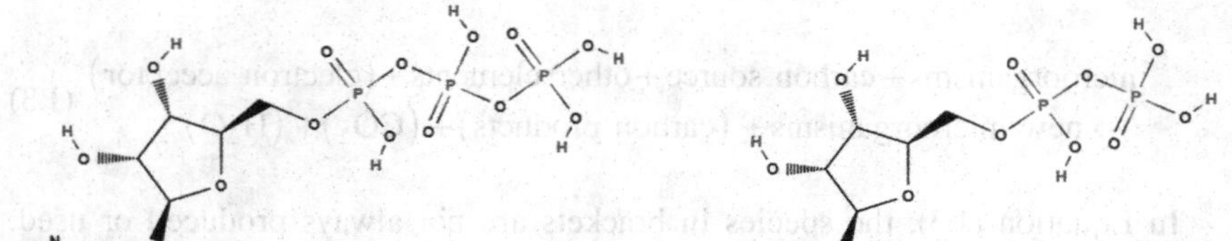

FIGURE 1.9 Adenosine triphosphate (ATP, left) and adenosine diphosphate (ADP, right). (Structures from molview.org.)

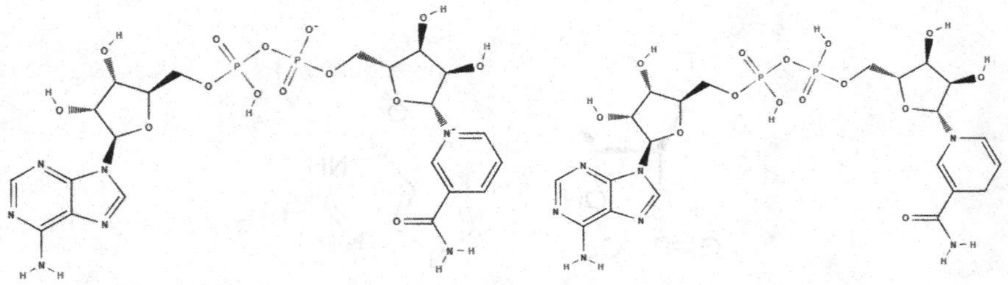

FIGURE 1.10 NAD^+ (left) and NADH (right). (Structures from molview.org.)

NAD^+, NADH (nicotinamide adenine dinucleotide). NAD^+ and NADH (Figure 1.10, NAD^+ is the oxidised form, NADH is the reduced form) are used in oxidation and reduction reactions in the cells. These molecules are made of two nucleotides (adenine and nicotinamide) joined through their phosphate groups. Each molecule of NAD^+ removes two hydrogen atoms from the biological molecule to be oxidised. One of these hydrogen atoms is removed as H^+ (proton), the other as H^- (hydride ion, hydrogen atom with one proton and two electrons). The hydride ion is transferred to NAD^+, which becomes NADH, the proton is released into the solution. The reverse reaction occurs when NADH is converted back to NAD^+: the hydride ion from NADH and a proton from the solution are transferred to the molecule to be reduced, regenerating the NAD^+. As the alternatives to NAD^+/NADH, the nucleotides FAD/$FADH_2$ (flavin adenine dinucleotide) are also used by cells with the same functions.

1.4 STOICHIOMETRY AND METABOLISM OF CELL GROWTH AND PRODUCT FORMATION

Overall, the growth and products generation by microorganisms can be schematised by the following general chemical stoichiometry:

$$\begin{aligned}&\text{microorganisms} + \text{carbon source} + \text{other elements} + (\text{electron acceptor})\\&\rightarrow \text{new microorganisms} + (\text{carbon products}) + (CO_2) + (H_2O)\end{aligned} \tag{1.3}$$

In Equation (1.3), the species in brackets are not always produced or used. Electron acceptors are molecules which remove electrons from the carbon source. The most common electron acceptor in fermentation processes is oxygen. Other electron acceptors can be nitrate (NO_3^-), sulphate (SO_4^{2-}) and others.

To understand the concept of electron acceptor, let us consider, for example, the aerobic fermentation of glucose into carbon dioxide (CO_2). In the carbon dioxide molecule, the two electrons in each bond between C and O atoms are more attracted by the O atom than by the C atom. Therefore, we say that each

O atom in the CO_2 molecule has oxidation state "–2". The C atom in the CO_2 molecule has oxidation state "+4" so that overall, the CO_2 molecule has oxidation state "0". In the O_2 molecule, the electrons are equally shared between the two O atoms, so the oxidation state of oxygen is 0. Therefore, in the aerobic fermentation of glucose into carbon dioxide each oxygen atom gains two electrons, so oxygen is called "electron acceptor". The substrate (glucose in this case) is called "electron donor".

To understand the cellular metabolism better, it is useful to introduce the concepts of anabolism and catabolism. Anabolism includes all the reactions which lead to the generation of new cell components, e.g. new proteins, cell membranes, etc. Catabolism includes all the reactions which generate the energy required by anabolic reactions. In aerobic fermentation processes (or in processes where there is an external electron acceptor different from oxygen, e.g. nitrate or sulphate), energy is obtained by the oxidation of the substrate to carbon dioxide and water. In aerobic processes, the substrate is the electron donor, and oxygen is the electron acceptor. In anaerobic processes, on the other hand, there is no external electron acceptor. In these cases, in the catabolic reactions a fraction of the substrate is used as an electron acceptor and another fraction is used as an electron donor, generating the required energy. The fraction of the substrate that is used as the electron acceptor is converted to reduced compounds (e.g. methane or hydrogen), while the fraction of the substrate that is used as the electron donor is converted to oxidised compounds (carbon dioxide). In any cases, transfer of energy from catabolic to anabolic reactions happens through ATP molecules.

As an example of anabolic and catabolic processes, let us consider the metabolism of glucose by microorganisms. The first reactions in this metabolism (Figure 1.11) are the same for aerobic and anaerobic microorganisms and involve the conversion of glucose into pyruvic acid (or pyruvate). This metabolic pathway forms, per each molecule of glucose converted, two molecules of ATP and two molecules of NADH. The stoichiometry for the conversion of glucose into pyruvate (called glycolysis) is:

$$\text{Glucose} + 2\text{NAD}^+ + 2\text{ADP} + 2\text{P}_i \rightarrow 2\text{pyruvate} + 2\text{NADH} + 2\text{ATP} \quad (1.4)$$

Part of the metabolites produced in this pathway, instead of being converted to pyruvate, can be used for anabolic pathways for the formation of the various cellular components. For example, the first reaction product, glucose 6-phosphate can be used for the formation of nucleotides (DNA and RNA) and polysaccharides for cell walls. P-glycerate can be used for the formation of amino acids and other cellular components.

While the pathways to produce pyruvate from glucose are the same in aerobic and anaerobic microorganisms, the fate of pyruvate depends on whether we have an electron acceptor or not. In the presence of an electron acceptor, pyruvate is converted to acetyl-coenzyme A, which is then oxidised to carbon dioxide in a sequence of reactions called the tricarboxylic acid cycle (also called Krebs

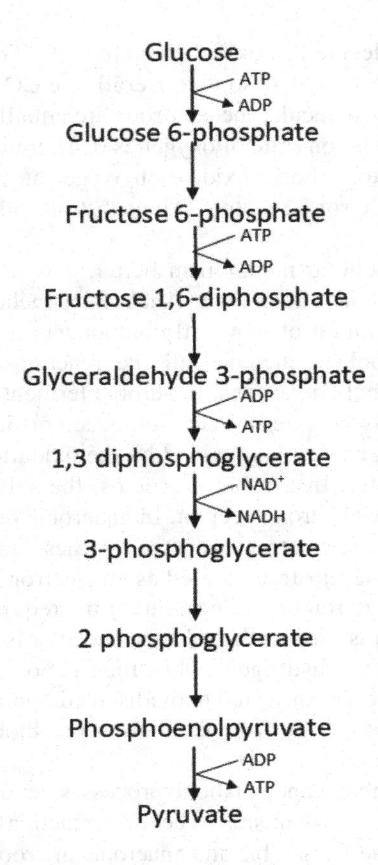

FIGURE 1.11 Steps of the conversion of glucose into pyruvate (glycolysis) according to the Embden-Meyerhof-Parnas pathway. From one molecule of fructose 1,6-diphosphate two molecules of glyceraldehyde 3-phosphate are produced, which both go through the subsequent steps; therefore, overall glycolysis produces 2 moles of ATP and 2 moles of NADH per mole of glucose. The molecules in this pathway can also be used for the anabolism to produce the biosynthetic components.

cycle, Figure 1.12). The oxidants in the tricarboxylic acid cycle are NAD^+ and FAD molecules, which are converted to NADH and $FADH^+$. The stoichiometry of pyruvate oxidation to acetyl-CoA and of the tricarboxylic acid cycle is reported in Equations (1.5) and (1.6), respectively:

$$\text{Pyruvate} + \text{CoA} + NAD^+ \rightarrow \text{acetyl-CoA} + CO_2 + \text{NADH} \tag{1.5}$$

$$\text{acetyl-CoA} + 3NAD^+ + \text{FAD} + \text{ADP} \rightarrow \text{CoA} + 2CO_2 + 3\text{NADH} + FADH^+ + \text{ATP} \tag{1.6}$$

Overall, in aerobic conditions, the oxidation of pyruvate to acetyl-CoA and the oxidation of acetyl-CoA in the tricarboxylic acid cycle yield, per mole of pyruvate,

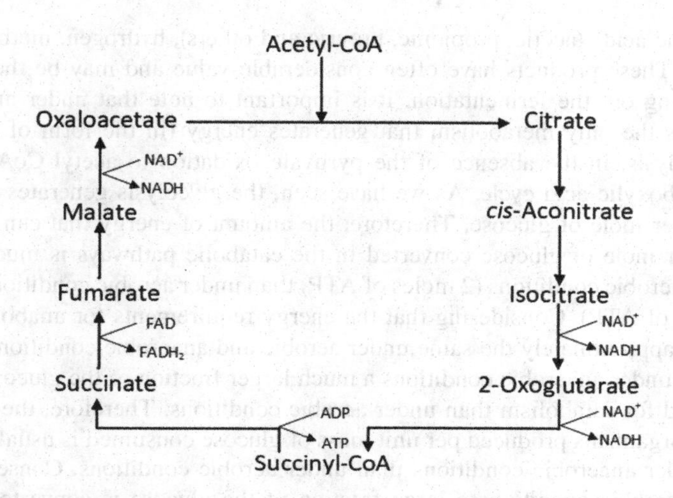

FIGURE 1.12 Tricarboxylic acid cycle. Only the main compounds in the cycle are shown. In this cycle, ADP and ATP are replaced, in many microorganisms, by guanosine diphosphate (GDP) and guanosine triphosphate (GTP) with the same functions.

4 moles of NADH, 1 mole of $FADH^+$ and 1 mole of ATP. Considering that 1 mole of glucose converted to pyruvate yields 2 moles of pyruvate, 2 moles of NADH and 2 moles of ATP, the overall catabolic reactions for the aerobic metabolism of glucose produce 10 moles of NADH, 2 moles of $FADH^+$ and 4 moles of ATP.

In aerobic conditions, the nucleotides NADH and $FADH^+$ are oxidised back to NAD^+ and FAD by oxygen. This process is coupled to the generation of ATP from ADP and is called oxidative phosphorylation. Ideally, although there is variability for different microorganisms and different environmental conditions, oxidative phosphorylation produces 3 moles of ATP per mole of NADH oxidised and 2 moles of ATP per mole of $FADH_2$ oxidised (Equations 1.7 and 1.8):

$$NADH + 3ADP + 3P + 0.5O_2 \rightarrow NAD^+ + 3ATP + H_2O \quad (1.7)$$

$$FADH_2 + 2ADP + 2P + 0.5O_2 \rightarrow FAD + 2ATP + H_2O \quad (1.8)$$

Therefore, the aerobic catabolism of glucose can produce up to 38 moles of ATP per mole of glucose. Of these 38 moles of ATP, 2 come from the glycolysis, 2 from the tricarboxylic acid cycle and 34 from the oxidative phosphorylation (of which 30 come from NADH and 4 from $FADH_2$). These ATP molecules are used as energy vectors for the anabolic reactions that produce the cell components.

Under anaerobic conditions, there is no external electron acceptor that oxidises the reduced nucleotide NADH to their NAD^+ form. Therefore, under anaerobic conditions the NADH, produced by the glycolysis reaction, is oxidised to NAD^+ by the glucose itself or by other metabolic intermediates, with the overall result that reduced products are generated and released into the medium. These products are organic acids (acetic, propionic, butyric and others), hydrogen, methane and alcohols. These products have often considerable value and may be the reason for carrying out the fermentation. It is important to note that under anaerobic conditions the only metabolism that generates energy (in the form of ATP) is the glycolysis, in the absence of the pyruvate oxidation to acetyl-CoA and of the tricarboxylic acid cycle. As we have seen, the glycolysis generates 2 moles of ATP per mole of glucose. Therefore, the amount of energy that can be generated per mole of glucose converted in the catabolic pathways is much lower under anaerobic conditions (2 moles of ATP) than under aerobic conditions (up to 38 moles of ATP). Considering that the energy requirements for anabolic reactions are approximately the same under aerobic and anaerobic conditions, it follows that under anaerobic conditions a much larger fraction of the glucose needs to be used for catabolism than under aerobic conditions. Therefore, the number of microorganisms produced per unit mass of glucose consumed is usually much lower under anaerobic conditions than under aerobic conditions. Consequently, under anaerobic conditions a large fraction of the glucose is converted to the reduced products. In most cases, these reduced products and not the microorganisms are the desired products of fermentation; therefore, the low growth yield of anaerobic microorganisms is a favourable outcome. However, the low generation of microorganisms under anaerobic conditions also makes fermentation slower, which is usually a negative consideration.

1.4.1 Metabolic Products

Products are produced during the cell metabolism for different reasons. One important class of products is those produced during cell growth (primary metabolites). Primary metabolites are essential products of cell metabolism and are produced during the growth of microorganisms. Primary metabolites can be produced during anaerobic or aerobic growth. Production of primary metabolites under anaerobic conditions is more common. As we have seen in Section 1.4, under anaerobic conditions there are no external electron acceptors and, therefore, the substrate acts both as an electron acceptor and electron donor, producing oxidised (carbon dioxide) and reduced (organic acids, ethanol, methane) compounds. Under aerobic conditions, oxygen is the external electron acceptor and normally the only final metabolic products are carbon dioxide and water. However, in some cases, e.g. lack of certain enzymes, primary metabolites may be produced under aerobic conditions.

Another important class of products is called secondary metabolites. These metabolites are normally not produced during normal cell metabolism and growth and are mainly produced when cell growth has stopped or has slowed down considerably. Secondary metabolites are produced when microorganisms are stressed by environmental factors such as deficiency in any nutrients, trace elements or an unbalanced mineral medium. For example, polyhydroxyalkanoates (PHAs) are biological polymers that are produced when the supply of carbon is more than the supply of nitrogen or phosphorus (or of other elements) or when the metabolic apparatus is not ready for a fast growth on the available carbon source. As another example, production of citric acid is carried out by the aerobic microorganism *Aspergillus niger* and is favoured using iron-deficient media, because iron is needed for the enzymes that convert citric acid. Many secondary metabolites have important applications, for example, penicillin and L-glutamic acid (Chapter 6). L-glutamic acid is produced when the mineral medium is unbalanced because of vitamin (biotin) limitation, the presence of surfactants or for other reasons.

1.4.2 Maintenance and Endogenous Metabolism

Equation (1.3) describes the use of the substrate to produce new microorganisms ("growth" process). In addition to growth, other types of metabolism take place in fermentation processes. One of these metabolisms is called maintenance. Maintenance metabolism is the use of substrate not related to microbial growth. This metabolism includes the substrate use to provide the energy needed to maintain the internal osmotic pressure, to repair cellular components, etc. The substrate used for maintenance is oxidised to generate energy for these processes. Another process which is non-growth related is endogenous metabolism. Endogenous metabolism includes the self-oxidation by the cells of internal components (e.g. enzymes) which are no longer required for the metabolism. This later metabolism does not involve the use of the substrate, but it causes a reduction in the concentration of microorganisms. Therefore, endogenous metabolism can be assimilated to a death process for the microorganisms.

1.5 QUALITATIVE TIME PROFILES OF MICROBIAL GROWTH AND PRODUCT FORMATION

Let us assume that we have a small number of microorganisms of the same species in suspension in water. We assume that all the elements necessary for growth (e.g. N, P, metals, etc.) and one biodegradable carbon source (e.g. glucose, ethanol, etc.) are present in the water solution. We also assume that the pH and temperature of the medium are adequate for the growth of microorganisms. What will happen to the microorganisms and substrate concentration over time in this system? Initially, the microorganisms will need to adapt to the carbon source (acclimation phase). In the acclimation phase, the microorganisms will develop the internal enzymes and the metabolic apparatus needed to degrade the substrate. The length of the acclimation phase might vary from several hours to days, depending on the microorganisms and on the substrate. If the microorganisms are already acclimated to the substrate, there will be no acclimation phase.

At the end of the acclimation phase, the microorganisms are adapted to the substrate and start growing on it. During the growth phase, the concentration of microorganisms increases and the substrate concentration decreases. When the substrate becomes limiting, the growth rate decreases until it stops completely when the substrate is completely removed. Figure 1.13 shows, qualitatively, the

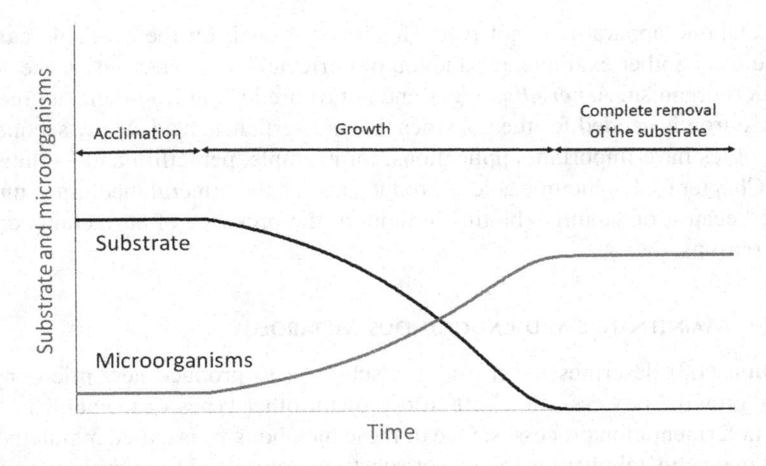

FIGURE 1.13 Qualitative profiles of microorganisms and substrate during batch growth.

profiles of substrate and biomass during this experiment (a process like this, with no continuous feeding of the substrate is called a batch process, see Section 1.6).

As we have seen, in many cases, fermentation processes are carried out to produce a certain product of interest. As we have seen in Section 1.4.1, products can be produced either during biomass growth on the substrate (primary metabolites) or at the end of biomass growth when the substrate has been (almost) completely removed (secondary metabolites). Figure 1.14 shows the qualitative profiles of substrate, microorganisms and products during batch growth.

1.6 PROCESSES AND REACTORS USED IN FERMENTATION

The three main fermentation processes used in the industry are batch, continuous and fed-batch. In batch fermentations, the batch reactor is inoculated with microorganisms, a substrate is added, microorganisms grow (and possibly produce a product) and when the maximum biomass concentration is reached the reactor is emptied to start a new batch. Batch fermentations are relatively easy to manage and to keep free of microbial contaminants as there is no continuous addition of the medium. Disadvantages of batch fermentations are the downtime required for cleaning and emptying, the need to add a fresh inoculum for each batch and the low initial reactor rates, due to the low biomass concentration at the start of the batch.

In continuous fermentation, the substrate and nutrients are continuously added to the reactor, and the medium which contains microorganisms and products is continuously removed. Continuous fermenters with a growth-limiting substrate are also called chemostats or, when the concentration of microorganisms is measured from the turbidity of the culture and controlled by manipulation of the residence time called turbidostats. Ideally, the continuous reactor's feed is sterile (does not contain microorganisms) and microorganisms are added only

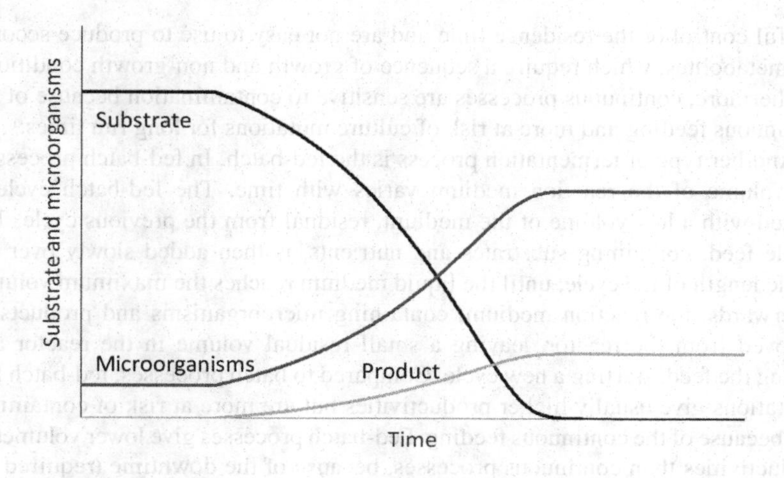

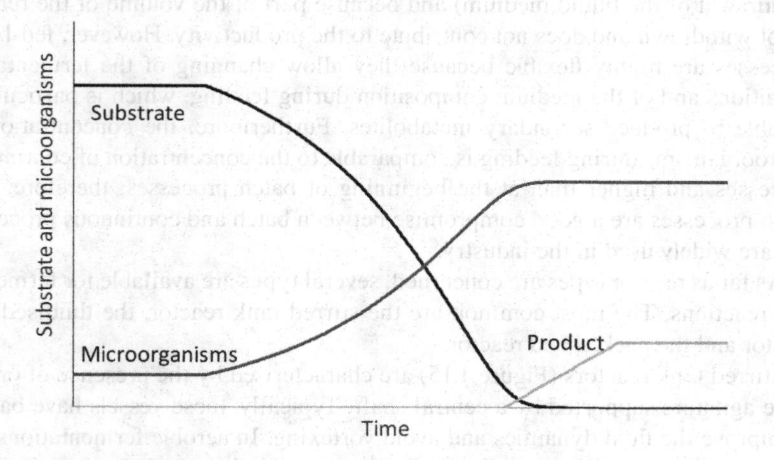

FIGURE 1.14 Qualitative profiles of substrate, microorganisms and product during batch growth. Top: Product produced during biomass growth (primary metabolite). Bottom: Product produced at the end of biomass growth (secondary metabolite).

at the start-up (inoculation) of the reactor. At the end of the start-up, typically continuous reactors reach a steady state, i.e. a condition where all the measured parameters (concentrations and temperature) are constant with time. Compared to batch processes, continuous processes have the advantages of having no downtime and higher biomass concentrations (because at steady state typically most of the substrate is removed and the biomass concentration is the highest), with consequent higher reaction rates and higher productivities. In addition, the fact that continuous processes are operated at a steady state with constant values of substrate, biomass and nutrient concentration makes it easier to study the effect of these parameters on the fermentation. However, continuous processes need careful control of the residence time and are not easy to use to produce secondary metabolites, which require a sequence of growth and non-growth conditions. Furthermore, continuous processes are sensitive to contamination because of the continuous feeding and more at risk of culture mutations for long run times.

Another type of fermentation process is the fed-batch. In fed-batch processes, the volume of the reaction medium varies with time. The fed-batch cycle is started with a low volume of the medium, residual from the previous cycle. The sterile feed, containing substrates and nutrients, is then added slowly over the whole length of the cycle, until the liquid medium reaches the maximum volume. Afterwards, the reaction medium, containing microorganisms and products, is removed from the reactor, leaving a small residual volume in the reactor and adding the feed, starting a new cycle. Compared to batch processes, fed-batch fermentations give usually higher productivities but are more at risk of contamination because of the continuous feeding. Fed-batch processes give lower volumetric productivities than continuous processes, because of the downtime (required for withdrawal of the liquid medium) and because part of the volume of the reactor is not withdrawn and does not contribute to the productivity. However, fed-batch processes are highly flexible because they allow changing of the fermentation conditions and of the medium composition during feeding, which is particularly suitable to produce secondary metabolites. Furthermore, the concentration of microorganisms during feeding is comparable to the concentration of continuous processes and higher than at the beginning of batch processes; therefore, fed-batch processes are a good compromise between batch and continuous processes and are widely used in the industry.

As far as reactor types are concerned, several types are available for fermentation reactions. The most common are the stirred tank reactor, the fluidised bed reactor and the packed bed reactor.

Stirred tank reactors (Figure 1.15) are characterised by the presence of one or more agitators supported by a central shaft. Typically, these vessels have baffles to improve the fluid dynamics and avoid vortexing. In aerobic fermentations, air or oxygen is typically supplied using spargers located in the bottom part of the vessel. Various types of agitators are available, depending on the characteristics of the fermentation medium and on the main tasks required by the agitation system. If the main task is dispersing and breaking the gas bubbles, ensuring high-mass transfer coefficient for oxygen, then small diameter agitators such as Ruston turbines are used. These agitators can rotate at high-speed ensuring good gas dispersion in relatively low viscosity fermentation media. For fermentation media with higher viscosity and higher cell concentrations, large low-speed agitators are used, such as the hydrofoil agitators. Ideally it can be assumed that agitated vessels are perfectly mixed, i.e. that the concentration of substrate, microorganisms and products is uniform in any points of the reactor.

Packed bed reactors are characterised by microorganisms growing attached to support media. The feed containing the substrate and nutrients flows downwards from the top of the bed. Microorganisms are retained on the support and, ideally, are not present in the liquid effluent, containing the products, which is collected

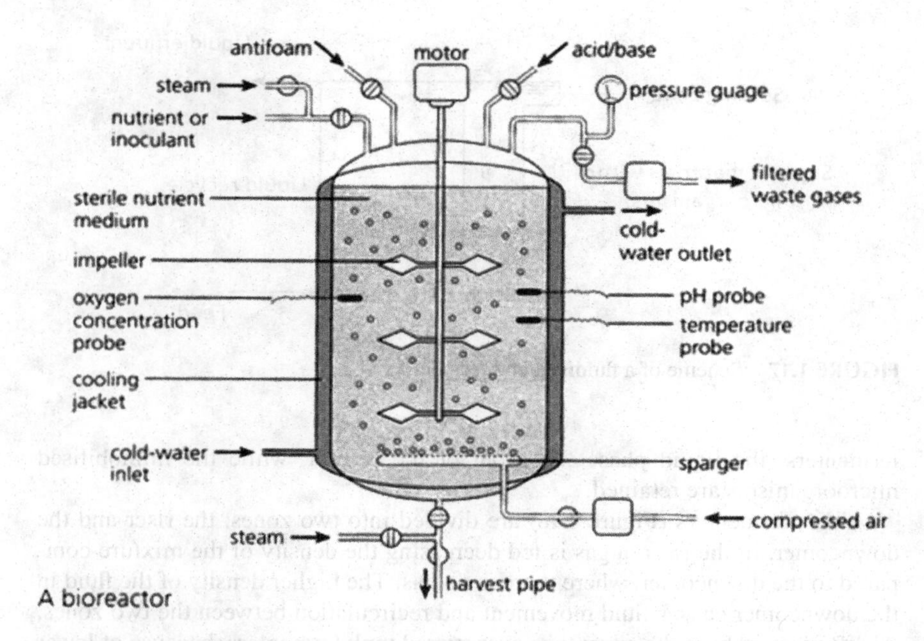

FIGURE 1.15 Scheme of the main components of a stirred tank fermenter. (From http://2010.igem.org/Team:UCL_London/Fermenter_Mechanics.)

from the bottom of the reactor. In packed bed reactors, the concentration of substrates and products varies along the vertical length of the column; however, the concentration gradients are lower at higher flow rates because of the increased mixing due to the higher turbulence.

Fluidised bed fermenters (Figure 1.17) are similar to packed beds, however, in this case the microorganisms, immobilised on support materials, are kept in suspension by circulation of the gas or of the liquid phase. As in packed bed

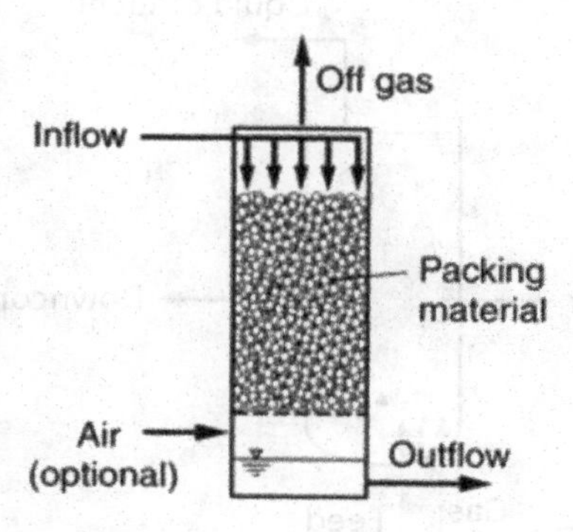

FIGURE 1.16 Scheme of a packed bed fermenter. (From https://sites.google.com/site/fermentersin/types.)

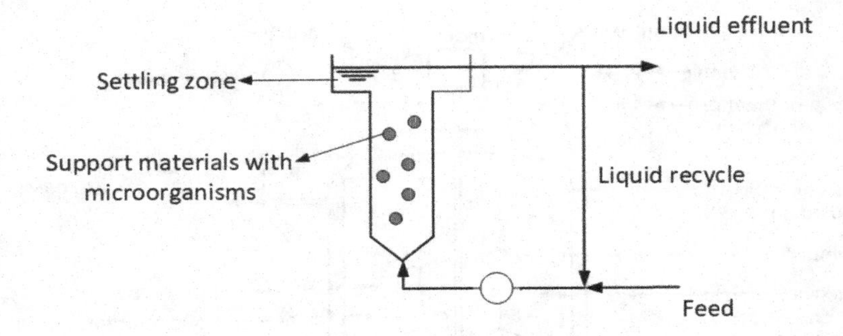

FIGURE 1.17 Scheme of a fluidised bed fermenter.

fermenters, the liquid phase flows out of the reactor, while the immobilised microorganisms are retained.

Airlift fermenters (Figure 1.18) are divided into two zones: the riser and the downcomer. In the riser, a gas is fed decreasing the density of the mixture compared to the downcomer, where there is no gas. The higher density of the fluid in the downcomer causes fluid movement and recirculation between the two zones. Airlift reactors have the advantage over stirred tank fermenters, because of lower shear rates (due to the absence of mechanical aerators) and are, therefore, suitable for shear-sensitive cultures.

The bubble column reactor (Figure 1.19) is the simplest type of fermenter. It is conceptually similar to stirred tank reactors, but the mixing is only provided by the flow of the gas phase, in the absence of mechanical agitators. The mass and heat transfer rates obtained in bubble column fermenters are lower, due to the absence of agitation, than the rates for stirred tank reactors; however, bubble columns are suitable for shear-sensitive cultures.

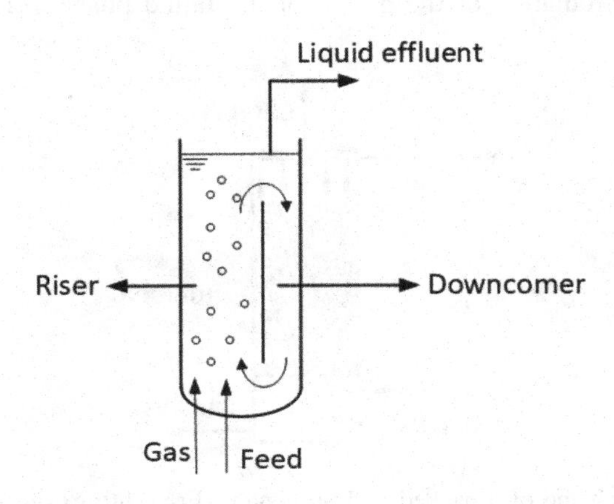

FIGURE 1.18 Scheme of an airlift fermenter.

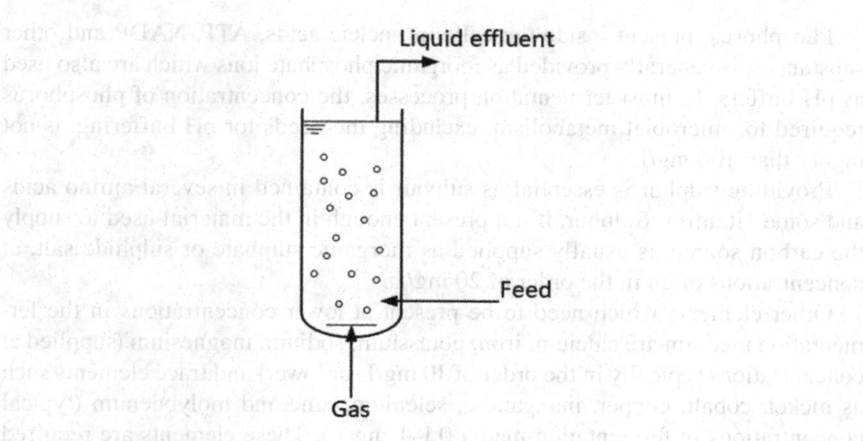

FIGURE 1.19 Scheme of a bubble column fermenter.

1.7 STARTING MATERIALS

Fermentation media must contain all the substances and nutrients required for the microorganisms to grow and to produce the desired products. Since microorganisms are made of carbon, hydrogen, oxygen, nitrogen, and phosphorus and of smaller amounts of many other elements, all these chemical elements must be present in the fermentation medium.

Water is usually the main component, by weight, of the fermentation medium and, in most cases must be purified, with removal of suspended solids and microorganisms before use. If the water has high c salts or calcium carbonate concentration (hard water), these salts must also be removed.

For most fermentation processes, the carbon source is organic and supplied at concentrations of 10 g/L or higher. Typical organic carbon sources are sugars such as glucose and sucrose. Sugars are sometimes used as purified substrates, however, in most cases, for cost reasons, sugar-rich by-products of other processes are used, e.g. molasses. Molasses is a by-product of sugar cane or sugar beet production and are rich in carbohydrates, mainly sucrose, nitrogen and other elements and vitamins. Other starting materials used in fermentations to provide organic carbon sources are whey, a lactose-rich by-product of the dairy industry, and methanol, typically produced from natural gas, vegetable oils and others. In some fermentations, the carbon source is inorganic, CO_2, e.g. in the microalgae fermentations to produce biofuels, however, these processes have not yet reached large-scale commercialisation.

Nitrogen must also be supplied in relatively large amounts, as the nitrogen is present in proteins and nucleic acids and its content in microorganisms is 10–15% (w/w). Nitrogen concentration in fermentation media depends on the type of fermentation and is often higher than 1 g/L. Nitrogen is supplied as ammonium salts or as nitrogen-rich substances such as urea, corn steep liquor (a by-product of starch extraction from corn), yeast extract (derived from the lysis of yeast cells) or others.

Phosphorus, present inside the cells in nucleic acids, ATP, NADP and other substances, is generally provided as inorganic phosphate ions which are also used as pH buffers. In most fermentation processes, the concentration of phosphorus required for microbial metabolism, excluding the needs for pH buffering, is not higher than 100 mg/L.

Providing sulphur is essential as sulphur is contained in several amino acids and some vitamins. Sulphur, if not present enough in the material used to supply the carbon source, is usually supplied as inorganic sulphate or sulphide salt, at concentrations often in the order of 20 mg/L.

Other elements which need to be present at lower concentrations in the fermentation medium are calcium, iron, potassium, sodium, magnesium (supplied at concentrations typically in the order of 10 mg/L or lower) and trace elements such as nickel, cobalt, copper, manganese, selenium, zinc and molybdenum (typical concentrations in fermentation media 0.1–1 mg/L). These elements are required for certain enzymes and for a variety of functions in the cells and are often present in the required amount in the water or in the material used to supply the carbon source. If they are not present at high enough concentrations, they need to be added with their salts.

Other substances such as vitamins, growth factors and precursors can also be added to the fermentation medium if required by specific microorganisms and processes.

1.8 CULTURE CULTIVATION AND MAINTENANCE

Regardless of the type of process, in many cases, fermentations are started in very small reactors where the microorganisms are inoculated from frozen stock cultures. In these reactors, microorganisms are supplied with the carbon source and with all the required nutrients and are left to acclimate to the substrate and to grow in numbers. Once microorganisms are acclimated and have reached a high enough concentration, they are inoculated in a larger reactor, in order to increase the cell mass and numbers. The process is repeated several times in larger reactors until the cell mass is large enough to be inoculated in the full-scale fermentation vessel.

The progress of the fermentation is usually monitored by measuring the growth of the microorganisms using direct or indirect methods. Direct methods include direct microscopic counts, electronic cell counters and plate counting techniques. Indirect methods include turbidimetric techniques, which measure the turbidity of the suspension, which is proportional to the concentration of microorganisms, and dry weight measurement. ATP can also be measured using a bioluminometry technique, giving an estimation of the concentration of viable microorganisms.

During the run of fermentation processes, the medium conditions need to be monitored to ensure maintenance of the optimum values of pH, dissolved oxygen and temperature. pH is maintained at the desired value by addition to the medium of a buffer and/or by controlled addition of acid or base in response to the readings of a pH sensor. Temperature is controlled by sensors in the medium that usually control the flow rate of a cooling or heating fluid in the vessel's jacket or coils. The concentration of dissolved oxygen is measured with dissolved oxygen sensors and is controlled by manipulating either the air (or oxygen) flow rate or the agitation speed. During fermentation processes, foam production can also be a problem, especially for aerobic fermentations. Foam formation is due to the generation by the microorganisms of surface-active agents, e.g. proteins. Foam is controlled by automatic addition of antifoam agents, e.g. vegetable oils or synthetic antifoams. Alternatively, the medium composition can be modified to avoid foam generation.

1.9 EXTRACTION AND SEPARATION OF METABOLITES

After the fermentation, the target product needs in most cases to be separated from the fermentation medium and purified. The sequence of unit operations required for the separation and purification of the product is called downstream processing. The processes used in downstream processing are different for different fermentations and different products and depend on the physical and chemical properties of the product and on its concentration.

In general, the first step of downstream processing is the separation of the microorganisms from the fermentation medium. This is achieved by a solid-liquid separation stage, which can be sedimentation, centrifugation or filtration.

The processes used to separate and purify the target product depend on whether the product is intracellular or extracellular. For intracellular products (e.g. many enzymes), the cells need to be disrupted to allow the release of the product. Cell disruption can be achieved with mechanical or non-mechanical methods. Mechanical methods include the use of pressure and liquid shear (high-pressure homogenizers) of high-speed bead mills or of ultrasounds. Non-mechanical methods cause cell disruption by permeabilisation of the cell membrane, which can be achieved by autolysis, osmotic or heat shock. When the cell membrane is made permeable by one of these methods, the products are released into the medium and can be purified.

Extracellular products are present in the liquid medium after separation of the microorganisms. These products can be separated and purified using a range of unit operations, some of them widely used in other areas of chemical engineering, others more typical of fermentation processes. Chromatography is a separation process used for the recovery of high-value products, e.g. proteins. In chromatography, products are separated based on their different retention when they flow through a stationary phase. The separation of the products in chromatography can occur via different mechanisms, e.g. size exclusion, ion-exchange. Other separation and purification processes include solvent extraction, evaporation, distillation, crystallisation and drying. In solvent extraction, separation is achieved using organic solvents to extract the product from the water-based fermentation medium. Evaporation and distillation require bringing the fermentation medium to the boiling point by supplying heat, using the different volatility of the product and of the other components of the medium. When crystallisation is used, the product is recovered as solid crystals, which can be obtained from the fermentation medium by decreasing the temperature or by adding a chemical which reacts with the product decreasing its solubility. Drying, when used, is the final purification stage and it involves the removal of any residual solvent or water by application of heat.

1.10 STERILISATION OF MEDIA AND EQUIPMENT

In many fermentation processes, it is necessary to avoid the growth of undesired microorganisms. Indeed, undesired microorganisms may compete with the desired species for the substrate, reducing the substrate available for the desired process. In addition, undesired microorganisms may metabolise the target product or contaminate it with other undesired metabolic products, whose removal in downstream processing may be difficult or expensive.

Undesired microorganisms are removed before the start of the fermentation by sterilising the reactor and the medium. Sterilisation is usually carried out by using heat because microorganisms cannot survive at high temperatures. In many cases, the vessel is heated using pressurised steam in the jacket or coils at temperatures up to 121°C, for an appropriate length. In some cases, the steam is injected into the head space of the vessel above the medium. The medium can also be sterilised by steam heating, however, some medium components, e.g. vitamins and glucose, can degrade when exposed to heat. In these cases, the medium is sterilised by filtration, rather than by heating. As an alternative, for heat-sensitive media heat can be applied for very short times at higher temperatures (e.g. flowing the media through heat exchangers) to avoid degradation of the components. Since microorganisms are also present in air, all the air inlets to the fermentation vessels need to be equipped with filters to remove undesired microorganisms, if sterile conditions are required. In addition, if contamination of the environment by the microorganisms in the fermentation is a concern, e.g. in the case of the cultivation of pathogenic microorganisms, the gas exhaust ports in the fermenter need also to be equipped with filters to remove any microorganisms from the outlet gases.

Sterilisation of the vessel and medium can also be achieved by means other than that of heat, e.g. irradiation with microwave, UV or X rays or with high voltage electric fields. Sterilisation by irradiation is used frequently to sterilise small fermentation accessories, e.g. Petri dishes.

1.11 GEOMETRY AND MATERIALS FOR FERMENTATION VESSELS

Fermentation vessels are typically of cylindrical shape, equipped, in case of agitated vessels, with agitators in the centre. Agitators are commercially available in many different types and manufacturers follow guidelines on the geometrical ratios (e.g. ratio between agitator and vessel diameter). Small lab-scale vessels (up to a few litres) are usually made of glass, while pilot and production scale vessels are made of mild or stainless steel. The construction materials should be non-toxic, resistant to corrosion and able to withstand repeated cycles of sterilisation, i.e. they need to be designed for pressures larger than the atmospheric due to the use of pressurised steam. The pipework should also be designed to be sterilisable and to avoid undesired microorganisms entering the process. For this reason, welded joints are usually preferable to flanges, as flanged joints may loosen due to the repeated sterilisation cycles with heating and cooling, favouring the contamination with air-borne microorganisms.

Questions and Problems

1.1 Describe the main components in cells and their function.
1.2 Explain the difference between primary and secondary products produced by microorganisms.
1.3 Explain why the mass of microorganisms produced per unit mass of substrate removed is usually much larger for aerobic reactions than for anaerobic ones.
1.4 Explain the advantages and disadvantages of the main types of fermentation processes: batch, continuous and fed-batch.
1.5 Explain the advantages and disadvantages of the main types of reactors for fermentation: stirred tank reactor, packed column, fluidised bed, airlift and bubble column.

2 Kinetics and Stoichiometry

This chapter presents the most common models used to describe the growth rate of microorganisms or cell cultures and the quantitative relationship between reagents and products (stoichiometry). Microorganisms grow at very different rates, depending on the nature of the microorganism and on the growth conditions (substrate, mineral elements, temperature and pH). The number of *Escherichia coli* cells can double every hour when growing aerobically on glucose under optimum conditions, while certain bacteria that oxidise ammonia under anaerobic conditions using nitrite only double their number every two days under their optimum conditions. Mathematical models are needed to describe the different growth rates of microorganisms and to design bioreactors that take the different microbial behaviours into account.

2.1 KINETICS OF FERMENTATION PROCESSES

In order to design fermentation processes, it is important to know how to express the rate of the various processes, i.e. biomass growth, substrate removal and product generation.

We will start by considering the rate of biomass growth. In quantitative terms, growth is defined as the increase in the number and mass of the microorganisms present in the medium. It is expected that the rate of biomass growth will be proportional to the number, or mass, of microorganisms. Indeed, in any given period of time, two microorganisms will produce more microorganisms than one microorganism, four microorganisms will produce more microorganisms than two of them, and so on. Conceptually, the same happens for the human population (and indeed for the population of any other living species), the more people there are on the planet, the faster the population will grow.

Based on these considerations, we will use the following rate equation for microorganisms' growth:

$$r_X\left(\frac{kg}{m^3d}\right) = \mu \cdot X \tag{2.1}$$

r_X is the microorganism growth rate and X is the microorganism concentration. In this book, we'll report the variables with their typical units, e.g. $\frac{kg}{m^3d}$ for r_X, but

DOI: 10.1201/9781003217275-2

equivalent units (e.g. $\frac{g}{L \cdot h}$) are of course also possible. Typically, the microorganism concentration is expressed as mass, rather than number, so the units of X are mass/volume, e.g. kg/m^3 or equivalent. Note that the unit of time for the rate of fermentation processes is often taken as days (d) or hours (h), rather than seconds which are more typical for non-fermentation reactions. Clearly, the microorganism growth rate does not depend only on the microorganism concentration, otherwise microorganisms would never stop growing. There must be something that limits microorganisms' growth. If all the other elements and nutrients, which the cells need, are more than the requirements for biomass growth, growth of the microorganisms will only be limited by the concentration of the substrate. The effect of the substrate concentration on the growth rate is included in the term μ, which is the specific growth rate (it's called 'specific' because it's the growth rate per unit mass of microorganisms). In general, we expect, and this has been experimentally observed in a great number of cases, that substrate concentration only limits biomass growth when the substrate concentration is low. Instead, when the substrate concentration is very high, changes in the substrate concentration don't affect the growth rate, which then reaches its maximum value for the given substrate and type of microorganisms. These considerations lead to the very frequent use of the equation for μ represented in Equation (2.2):

$$\mu\left(d^{-1}\right)=\frac{\mu_{\max} S}{K_S+S} \tag{2.2}$$

S is the substrate concentration, which is often expressed as kg/m^3, even though mol/m^3 or other equivalent units can also be used. μ_{max} (maximum growth rate, d^{-1} or equivalent units, Table 2.1) and K_S (half-saturation constant, same units as the substrate) are parameters, which depend on the species of microorganism and on the substrate. They also depend on the environmental conditions, mainly temperature and pH. Equation (2.2) is called the 'Monod' equation, named after the French researcher who first proposed it.

TABLE 2.1
μ_{max} Values Obtained for Various Microorganisms

Microorganism	Limiting Substrate	μ_{max} (d^{-1})	Ks (kg/m^3)
Pseudomonas sp.	Benzoate	10.4	4.5×10^{-4}
Pseudomonas putida F1	Toluene	20.6	1.4×10^{-2}
Pseudomonas putida F1	Benzene	17.5	1.2×10^{-4}
Mixed culture (activated sludge)	Sodium dodecyl sulphate	22.2	9.6×10^{-2}
Streptococcus zooepidemicus	Glucose	10.8–19.2	-

Combining Equations (2.1) and (2.2), the biomass growth rate is:

$$r_X\left(\frac{kg}{m^3d}\right)=\frac{\mu_{max}S}{K_S+S}X \tag{2.3}$$

Equation (2.2) is a shifting order rate equation. When the substrate concentration is very large, i.e. $S \gg K_S$, the specific growth rate does not depend on the substrate concentration and $\mu = \mu_{max}$. When the substrate concentration is very low, i.e. $S \ll K_S$ and $\mu = \frac{\mu_{max}}{K_S}$, the specific growth rate shows a first-order dependence on the substrate concentration. Note that numerically the value of K_S coincides with the substrate concentration for which $\mu = \frac{\mu_{max}}{2}$. The profile of μ vs S given by the Monod equation is shown in Figure 2.1.

Example 2.1

The kinetic parameters of a microorganism on a certain substrate are $\mu_{max} = 4.9\ d^{-1}$ and $K_S = 0.1\ kg/m^3$.

Calculate:

a. The specific growth rate *m* when the substrate concentration is 10, 5, 0.002, 0.001 kg/m³.
b. The growth rate r_X when the substrate concentration is 5 kg/m³ and the microorganisms' concentration is 2.5 kg/m³.

SOLUTION

a. We use Equation (2.2). For $S = 10\ kg/m^3$, $\mu = 4.85\ d^{-1}$; for $S = 5\ kg/m^3$, $\mu = 4.80\ d^{-1}$; for $S = 0.002\ kg/m^3$, $\mu = 0.096\ d^{-1}$; for $S = 0.001\ kg/m^3$, $\mu = 0.049\ d^{-1}$.

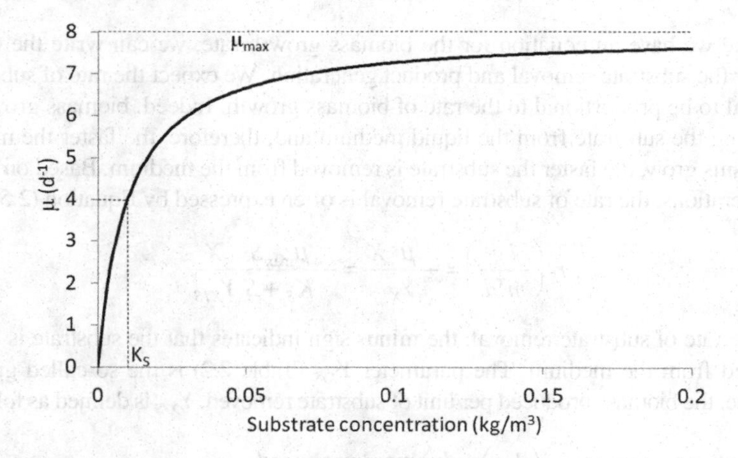

FIGURE 2.1 Typical shape of the rate equation for microbial growth on the limiting substrate S. This curve has been obtained with $\mu_{max} = 8\ d^{-1}$ and $K_S = 0.010\ kg/m^3$.

Note that for high concentrations of the substrate, 10 and 5 kg/m^3, μ is very close to μ_{max} and almost independent of the substrate concentration, while for very low concentrations μ is approximately linearly dependent on S.

b. For $X = 2.5$ kg/m^3 and $\mu = 4.80$ d^{-1}, using Equation (2.1) $r_X = 12.0 \frac{kg}{m^3 d}$.

The Monod equation shown in Equation (2.2) assumes that the specific growth rate is only limited by the concentration of the carbon source. This is true if all other elements and nutrients, needed by the microorganisms, are present in excess in the growth medium. If any other elements are limiting, Equation (2.2) can be modified to include limitation by multiple substances, e.g.:

$$\mu\left(d^{-1}\right) = \mu_{max} \frac{S_1}{K_{S1} + S_1} \cdot \frac{S_2}{K_{S2} + S_2} \cdot \ldots \quad (2.4)$$

In Equation (2.4), S_1, S_2, etc. are the concentration of the any limiting substances, e.g. substrate, oxygen, nitrogen, etc., and K_{S1}, K_{S2}, etc. are the corresponding half-saturation constants.

Example 2.2

With the kinetic parameters of Example 2.1, assume that there is a second limiting substrate, i.e. oxygen, for which $K_{SO2} = 0.0005$ kg/m^3. Calculate the specific growth rate when $S = 5$ kg/m^3 and $O_2 = 0.001$ kg/m^3 or 0.008 kg/m^3.

SOLUTION

Using Equation (2.4), for $O_2 = 0.001$ kg/m^3, $\mu = 3.20$ d^{-1} and for $O_2 = 0.008$ kg/m^3, $\mu = 4.52$ d^{-1}.

It's important to observe that Equations (2.1–2.3) do not describe the acclimation phase, but only apply to completely acclimated microorganisms.

Once we have an equation for the biomass growth rate, we can write the equation for the substrate removal and product generation. We expect the rate of substrate removal to be proportional to the rate of biomass growth. Indeed, biomass grows by removing the substrate from the liquid medium and, therefore, the faster the microorganisms grow, the faster the substrate is removed from the medium. Based on these considerations, the rate of substrate removal is often expressed by Equation (2.5):

$$r_S\left(\frac{kg}{m^3 d}\right) = -\frac{\mu \cdot X}{Y_{X/S}} = -\frac{\mu_{max} S}{K_S + S} \frac{X}{Y_{X/S}} \quad (2.5)$$

r_S is the rate of substrate removal, the minus sign indicates that the substrate is being removed from the medium. The parameter $Y_{X/S}$ (Table 2.2) is the so-called growth yield, i.e. the biomass produced per unit of substrate removed. $Y_{X/S}$ is defined as follows:

$$Y_{X/S}\left(\frac{kg}{kg}\right) = \frac{biomass\ produced}{substrate\ removed} = \frac{r_X}{(-r_S)} \quad (2.6)$$

TABLE 2.2
$Y_{X/S}$ Values Obtained for Various Microorganisms

Microorganism	Substrate	$Y_{X/S}$ (kg/kg)
Klebsiella sp.	Methanol	0.38
Enterobacter aerogenes	Maltose	0.46
Candida utilis	Glucose	0.51
Pseudomonas fluorescens	Acetate	0.28
Pseudomonas methanica	Methane	0.56

The parameter $Y_{X/S}$ depends on the microbial species, on the substrate nature and also on the environmental conditions, e.g. temperature and pH.

As far as the rate of product production is concerned, we need to distinguish whether the product is a primary or secondary metabolite. If the product is a primary metabolite (kinetics for secondary metabolites production will be covered in Section 2.6), i.e. it is produced during microbial growth, then the rate of product production will be proportional to the biomass growth rate:

$$r_P\left(\frac{kg}{m^3d}\right) = Y_{P/X}r_X = Y_{P/X}\mu \cdot X \tag{2.7}$$

$Y_{P/X}$ is the product yield, i.e. the product produced per unit mass of biomass produced:

$$Y_{P/X}\left(\frac{kg}{kg}\right) = \frac{product\ produced}{biomass\ produced} = \frac{r_P}{r_X} \tag{2.8}$$

Instead of using the parameter $Y_{P/X}$, the parameter $Y_{P/S}$ (Table 2.3) can be introduced, with the following definition:

$$Y_{P/S}\left(\frac{kg}{kg}\right) = \frac{product\ produced}{substrate\ removed} = \frac{\Delta P}{(-\Delta S)} = \frac{r_P}{(-r_S)} \tag{2.9}$$

TABLE 2.3
$Y_{P/S}$ Values Obtained for Various Microorganisms

Microorganism	Substrate	Product	$Y_{P/S}$ (kg/kg)
Saccharomyces cerevisiae	Pre-treated rice husks	Ethanol	0.30–0.40
Scheffersomyces stipitis	Pre-treated rice husks	Ethanol	0.24–0.36
Yeast culture	Pre-treated orange peels	Ethanol	0.36
Streptococcus zooepidemicus	Glucose	Hyaluronic acid	0.070–0.085

By using $Y_{P/S}$, the rate of product production can be expressed as:

$$r_P\left(\frac{kg}{m^3d}\right)=Y_{P/S}\left(-r_S\right)=Y_{P/S}\frac{\mu X}{Y_{X/S}} \tag{2.10}$$

Comparing Equations (2.6) and (2.9) it is also evident that:

$$Y_{P/X}=\frac{Y_{P/S}}{Y_{X/S}} \tag{2.11}$$

We have seen that the parameters $Y_{X/S}$, $Y_{P/S}$ and $Y_{P/X}$ link biomass growth, substrate consumption and product generation. In fermentations, it is often also important to express the rate of oxygen and nutrients consumption by the microorganisms. These can be expressed by using the same approach, e.g. the biomass produced per unit of oxygen consumed can be expressed by the parameter $Y_{X/O2}$:

$$Y_{X/O2}\left(\frac{kg}{kg}\right)=\frac{Biomass\ produced}{Oxygen\ consumed}=\frac{r_X}{\left(-r_{O2biom}\right)} \tag{2.12}$$

In Equation (2.12), r_{O2biom} is the rate of oxygen consumption by the microorganisms per unit volume of the reactor. r_{O2biom} is negative, because oxygen is consumed, so in the definition of the parameter $Y_{X/O2}$, which is positive, the term $-r_{O2biom}$ is used. From Equation (2.12), the equation for the calculation of the rate of oxygen consumption follows immediately:

$$r_{O_2biom}\left(\frac{kg}{m^3d}\right)=-\frac{r_X}{Y_{X/O_2}}=-\frac{\mu_{\max}S}{K_S+S}\frac{X}{Y_{X/O_2}} \tag{2.13}$$

As an example of the rate of nutrients consumption, let us write an equation for the rate of nitrogen uptake by the microorganisms. This can be written in the same way as done for oxygen, by defining the parameter $Y_{X/N}$ which represents the biomass produced per unit of nitrogen consumed:

$$Y_{X/N}\left(\frac{kg}{kg}\right)=\frac{Biomass\ produced}{Nitrogen\ consumed}=\frac{r_X}{\left(-r_N\right)} \tag{2.14}$$

In Equation (2.14), r_N is the volumetric rate of nitrogen consumption by the microorganisms. r_N is negative because nitrogen is consumed, so in the definition of the parameter $Y_{X/N}$, which is positive, the term $-r_N$ is used. From Equation (2.14), the equation for the calculation of the rate of oxygen consumption follows immediately:

$$r_N\left(\frac{kgN}{m^3d}\right)=-\frac{r_X}{Y_{X/N}}=-\frac{\mu_{\max}S}{K_S+S}\frac{X}{Y_{X/N}} \tag{2.15}$$

The same approach can be used to express and calculate the production and consumption rate of any products and metabolites generated or consumed during biomass growth.

Example 2.3

A certain species of microorganisms (assume empirical formula $C_5H_7O_2N$) is growing on glucose ($C_6H_{12}O_6$) at a rate of 0.1 kg/m³·min. The rate of substrate removal is 0.5 kg/m³·min. Calculate the growth yield $Y_{X/S}$ in units of kg/kg and mol/mol.

SOLUTION

From the definition of $Y_{X/S}$:

$$Y_{X/S} = \frac{r_X}{(-r_S)} = \frac{0.1\frac{kg}{m^3 min}}{0.5\frac{kg}{m^3 min}} = 0.2\frac{kg}{kg}$$

On a molar basis, 0.1 kg of microorganisms corresponds to 0.88 mol, and 0.5 kg of glucose corresponds to 2.78 mol. So, on a molar basis the growth yield is:

$$Y_{X/S} = \frac{r_X}{(-r_S)} = \frac{0.88\frac{mol}{m^3 min}}{2.77\frac{mol}{m^3 min}} = 0.32\frac{mol}{mol}$$

This example shows that it is important to express the growth yield $Y_{X/S}$ with the appropriate units (kg/kg or mol/mol) rather than in the dimensional form without specifying any units.

Example 2.4

A microorganism grows on a certain substrate at a rate of 1.5 $\frac{kg}{m^3h}$. The yield parameters are $Y_{X/S} = 0.3$ kg/kg, $Y_{P/S} = 0.2$ kg/kg, $Y_{X/O2} = 0.5$ kg/kg and $Y_{X/N} =$ 10 kg/kg. Calculate the rate of substrate, oxygen and nitrogen removal and product production. Also, calculate the parameter $Y_{P/X}$.

SOLUTION

$$r_S = -\frac{r_X}{Y_{X/S}} = -\frac{1.5\frac{kg}{m^3 hr}}{0.3\frac{kg}{kg}} = -5\frac{kg}{m^3h}$$

$$r_{O_{2biom}} = -\frac{r_X}{Y_{X/O_2}} = -3\frac{kg}{m^3h}$$

$$r_N = -\frac{r_X}{Y_{X/N}} = -0.15\frac{kgN}{m^3h}$$

$$Y_{P/X} = \frac{Y_{P/S}}{Y_{X/S}} = 0.67\frac{kg}{kg}$$

2.2 EFFECT OF INHIBITORS, TEMPERATURE AND pH ON GROWTH RATE

The rate of microorganism growth and the kinetic parameters used to describe it depend on many environmental factors, the most important of which are the presence of inhibitors, temperature and pH.

Inhibition may be due to the substrate, to the fermentation products or to other chemicals in the fermentation medium. The Haldane or Andrews Equation (2.16) includes substrate inhibition:

$$\mu = \frac{\mu_{max} S}{K_S + S + \frac{S^2}{K_I}} \tag{2.16}$$

Equation (2.16) is valid for those substrates which inhibit biomass growth when they are present above a certain concentration. In Equation (2.16) in addition to μ_{max} and K_S, we have an additional parameter K_I. The lower the K_I value, the larger the inhibiting effect of the substrate or the lower the substrate concentration which is required to give a decrease in μ. A typical plot of the Haldane equation is shown in Figure 2.2.

If microorganisms are growing on substrate S but are inhibited by another substance I (which may be either a product of the fermentation or another chemical), the specific growth rate can have the following form:

$$\mu = \frac{\mu_{max} S}{K_S + S} \frac{1}{1 + \frac{I}{K_I}} \tag{2.17}$$

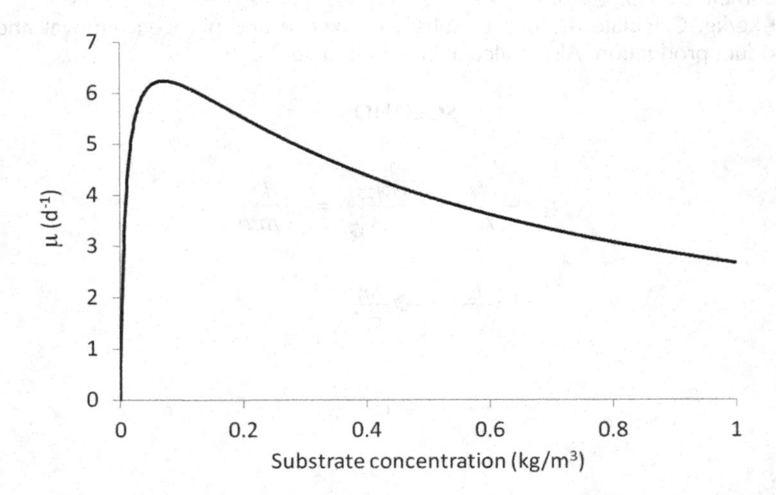

FIGURE 2.2 Typical shape of the rate equation for microbial growth on an inhibiting substrate S. This curve has been obtained with $\mu_{max} = 8.0\ d^{-1}$, $K_S = 0.01\ kg/m^3$ and $K_I = 0.5\ kg/m^3$.

Another important effect on the specific growth rate is pH. Microorganisms are typically only able to grow in a restricted pH range and, within this range, there is typically a pH value for which the specific growth rate is maximum. The effect of pH on the specific growth rate is often described by equations such as Equation (2.18):

$$\mu = \frac{\mu_{\max} S}{K_S + S} \frac{1 + 2 \cdot 10^{0.5(a-b)}}{1 + 10^{(pH-b)} + 10^{(a-pH)}} \tag{2.18}$$

where a and b are two parameters which represent, respectively, the low and high extremes of pH for which there is still microbial activity. A plot of the pH inhibition factor, $\frac{1+2\cdot 10^{0.5(a-b)}}{1+10^{(pH-b)}+10^{(a-pH)}}$, is shown in Figure 2.3.

Another important effect on microbial growth rate is the temperature. Typically, the rate of microbial growth increases with temperature. When the temperature increases, from room temperature to approximately 35–40°C, microbial growth increases even though many microbial species may have an optimum temperature outside this range. At higher temperatures, denaturation of the cellular enzymes may start to occur, with a consequent decrease in microbial activity and in the growth rate. In the region where the reaction rate increases with temperature, it is usually assumed that the temperature dependence can be included in the $\mu_{\max}$ parameter, e.g. with a conventional Arrhenius equation such as:

$$\mu_{\max} = Ae^{-\frac{Ea}{RT}} \tag{2.19}$$

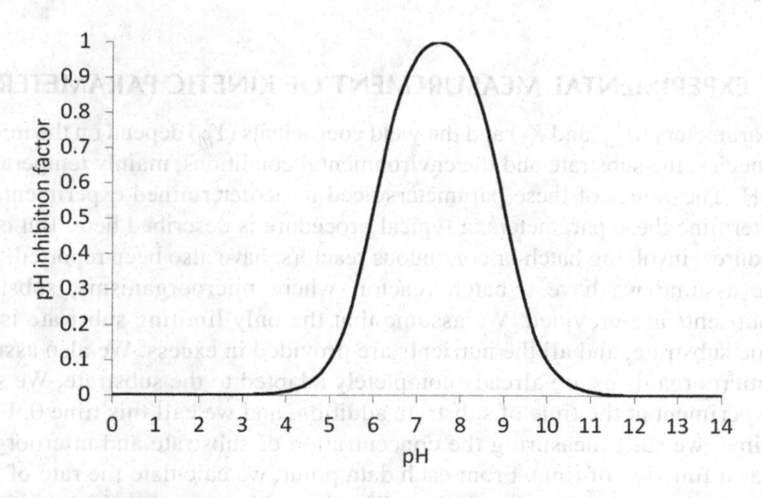

FIGURE 2.3 Typical shape of the pH inhibition factor. This curve has been obtained with $a = 6$, $b = 9$.

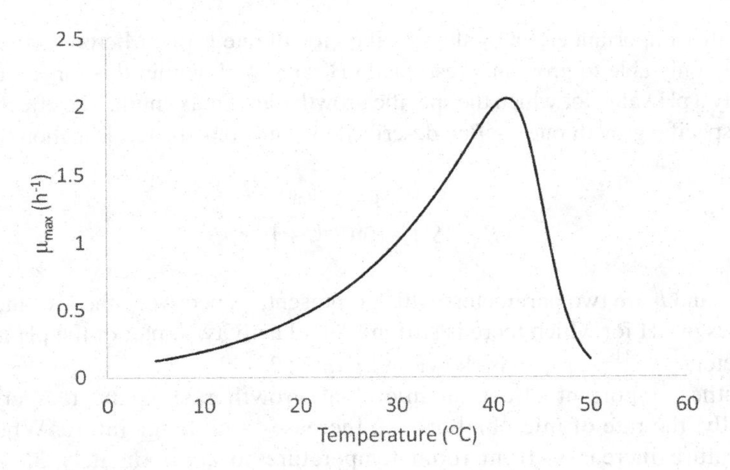

FIGURE 2.4 Plot of μ_{max} vs T according to Equation (2.20), using $A = 10^{10}$ h^{-1}, $B = 3.0 \times 10^{90}$, $Ea = 58{,}000$ J/mol and $\Delta G_d = 550{,}000$ J/mol.

An equation that accounts for the effect of temperature in a wider range, including both the increase and decrease in microbial activity is the following:

$$\mu_{max} = \frac{Ae^{-\frac{Ea}{RT}}}{1 + Be^{-\frac{\Delta G_d}{RT}}} \tag{2.20}$$

In Equation (2.20), the parameters A, B, Ea and ΔG_d (which physically corresponds to the free energy change of the deactivation reaction) must be determined experimentally. A typical plot of μ_{max} vs T using Equation (2.20) is shown in Figure 2.4.

2.3 EXPERIMENTAL MEASUREMENT OF KINETIC PARAMETERS

The parameters (μ_{max} and K_S) and the yield coefficients ($Y_{i/j}$) depend on the microbial species, the substrate and the environmental conditions, mainly temperature and pH. The values of these parameters need to be determined experimentally. To determine these parameters, a typical procedure is described here. But other procedures, involving batch or continuous reactors, have also been reported.

We assume we have a batch reactor, where microorganisms, substrate and nutrients are provided. We assume that the only limiting substrate is the organic substrate, and all the nutrients are provided in excess. We also assume that microorganisms are already completely adapted to the substrate. We start the experiment at the time of substrate addition, and we call this time 0. From that time, we start measuring the concentration of substrate and microorganisms as a function of time. From each data point, we calculate the rate of biomass growth, r_X, as a function of time. The data will be arranged in a table, as shown in Table 2.4.

TABLE 2.4
Example of Experimental Data to Be Collected for the Determination of Kinetic Parameters

Time	X	S	r_X
0	X_0	S_0	
t_1	X_1	S_1	$\frac{X_1 - X_0}{t_1 - t_0}$
t_2	X_2	S_2	$\frac{X_2 - X_1}{t_2 - t_1}$
....	....	....	...

The parameters μ_{max} and K_S can be obtained by re-arranging Equation (2.3) as:

$$\frac{X}{r_X} = \frac{K_S}{\mu_{max}}\frac{1}{S} + \frac{1}{\mu_{max}} \tag{2.21}$$

Therefore, the experimental data collected in Table 2.4 can be re-arranged to plot $\frac{X}{r_X}$ vs $\frac{1}{S}$ (Figure 2.5). The intercept of the graph corresponds to $\frac{1}{\mu_{max}}$ and the slope to $\frac{K_S}{\mu_{max}}$, allowing, therefore, the calculation of μ_{max} and K_S.

From the same experiment, the parameter $Y_{X/S}$ can also be obtained by calculating the ratio between the biomass produced, $\Delta X = X_{final} - X_0$, and the substrate removed, $(-\Delta S) = -[S_{final} - S_0]$, over the whole length of the experiment. Alternatively, $Y_{X/S}$ can be calculated as the average of the ratios $\frac{\Delta X}{(-\Delta S)}$ over the length of the experiment.

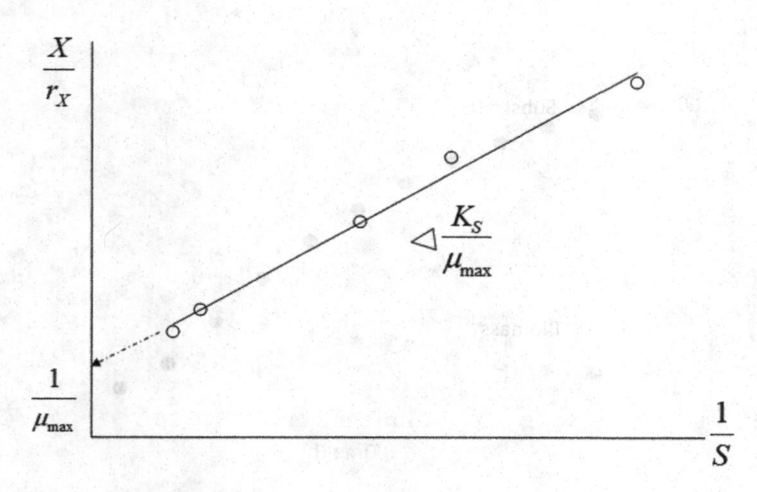

FIGURE 2.5 Plot of $\frac{X}{r_X}$ vs $\frac{1}{S}$ for the calculation of the parameters μ_{max} and K_S.

TABLE 2.5
Data for Example 2.5

Time (h)	Substrate (S, mg/L)	Biomass (X, mg/L)
0	2,000	100
2	1,930	122
4	1,840	148
6	1,730	180
8	1,600	219
10	1,450	265
12	1,260	321
14	1,040	388
16	780	466
18	480	556
20	180	646
22	0	696

Example 2.5

The following data (Table 2.5) have been obtained during the batch growth of a certain microorganism on an organic substrate.

Calculate μ_{max}, K_S and $Y_{X/S}$.

SOLUTION

The plot of the experimental profiles of S and X vs time is given in Figure 2.6.

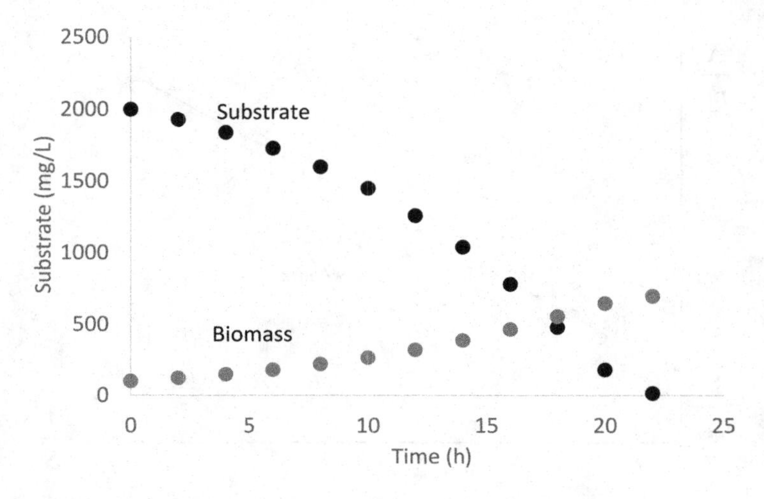

FIGURE 2.6 Plot of experimental data for Example 2.5.

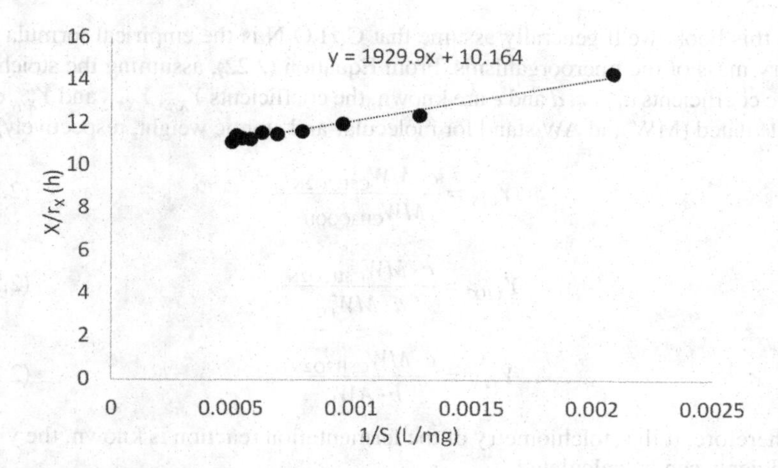

FIGURE 2.7 Linearisation of experimental data for Example 2.5.

In order to calculate μ_{max} and K_S, we need to generate the plot of X/r_X vs $1/S$. This plot is reported in Figure 2.7. Note that in the plot of Figure 2.7, we have excluded the last data point, at $t = 22$ h, because at that time there was no substrate left and we don't know at what time the substrate had been totally removed.

The intercept of the X/r_X vs $1/S$ graph is equal to 10.164 h. This intercept is equal to $1/\mu_{max}$ (Equation 2.21). This means that:

$$\mu_{max} = 0.098\ h^{-1}$$

The intercept is 1929.9 mg/L·h. This corresponds to $\frac{K_S}{\mu_{max}}$. With the known value of μ_{max}, we obtain:

$$K_S = 189.1\frac{mg}{L}$$

To calculate $Y_{X/S}$, we can use the total biomass formation and substrate removal during the experiment:

$$Y_{X/S} = \frac{(696-100)\,mg/L}{2{,}000\,mg/L} = 0.3\frac{kg}{kg}$$

2.4 YIELD COEFFICIENTS AND GROWTH STOICHIOMETRY

The yield coefficients introduced in the previous sections are linked by the overall stoichiometry that describes the fermentation reaction. For example, let us consider the generic stoichiometry for growth of microorganisms on acetic acid (CH_3COOH):

$$CH_3COOH + aO_2 + bNH_3 \rightarrow cC_5H_7O_2N + dCO_2 + eH_2O \qquad (2.22)$$

In this book, we'll generally assume that $C_5H_7O_2N$ is the empirical formula for the dry mass of the microorganisms. From Equation (2.22), assuming the stoichiometric coefficients *a, b, c, d* and *e* are known, the coefficients $Y_{X/S}$, $Y_{X/O2}$ and $Y_{X/N}$ can be calculated (MW and AW stand for molecular and atomic weight, respectively):

$$Y_{X/S} = \frac{c \cdot MW_{\mathrm{C5H7O2N}}}{MW_{\mathrm{CH3COOH}}} \tag{2.23}$$

$$Y_{X/O2} = \frac{c \cdot MW_{\mathrm{C5H7O2N}}}{a \cdot MW_{\mathrm{O2}}} \tag{2.24}$$

$$Y_{X/N} = \frac{c \cdot MW_{\mathrm{C5H7O2N}}}{b \cdot AW_{\mathrm{N}}} \tag{2.25}$$

Therefore, if the stoichiometry of the fermentation reaction is known, the yield coefficients can be calculated.

The procedure can also be applied in reverse, i.e. the knowledge of at least one of the yield coefficients can be used to calculate the fermentation stoichiometry if this is not known. For example, let us consider the fermentation reaction of glucose under aerobic conditions. In this case, we assume that the only products are microorganisms, water and carbon dioxide. Nitrogen is supplied as ammonia. The generic reaction stoichiometry is, therefore, written as:

$$\mathrm{C_6H_{12}O_6 + aO_2 + bNH_3 \rightarrow cC_5H_7O_2N + dCO_2 + eH_2O} \tag{2.26}$$

How can we calculate the stoichiometric coefficients *a, b, c, d* and *e*? These coefficients are linked by the elemental balances for C, H, N and O. These constitute four equations; however, a fifth equation is needed to calculate the five coefficients. The fifth equation can be obtained by knowledge of an experimental value for one of the yield coefficients. For example, if we assume the growth yield ($Y_{X/S}$) is known, the fifth equation is:

$$Y_{X/S} = \frac{c \cdot 113}{180} \tag{2.27}$$

In Equation (2.27), 113 and 180 are the molecular weights of microorganisms and glucose, respectively. Equation (2.27) can be coupled with the elemental balances for C, O, N and H to calculate all the stoichiometric coefficients. In summary, the five equations that give the stoichiometric coefficients are:

$$\begin{cases} C-balance\colon 5c+d=6 \\ O-balance\colon 2c+2d+e=2a+6 \\ H-balance\colon 7c+2e=3b+12 \\ N-balance\colon b=c \\ Y_{X/S}=\dfrac{c\cdot 113}{180} \end{cases} \tag{2.28}$$

The equations in (2.28) constitute a system of five equations in the five unknowns *a*, *b*, *c*, *d* and *e*.

The concepts expressed in this section can be applied in the same way if the fermentation produces a product of interest. For example, let us consider the generic stoichiometry for ethanol production from glucose under anaerobic conditions:

$$C_6H_{12}O_6 + aNH_3 \rightarrow bC_5H_7O_2N + cC_2H_5OH + dCO_2 + eH_2O \quad (2.29)$$

If the coefficients *a*, *b*, *c*, *d* and *e* are known, the product yields can be calculated:

$$Y_{P/X} = \frac{kg\ product}{kg\ biomass} = \frac{cMWC_2H_5OH}{bMWC_5H_7O_2N} \quad (2.30)$$

$$Y_{P/S} = \frac{kg\ product}{kg\ substrate} = \frac{cMWC_2H_5OH}{MWC_6H_{12}O_6} \quad (2.31)$$

Similarly, if one of the yield coefficients ($Y_{X/S}$, $Y_{X/P}$, $Y_{P/S}$) is known, the stoichiometric coefficients for Equation (2.29) can be calculated by combining the equation for the known coefficient with the elemental balances for C, H, N and O.

Example 2.6

Consider the production of citric acid by the aerobic fermentation of glucose by the microorganism *Aspergillus niger*. Assume the stoichiometry of the reaction reported below:

$$C_6H_{12}O_6 + 2.35O_2 + 0.1NH_3 \rightarrow 0.1C_5H_7O_2N$$
$$+ 0.7C_6H_8O_7 + 1.3CO_2 + 3H_2O$$

Calculate the yields $Y_{X/S}$, $Y_{P/X}$, $Y_{P/S}$, $Y_{X/O2}$ and $Y_{X/N}$.

SOLUTION

$$Y_{X/S} = \frac{0.1 \cdot MW_{C5H7O2N}}{MW_{C6H12O6}} = 0.063\frac{kg}{kg}$$

$$Y_{P/X} = \frac{0.7 \cdot MW_{C6H8O7}}{0.1 \cdot MW_{C5H7O2N}} = 11.89\frac{kg}{kg}$$

$$Y_{P/S} = \frac{0.7 \cdot MW_{C6H8O7}}{MW_{C6H12O6}} = 0.75\frac{kg}{kg}$$

$$Y_{X/O2} = \frac{0.1 \cdot MW_{C5H7O2N}}{2.35 \cdot MW_{O2}} = 0.15 \frac{kg}{kg}$$

$$Y_{X/N} = \frac{0.1 \cdot MW_{C5H7O2N}}{0.1 \cdot AW_N} = 8.07 \frac{kg}{kg}$$

Example 2.7

Consider the fermentation of glucose ($C_6H_{12}O_6$) to produce ethanol (C_2H_6O) by an anaerobic microorganism. The reaction consumes glucose and ammonia and produces ethanol, water, carbon dioxide and microorganisms. It is known that 0.10 kg of microorganisms are produced per kg of glucose. Determine the stoichiometry of the fermentation reaction.

SOLUTION

The general form of the unbalanced stoichiometry is:

$$C_6H_{12}O_6 + aNH_3 \rightarrow bC_5H_7O_2N + cC_2H_6O + dCO_2 + eH_2O$$

There are five stoichiometric coefficients to determine. We have the four elemental balances for C, H, N and O and the information on the product yield:

$$Y_{X/S} = \frac{b \cdot MW_{C5H7O2N}}{MW_{C6H12O6}} = 0.627b = 0.10 \frac{kg}{kg}$$

From this equation, we get that $b = 0.16$.
The elemental balances are:

$$\begin{cases} C-balance\text{: } 6 = 5b + 2c + d \\ O-balance\text{: } 6 = 2b + c + 2d + e \\ H-balance\text{: } 12 + 3a = 7b + 6c + 2e \\ N-balance\text{: } a = b \end{cases}$$

From the nitrogen balance:

$$a = b = 0.16$$

From the carbon balance:

$$c = \frac{6 - 5b - d}{2} = 2.6 - 0.5d$$

From the oxygen balance:

$$e = 6 - 2 \cdot 0.16 - 2.6 + 0.5d - 2d = 3.08 - 1.5d$$

From the hydrogen balance:

$$12+3\cdot 0.16=7\cdot 0.16+6\cdot(2.6-0.5\,d)+2\cdot(3.08-1.5\,d)$$

which gives:

$$d=1.73$$

Therefore, e = 0.49 and c = 1.73.
And the balanced stoichiometry is:

$$C_6H_{12}O_6+0.16NH_3\rightarrow 0.16C_5H_7O_2N+1.73C_2H_6O+1.73CO_2+0.49H_2O$$

2.5 RATE EQUATIONS FOR SECONDARY METABOLITES, MAINTENANCE AND ENDOGENOUS METABOLISM

The rate equations developed in previous sections apply to microorganism's growth on the substrate with simultaneous production of the product (primary metabolite). However, in some cases, the product is not synthesised during microbial growth but rather when the substrate is (almost) completely removed and growth has stooped (secondary metabolite). For secondary metabolites, the rate of product production is no longer proportional to the growth rate according to Equation (2.7), but instead it will be proportional to the biomass concentration:

$$r_P=\alpha X \tag{2.32}$$

In Equation (2.32), α is a parameter to be determined from experiments. In the most general case, the rate of product production can be expressed as a function of both the biomass growth rate and the biomass concentration:

$$r_P=Y_{P/X}r_X+\alpha X \tag{2.33}$$

During the production of secondary metabolites, there will be some oxygen consumption and change in biomass concentration; however, these are usually small and can be ignored in the development of kinetic models.

Other modifications of the growth kinetics are used to account for the phenomena of maintenance and endogenous metabolism (Chapter 1). Maintenance (the use of substrate for non-growth purposes) gives a consumption of the substrate in the absence of growth and is usually described by the following rate equation:

$$r_{Sm}=-m_S X \tag{2.34}$$

In Equation (2.34), r_{Sm} is the rate of substrate removal due to the maintenance metabolism. The total rate of substrate consumption is the sum of the substrate removed due to growth and due to maintenance:

$$r_S\left(\frac{kgsubstrate}{m^3d}\right) = -\frac{\mu \cdot X}{Y_{X/S}} - m_S X \tag{2.35}$$

The substrate used for maintenance is oxidised for energy generation; therefore, there is an oxygen consumption associated with maintenance ($r_{O2biom,m}$):

$$r_{O2biom,m} = \frac{-m_S X}{Y_{S/O2,m}} \tag{2.36}$$

The coefficient $Y_{S/O2,m}$ can be calculated from the stoichiometry of total oxidation of the substrate. The total oxygen consumption is the sum of the consumption due to growth and maintenance:

$$r_{O2biom,m} = -\frac{r_X}{Y_{X/O_2}} - \frac{m_S X}{Y_{S/O2,m}} \tag{2.37}$$

Table 2.6 reports values of the maintenance coefficient measured for various microorganisms.

In summary, if we assume that biomass growth and substrate removal are described by the maintenance model, the equations that describe them are Equations (2.3) and (2.38):

$$r_X = \frac{\mu_{max} S}{K_S + S} X \tag{2.3}$$

$$r_S = -\frac{\mu_{max} S}{(K_S + S)} \frac{X}{Y_{X/S}} - m_S X \tag{2.38}$$

In order to determine the values of the kinetic parameters μ_{max}, K_S, $Y_{X/S}$ and m_S, we can use the same experiment as described in Section 2.4, with measurement of the microorganisms and substrate concentrations over time in a batch experiment.

TABLE 2.6
m_s Values Obtained for Various Microorganisms

Microorganism	Substrate	m_s (d^{-1})
Candida utilis	Glucose	0.74
Escherichia coli	Maltose	1.4
Penicillium chrysogenum	Glucose	0.50
Klebsiella aerogenes	Glycerol	1.78

The parameters μ_{max} and K_S can be determined by the same procedure described in Section 2.4, by plotting $\frac{X}{r_X}$ vs $\frac{1}{S}$ (Equation 2.21). $Y_{X/S}$ and m_S can be determined from the ratio $\frac{r_S}{r_X}$:

$$\frac{r_S}{r_X} = \frac{-\frac{\mu_{max} SX}{(K_S+S)Y_{X/S}} - m_S X}{\frac{\mu_{max} SX}{(K_S+S)}} = -\frac{1}{Y_{X/S}} - \frac{m_S}{\mu_{max}} - \frac{m_S K_S}{\mu_{max}} \frac{1}{S} \tag{2.39}$$

Therefore, according to Equation (2.39), plotting of $\frac{r_S}{r_X}$ vs $\frac{1}{S}$ should give a straight line with slope equal to $\frac{m_S K_S}{\mu_{max}}$ and intercept equal to $-\frac{1}{Y_{X/S}} - \frac{m_S}{\mu_{max}}$. Once μ_{max} and K_S are known, the slope of the line $\frac{r_S}{r_X}$ vs $\frac{1}{S}$ will give m_S and the intercept will give $Y_{X/S}$.

Example 2.8

The following data (Table 2.7) have been obtained during the batch growth of a certain microorganism on an organic substrate. Using the Monod model with maintenance, calculate μ_{max}, K_S, $Y_{X/S}$ and m_S.

SOLUTION

According to Equations (2.21) and (2.39), the parameters μ_{max}, K_S, $Y_{X/S}$ and m_S can be calculated by plotting $\frac{X}{r_X}$ vs $\frac{1}{S}$ and $\frac{r_S}{r_X}$ vs $\frac{1}{S}$. From the intercept of the $\frac{X}{r_X}$ vs $\frac{1}{S}$ plot (Figure 2.8), we calculate μ_{max}:

$$\mu_{max} = \frac{1}{8.85} = 0.11\, h^{-1}$$

TABLE 2.7
Data for Example 2.8

Time (h)	Substrate (*S*, mg/L)	Biomass (*X*, mg/L)
0	3,000	120
2	2,885	152
4	2,744	193
6	2,561	245
8	2,333	310
10	2,048	391
12	1,688	493
14	1,247	619
16	719	769
18	175	921
20	0	966

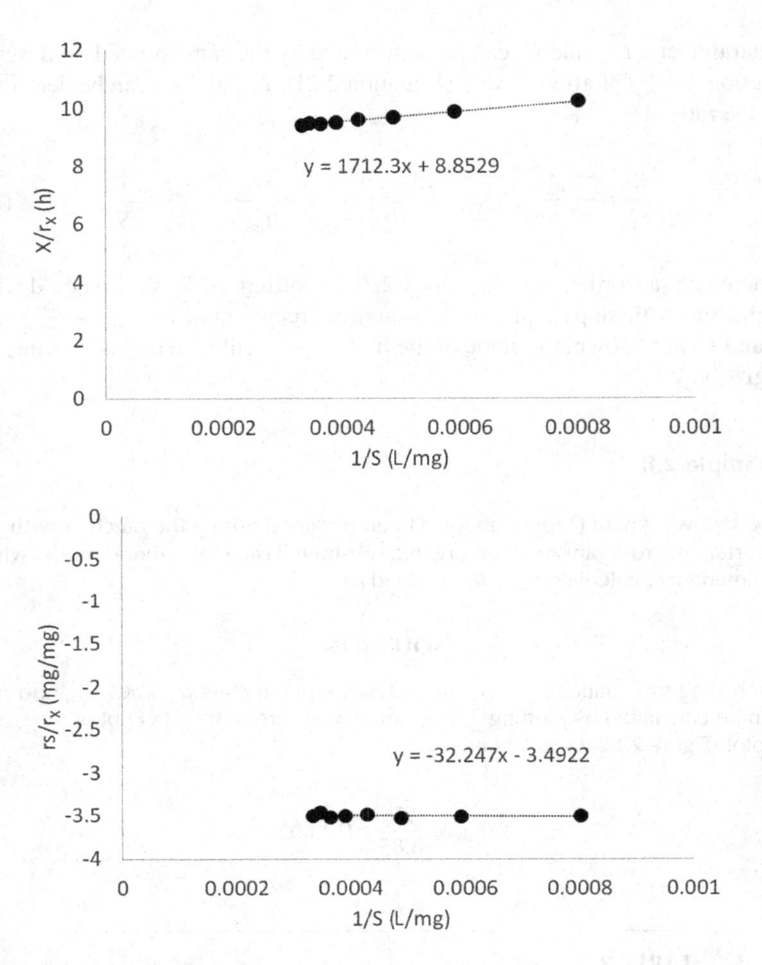

FIGURE 2.8 Linearisation of experimental data for Example 2.10.

From the slope of the plot, we calculate K_S:

$$K_S = 1712.3 \cdot \mu_{\max} = 188 \frac{mg}{L}$$

From the slope of the $\frac{r_S}{r_X}$ vs $\frac{1}{S}$ plot, we calculate m_S:

$$m_S = \frac{\mu_{\max}}{K_S} 32.25 = 0.019\ h^{-1}$$

From the intercept of the plot, we calculate $Y_{X/S}$:

$$Y_{X/S} = \frac{1}{3.49 - \frac{m_S}{\mu_{\max}}} = 0.30 \frac{kg}{kg}$$

Example 2.9

Calculate the coefficient $Y_{S/O2,m}$, which gives the ratio between substrate and oxygen consumed during maintenance metabolism, if the substrate is glucose.

SOLUTION

The coefficient $Y_{S/O2,m}$ can be calculated from the complete oxidation of glucose into carbon dioxide and water:

$$C_6H_{12}O_6 + 6O_2 \rightarrow 6CO_2 + 6H_2O$$

From this stoichiometry: $Y_{S/O2,m} = \frac{180}{6 \cdot 32} = 0.94 \frac{kg}{kg}$.

Another phenomenon that occurs during fermentations is endogenous metabolism. Endogenous metabolism is the consumption of cellular components by the microorganisms. Endogenous metabolism doesn't directly affect the rate of substrate removal but affects the rate of biomass growth:

$$r_{\text{end}}\left(\frac{kg}{m^3 d}\right) = -bX \tag{2.40}$$

In Equation (2.40), r_{end} is the rate of endogenous metabolism, i.e. the rate of consumption of the cellular components (which can be considered a death rate of the microorganisms), and b is an empirical parameter, dependent on temperature and on the type of microorganisms. The net rate of biomass growth is the algebraic sum of the biomass produced due to growth and the biomass consumed due to endogenous metabolism:

$$r_X = \mu X - bX \tag{2.41}$$

Under aerobic conditions, we can assume that endogenous metabolism gives the total oxidation of the biomass to carbon dioxide and water. Therefore, under aerobic conditions endogenous metabolism gives an oxygen consumption, written as

$$r_{\text{O2biom,end}} = \frac{-bX}{Y_{X/O2,\text{end}}} \tag{2.42}$$

The coefficient $Y_{X/O2,\text{end}}$ can be calculated from the stoichiometry of the total oxidation of the microorganisms. The total oxygen consumption is therefore given as the sum of oxygen consumption due to growth and due to endogenous metabolism:

$$r_{\text{O2biom}} = -\frac{\mu X}{Y_{X/O2}} - \frac{bX}{Y_{X/O2,\text{end}}} \tag{2.43}$$

In summary, if we assume that biomass growth and substrate removal are described by the endogenous metabolism model, the rate equations are:

$$r_X = \frac{\mu_{max} S}{K_S + S} X - bX \tag{2.44}$$

$$r_S = -\frac{\mu_{max} S}{(K_S + S)} \frac{X}{Y_{X/S}} \tag{2.5}$$

The kinetic parameters to be determined are now μ_{max}, K_S, $Y_{X/S}$ and b. They can be obtained by linearisation of $\frac{X}{r_S}$ vs $\frac{1}{S}$ and $\frac{r_X}{r_S}$ vs $\frac{1}{S}$.

The linearisation of $\frac{X}{r_S}$ vs $\frac{1}{S}$ gives:

$$\frac{X}{r_S} = -\frac{Y_{X/S}}{\mu_{max}} - \frac{Y_{X/S}}{\mu_{max}} K_S \frac{1}{S} \tag{2.45}$$

Plotting $\frac{X}{r_S}$ vs $\frac{1}{S}$ will give the values of $\frac{Y_{X/S}}{\mu_{max}}$ (intercept of the plot) and K_S (from the slope).

The linearisation of $\frac{r_X}{r_S}$ vs $\frac{1}{S}$ gives:

$$\frac{r_X}{r_S} = -Y_{X/S} + b\frac{Y_{X/S}}{\mu_{max}} + \frac{bK_S Y_{X/S}}{\mu_{max}} \frac{1}{S} \tag{2.46}$$

Once the values of $\frac{Y_{X/S}}{\mu_{max}}$ and K_S are known, the plot of $\frac{r_X}{r_S}$ vs $\frac{1}{S}$ will give the values of b (from the slope) and $Y_{X/S}$ (from the intercept). μ_{max} can then be calculated from $\frac{Y_{X/S}}{\mu_{max}}$.

Example 2.10

The following data (Table 2.8) have been obtained during the batch growth of a certain microorganism on an organic substrate. Using the Monod model with endogenous metabolism, calculate μ_{max}, K_S, $Y_{X/S}$ and b.

SOLUTION

According to Equations (2.45) and (2.46), the parameters μ_{max}, K_S, $Y_{X/S}$ and b can be calculated by plotting $\frac{X}{r_S}$ vs $\frac{1}{S}$ and $\frac{r_X}{r_S}$ vs $\frac{1}{S}$. The plots are shown in Figure 2.9.

From the intercept and slope of the $\frac{X}{r_S}$ vs $\frac{1}{S}$ plot, we obtain:

$$\frac{Y_{X/S}}{\mu_{max}} = 2.967 \frac{mg}{mg \cdot h} \text{ and } \frac{Y_{X/S}}{\mu_{max}} K_S = 567.28 \frac{mg}{L \cdot h}$$

From the intercept and slope of the $\frac{r_X}{r_S}$ vs $\frac{1}{S}$ plot, we obtain:

$$\frac{bK_S Y_{X/S}}{\mu_{max}} = 14.411 \text{ and } -Y_{X/S} + b\frac{Y_{X/S}}{\mu_{max}} = -0.359$$

TABLE 2.8
Data for Example 2.10

Time (h)	Substrate (*S*, mg/L)	Biomass (*X*, mg/L)
0	2,500	150
2	2,382	192
4	2,230	246
6	2,032	314
8	1,788	401
10	1,476	510
12	1,086	647
14	611	812
16	113	979
18	0	985

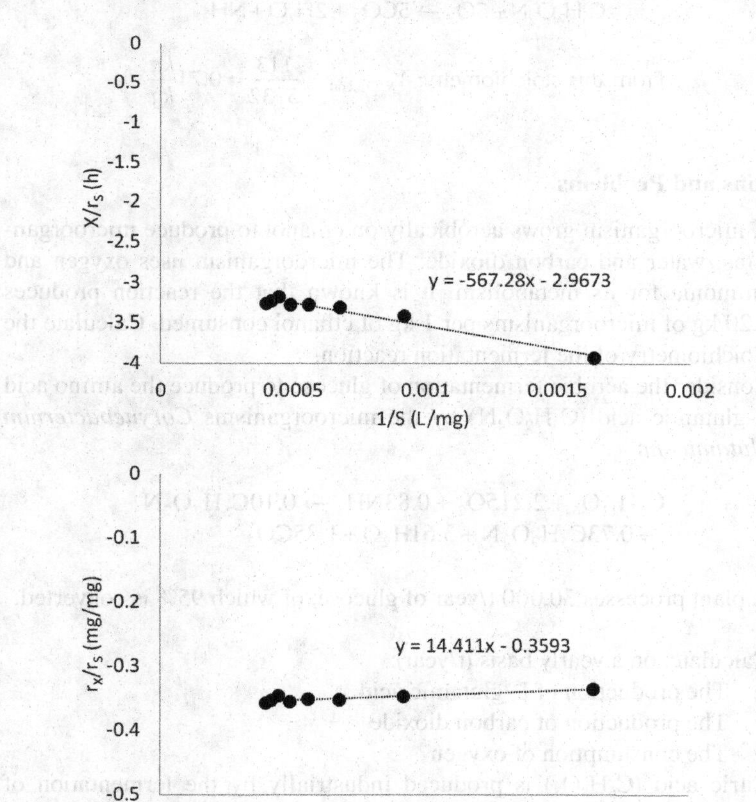

FIGURE 2.9 Linearisation of experimental data for Example 2.11.

Solving these four equations in the four unknowns μ_{max}, K_S, $Y_{X/S}$ and b, we obtain:

$$K_S = 191.2\ mg/L \quad b = 0.025\ h^{-1} \quad Y_{X/S} = 0.43\ kg/kg \quad \mu_{max} 0.14\ h^{-1}$$

Example 2.11

Calculate the coefficient $Y_{X/O2,end}$, which gives the ratio between biomass and oxygen consumed during endogenous metabolism. Assume that the empirical formula for the microorganisms is $C_5H_7O_2N$.

SOLUTION

The coefficient $Y_{X/O2,end}$ can be calculated from the complete oxidation of microorganisms into carbon dioxide and water:

$$C_5H_7O_2N + 5O_2 \rightarrow 5CO_2 + 2H_2O + NH_3$$

From this stoichiometry: $Y_{X/O2,end} = \dfrac{113}{5 \cdot 32} = 0.71 \dfrac{kg}{kg}$

Questions and Problems

2.1 A microorganism grows aerobically on ethanol to produce microorganisms, water and carbon dioxide. The microorganism uses oxygen and ammonia for its metabolism. It is known that the reaction produces 0.20 kg of microorganisms per 1 kg of ethanol consumed. Calculate the stoichiometry of the fermentation reaction.

2.2 Consider the aerobic fermentation of glucose to produce the amino acid L-glutamic acid ($C_5H_9O_4N$) by the microorganisms *Corynebacterium glutamicum*:

$$C_6H_{12}O_6 + 2.215O_2 + 0.83NH_3 \rightarrow 0.10C_5H_7O_2N + 0.73C_5H_9O_4N + 3.61H_2O + 1.85CO_2$$

A plant processes 30,000 t/year of glucose, of which 95% is converted.

Calculate, on a yearly basis (t/year):

- The production of L-glutamic acid
- The production of carbon dioxide
- The consumption of oxygen

2.3 Citric acid ($C_6H_8O_7$) is produced industrially by the fermentation of glucose using the microorganism *Aspergillus niger*. A simplified stoichiometry of the fermentation reaction is reported below:

$$C_6H_{12}O_6 + 2.35O_2 + 0.1NH_3 \rightarrow 0.1C_5H_7O_2N + 0.7C_6H_8O_7 + 1.3CO_2 + 3H_2O$$

It is desired to produce 10,000 tonnes per year of citric acid.

- Assuming that ammonia is provided as ammonium chloride (NH_4Cl), calculate the amount of ammonium chloride required per year.
- Calculate the amount of oxygen consumed by the microorganisms per year.
- Calculate the amount of carbon dioxide produced per year.

2.4 The microorganism *Bacillus subtilis* is used to produce the enzyme protease ($C_3H_5O_2N$). The fermentation reaction is aerobic and uses glucose as a substrate and ammonia as a nitrogen source. The products are microorganisms, the enzyme, water and carbon dioxide. It is known that 60 g of carbon dioxide and 20 g of microorganisms are produced per 100 g of glucose consumed. Calculate the stoichiometry of the fermentation.

2.5 The growth of an anaerobic microorganisms which produces ethanol on glucose is described by the following stoichiometry:

$$C_6H_{12}O_6 + 0.15NH_3 \rightarrow 0.15C_5H_7O_2N + xC_2H_6O + 2.1CO_2 + 3H_2O$$

Determine the stoichiometric coefficient for ethanol, x.

2.6 A microorganism produces lactic acid ($C_3H_6O_3$) from the sugar lactose ($C_{12}H_{22}O_{11}$) present in liquid cheese whey (the fermentation is anaerobic). Assume that cheese whey contains 50 kg/m^3 of lactose and that the fermentation started with 80% of the working volume made of cheese way (the remaining 20% is for the inoculum and nutrients). During the fermentation, when 70% of the initial lactose has been consumed, 2 g/L of microorganisms have been produced. Calculate the rate of lactic acid production when the rate of lactose removal is 0.4 g/L·h. Assume that the reactants and products in this reaction are lactose, ammonia, microorganisms, lactic acid and water with no carbon dioxide production.

2.7 Assuming that, in the considered temperature range, the temperature dependency of μ_{max} is described by Equation (2.19), describe a series of experiments to determine the values of the parameters A and Ea in Equation (2.19), showing the equation used to obtain the parameter values.

2.8 The growth of a microorganism on a substrate is described by the Monod model with maintenance metabolism, with $Y_{X/S}$ = 0.2 kg/kg and m_S = 0.4 d^{-1}. Calculate the rate of substrate removal when the biomass concentration is 2.5 g/L and the rate of biomass formation is 1.7 g/L·h.

2.9 It is suspected that a substance present in the fermentation medium may inhibit the growth of the microorganisms. Describe a set of experiments to determine the K_I parameter in Equation (2.17). Assume that the parameters μ_{max}, K_S and $Y_{X/S}$ are known.

2.10 The maximum growth rate of a microorganism on a substrate has been measured at different temperatures (Table 2.9). Assuming that, in this temperature range, μ_{max} depends on temperature according to Equation (2.19), calculate the parameters A and Ea and calculate the μ_{max} when the microorganism is at 28°C.

TABLE 2.9
Data for Problem 2.10

Temperature (°C)	μ_{max} (d^{-1})
20	4.5
25	7.5
30	12.8
35	22.1

2.11 The initial rate of substrate removal and microorganisms' growth was measured (Table 2.10) at various initial concentrations. Assuming that the microorganisms' growth can be described by the Monod model with endogenous metabolism, use the data to calculate the parameters μ_{max}, K_S, $Y_{X/S}$ and b.

TABLE 2.10
Data for Problem 2.11

Initial X (kg/m³)	Initial S (kg/m³)	Initial r_X (kg/m³·d)	Initial ($-r_S$) (kg/m³·d)
1.0	0.5	6.1	17.9
1.5	0.8	9.5	27.9
2.0	1.2	12.8	37.7
2.5	0.2	13.5	39.7
1.5	0.2	8.1	23.8
1.5	0.05	5.1	15.2
1.5	0.01	1.6	5.6
0.5	1.0	3.2	9.3

2.12 It is known that the growth of a microorganism on a substrate can be described by the Monod model with endogenous metabolism. Calculate the total oxygen consumption by the microorganisms when the biomass concentration is 1.2 kg/m³ and the net growth rate is 0.3 g/L·h. Assume that the coefficient $Y_{X/O2} = 0.8$ kg/kg, $b = 0.15$ d^{-1} and $Y_{X/O2,end} = 0.71$ kg/kg.

3 Mass Balances and Design for Batch, Continuous and Fed-Batch Reactors

In this chapter, we will focus on the design of batch, continuous and fed-batch fermentation processes. We will cover how to calculate the required volume or residence time required for fermentation processes, and how to calculate the composition of the fermentation effluents. Mass balances are the key equations for the design of these processes.

Many processes can be carried out in any fermentation mode: batch, continuous or fed batch. Examples of batch processes are the fermentation to produce penicillin and other antibiotics and the production of citric acid by *Aspergillus* or *Candida* microorganisms. An example of continuous fermentation is the production of bioethanol by *Saccharomyces cerevisiae* strains, while the fermentative production of amino acids uses most commonly fed-batch processes (e.g. production of L-lysine with *Escherichia coli*).

3.1 GENERAL EQUATION FOR MASS BALANCES

Mass balances have the general form:

$$\text{Accumulation} = \text{Input} - \text{Output} + \text{Generation} - \text{Consumption} \tag{3.1}$$

Mass balances can be written for each component in the reaction medium or in the system under consideration. All the terms in the mass balance should have the units of a mass flow rate (e.g. kg/d). Accumulation represents the rate at which the component's mass in the system changes. Input and output represent the mass flow rate of the component that enters or leaves the system. Generation represents the rate of production of the component in the system and consumption the rate at which the component is consumed in the system. With the signs used in Equation (3.1), all the terms will be taken as positive.

This general equation takes different forms depending on whether we are considering a batch or a continuous reactor.

DOI: 10.1201/9781003217275-3

3.2 MASS BALANCES FOR BATCH REACTORS

In batch reactors (Figure 3.1), the inoculum of microorganisms, the substrate and the required nutrients are added at time 0. No continuous addition of feed and continuous removal of the products is carried out. The only exception is oxygen, in the case of aerobic fermentations, which is continuously sparged in the reactor.

For the batch reactor, the general equation for mass balances (3.1) becomes:

$$\text{Accumulation} = \text{Generation} - \text{Consumption} \tag{3.2}$$

With the rate equations introduced in Section 2.1 (Chapter 2), the mass balances for microorganisms, substrate and products in a batch reactor are:

$$\frac{dX}{dt}\left(\frac{kg}{m^3d}\right) = r_X = \mu \cdot X = \frac{\mu_{\max} S}{K_S + S} X \tag{3.3}$$

$$\frac{dS}{dt}\left(\frac{kg}{m^3d}\right) = r_S = -\frac{\mu \cdot X}{Y_{X/S}} = -\frac{\mu_{\max} S}{K_S + S}\frac{X}{Y_{X/S}} \tag{3.4}$$

$$\frac{dP}{dt}\left(\frac{kg}{m^3d}\right) = r_P = \mu X \cdot Y_{P/X} = \frac{\mu_{\max} S}{K_S + S} X \cdot Y_{P/X} \tag{3.5}$$

Equations (3.3)–(3.5) represent a system of three differential equations in the three unknowns $S(t)$, $X(t)$ and $P(t)$ and need to be solved simultaneously in order to calculate the profiles of microorganisms, substrate and products in a batch fermentation. In order to solve these equations, the parameters $\mu_{\max}$, K_S, $Y_{X/S}$ and $Y_{P/X}$ need to be known and the initial conditions $X = X_0$, $S = S_0$ and $P = P_0$ need to be specified, where X_0, S_0 and P_0 are the concentrations of microorganisms, substrate and products, respectively, at the start of the fermentation.

The general method to solve systems of differential equations using Microsoft Excel is shown in Appendix B.

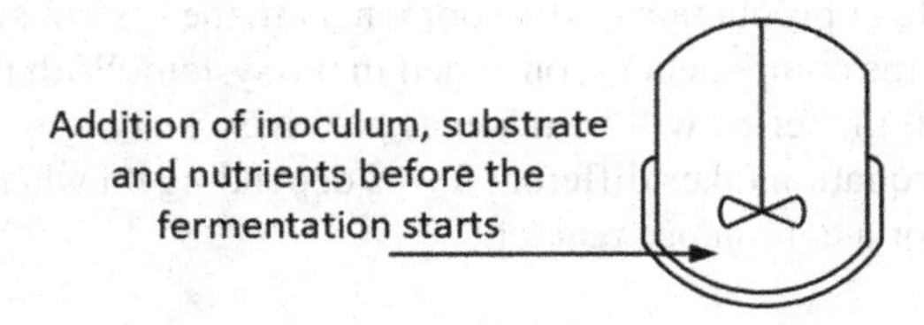

FIGURE 3.1 Batch reactor.

Example 3.1

Calculate the time profiles of substrate, microorganisms and product for a fermentation process characterised by the following parameters:

$$\mu_{max} = 5.2d^{-1};\ K_S = 0.05\ kg/m^3;\ Y_{X/S} = 0.15\ kg/kg;\ Y_{P/X} = 2.70\ kg/kg$$

The initial biomass, substrate and product concentrations are 0.1 kg/m³, 10 kg/m³ and 0 kg/m³, respectively.

SOLUTION

In this case, Equations (3.2)–(3.5) become (using units of kg for microorganisms, substrate and product, m³ for volume and d for time):

$$\frac{dX}{dt} = \frac{5.2S}{0.05+S}X$$

$$\frac{dS}{dt} = -\frac{5.2S}{0.05+S}\frac{X}{0.15}$$

$$\frac{dP}{dt} = \frac{5.2S}{0.05+S}X \cdot 2.70$$

The solution of these differential equations with the given initial conditions is reported in Figure 3.2.

In a batch process, we can also use the rate equations and mass balances to calculate the doubling time of microorganisms. The doubling time

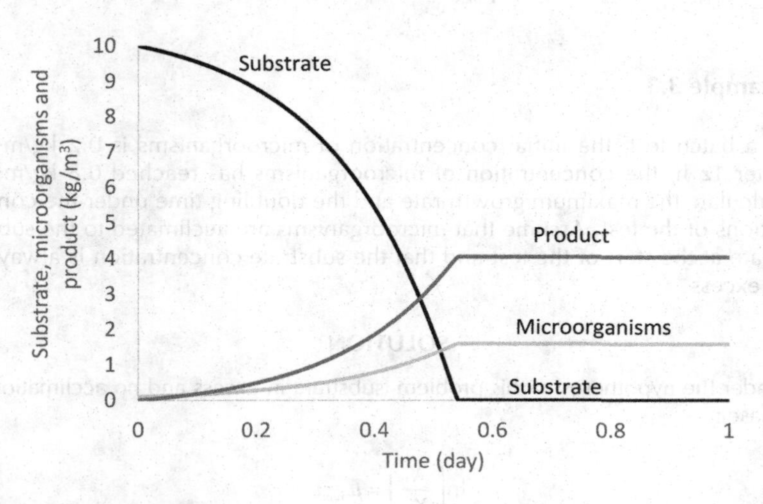

FIGURE 3.2 Solution of Example 3.1.

is the time the microorganisms need to double their concentration. The doubling time only refers to the condition where biomass is completely acclimated to the substrate and the substrate is in excess (i.e. $S \gg K_S$ and $\mu = \mu_{max}$). Under these hypotheses, the mass balance for the microorganisms is:

$$\frac{dX}{dt} = \mu_{max} X \tag{3.6}$$

Equation (3.6) can be integrated between X_0 and X and between 0 and t:

$$\int_{X_0}^{X} \frac{dX}{X} = \mu_{max} \int_{0}^{t} dt \rightarrow \ln\left(\frac{X}{X_0}\right) = \mu_{max} t \tag{3.7}$$

The doubling time (t_d) can be calculated from Equation (3.7), by imposing that $X = 2X_0$:

$$t_d = \frac{\ln 2}{\mu_{max}} \tag{3.8}$$

Example 3.2

A certain microorganism has a doubling time on a substrate of 2.5 h. Calculate the maximum growth rate of the microorganisms under these conditions.

SOLUTION

Re-arranging Equation (3.8):

$$\mu_{max} = \frac{\ln 2}{t_d} = 0.28\ h^{-1}$$

Example 3.3

In a batch test, the initial concentration of microorganisms is 0.2 kg/m^3. After 12 h, the concentration of microorganisms has reached 0.8 kg/m^3. Calculate the maximum growth rate and the doubling time under the conditions of the test. Assume that microorganisms are acclimated to the substrate at the start of the test and that the substrate concentration is always in excess.

SOLUTION

Under the hypotheses of this problem (substrate in excess and no acclimation phase):

$$\ln\left(\frac{X}{X_0}\right) = \mu_{max} t$$

Therefore:

$$\mu_{max} = \frac{1}{t}\ln\left(\frac{X}{X_0}\right) = \frac{1}{12}\ln\left(\frac{0.8}{0.2}\right) = 0.1155\ h^{-1}$$

and

$$t_d = \frac{\ln 2}{\mu_{max}} = 6\ h$$

3.3 MASS BALANCES FOR CONTINUOUS REACTORS

Continuous reactors are characterised by a continuous flow of feed and effluent. Continuous reactors are only inoculated at the start-up, but they require no further addition of microorganisms (the feed is sterile). Similarly as for batch reactors, we assume that continuous reactors are perfectly mixed (CSTR). This also means that the composition of the effluent is the same as inside the reactor (Figure 3.3).

Assuming we know the values of all the kinetic parameters that we need, in the design of continuous fermenters the main task is to determine the residence time that is required to achieve the desired conversion of the substrate and the desired concentration of microorganisms and products. The calculation of the residence time can be obtained by solving the mass balances for substrate, microorganisms and products.

Assuming that the fermenter is at steady state, the general form of the mass balance Equation (3.1) takes the form:

$$\text{Input-Output} = \text{Generation-Consumption} \tag{3.9}$$

Assuming the rate equations for substrate removal, microorganisms and product production seen in Chapter 2, the mass balances in a continuous fermenter are:

$$r_X \cdot V = Q \cdot X\ (\text{biomass balance}) \tag{3.10}$$

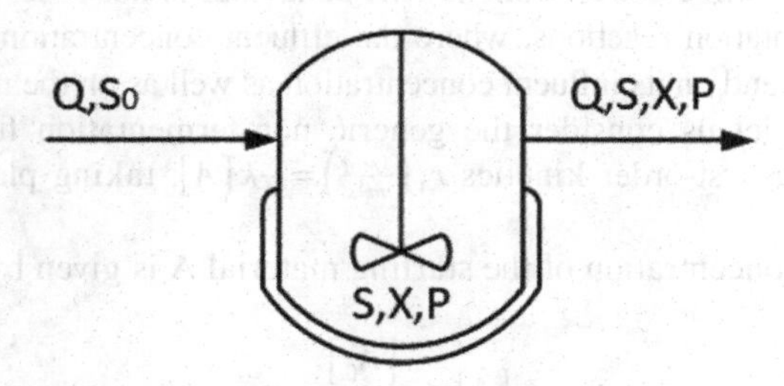

FIGURE 3.3 Scheme of a continuous fermenter.

$$Q \cdot S_0 = Q \cdot S + (-r_S) \cdot V \text{ (substrate balance)} \tag{3.11}$$

$$r_P \cdot V = Q \cdot P \text{ (product balance)} \tag{3.12}$$

Equations (3.10)–(3.12) can be re-arranged by introducing the dilution rate D, which is the reciprocal of the residence time τ:

$$D\left(\frac{1}{d}\right) = \frac{Q}{V} = \frac{1}{\tau} \tag{3.13}$$

Introducing Equation (3.13), Equations (3.10)–(3.12) become:

$$r_X = D \cdot X \tag{3.14}$$

$$D(S_0 - S) = (-r_S) \tag{3.15}$$

$$r_P = D \cdot P \tag{3.16}$$

Equations (3.14)–(3.16) can be used to calculate the dilution rate, or the residence time, that is required to obtain the desired concentrations of substrate, microorganisms and product in the effluent stream. Using the Monod kinetics defined in Chapter 2, Equation (3.14) becomes:

$$\mu = \frac{\mu_{\max} S}{K_S + S} = D \tag{3.17}$$

Equation (3.17) can be re-arranged as:

$$S = \frac{DK_S}{\mu_{\max} - D} \tag{3.18}$$

Equation (3.18) allows the calculation of the substrate concentration in a continuous fermenter and in the effluent stream as a function of the dilution rate if the kinetic parameters of the Monod equation are known. Equation (3.18) shows that, interestingly, the effluent substrate concentration in a continuous fermenter does not depend on the influent substrate concentration. This behaviour is not observed for conventional non-fermentation reactions, where the effluent concentration of the starting material does depend on its influent concentration as well as on the residence time.

For example, let us consider the generic non-fermentation first-order reaction $A \rightarrow B$, with first-order kinetics $r_A\left(\frac{kmol}{m^3 s}\right) = -k[A]$, taking place in a CSTR (Figure 3.4).

The effluent concentration of the starting material A is given by:

$$[A] = \frac{[A_0]}{1 + k\frac{V}{Q}} \tag{3.19}$$

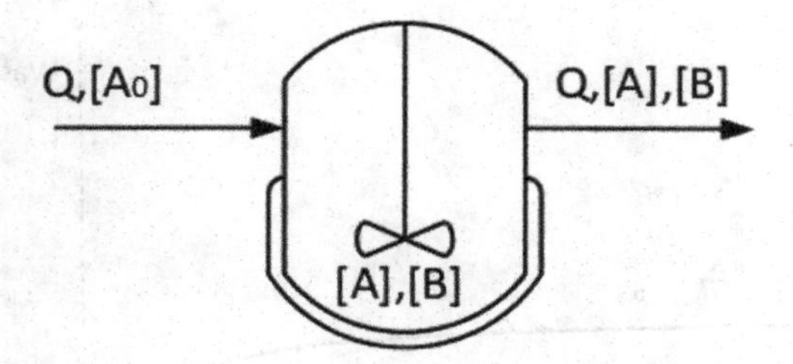

FIGURE 3.4 Scheme of a non-fermentation reaction taking place in a CSTR.

Equation (3.19), which is obtained from the mass balance of species A, can be compared with Equation (3.18), which is the analogous equation for fermentation reactions, showing that for non-fermentation reactions the effluent concentration of the starting material depends on its influent concentration, while for fermentation reactions it doesn't. The reason for this different behaviour is that in fermentation processes the rate of substrate removal is proportional to the concentration of the product, i.e. the microorganisms. If the substrate concentration in the feed increases, so does the concentration of microorganisms in the reactor (see also Equation 3.20), giving higher substrate removal rates and maintaining the same substrate concentration in the reactor.

The substrate and the product balances Equations (3.15) and (3.16) can also be re-arranged to calculate the biomass and the product concentration in the reactor and effluent:

$$X = Y_{X/S}\left(S_0 - S\right) \tag{3.20}$$

$$P = X \cdot Y_{P/X} \tag{3.21}$$

In summary, Equations (3.18), (3.20) and (3.21) give the concentrations *S*, *X* and *P* as a function of the dilution rate *D* (or of its reciprocal, the residence time) if the kinetic parameters are known.

The typical profiles of *S*, *X* and *P* as a function of *D* are shown in Figure 3.5.

Figure 3.5 can be explained as follows. Low values of *D* correspond to large values of the residence time and the substrate concentration is low because the microorganisms have a long time to grow on the substrate. Consequently, low values of *D* also correspond to the highest values of *X* and *P*. As *D* increases, and correspondingly the residence time decreases, *S* increases and *X* and *P* decrease. There is a maximum value of *D* for which the substrate concentration becomes equal to the influent concentration with no production of microorganisms or products. This condition ($D = D_{max}$) corresponds to the wash-out of the microorganisms, which cannot grow fast enough to be maintained in the reactor. The condition for the wash-out is obtained by using $S = S_0$ in Equation (3.17):

$$D_{max} = \frac{\mu_{max}}{K_S + S_0} \tag{3.22}$$

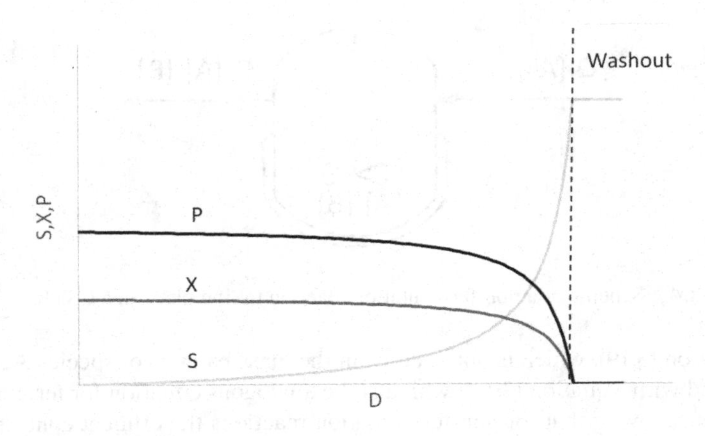

FIGURE 3.5 Typical profiles of X, P and S as a function of D.

A continuous fermenter needs to be operated with a value of D lower than D_{max} in order to avoid wash-out of the microorganisms. Since in many cases $S_0 \gg K_S$, then D_{max} is often very close to μ_{max}.

The mass balances Equations (3.18), (3.20) and (3.21) can also be used for the calculation of the biomass and product productivities as a function of the dilution rate. The productivity is the biomass or product produced per unit of reactor volume per unit of time.

$$\textit{Biomass productivity}\left(\frac{kg}{m^3 d}\right) = \frac{QX}{V} = DX = DY_{X/S}\left(S_0 - S\right) \tag{3.23}$$

$$\textit{Product productivity}\left(\frac{kg}{m^3 d}\right) = \frac{QP}{V} = DP = DXY_{P/X} \tag{3.24}$$

The productivity of biomass and product depends on the dilution rate. When the dilution rate is low, the productivity will also be low because the rate of biomass and product generation per unit volume is low. As the dilution rate increases, the productivity increases because the biomass and product production per unit volume increases. However, when the dilution rate approaches D_{max}, the productivity falls again because we are approaching wash-out of the microorganisms. It follows that the biomass and product productivities will have a maximum for a certain value of the dilution rate D. Figure 3.6 shows the typical profile of the biomass and product productivities as a function of the dilution rate.

Plots like the one in Figure 3.6 can be obtained by plotting Equations (3.23) and (3.24) as a function of D, once S, X and P have been obtained by the solution of Equations (3.18), (3.20) and (3.21).

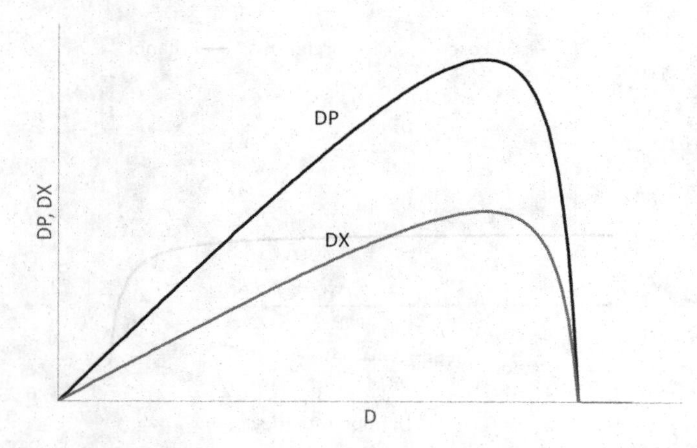

FIGURE 3.6 Typical profiles of biomass and product productivities in a continuous fermenter as a function of the dilution rate.

Example 3.4

The anaerobic fermentation of glucose to produce ethanol is carried out in a CSTR. The influent glucose concentration is 150 g/L. Calculate the profiles of microorganisms, glucose and ethanol as a function of the dilution rate. Also calculate the value of D which gives the highest ethanol productivity and the value of the maximum productivity. What is the volume of the reactor that gives the highest ethanol productivity if the feed flow rate is 1,000 m^3/d? What is the value of this productivity?

Stoichiometry of the reaction:

$$C_6H_{12}O_6 + 0.31NH_3 \rightarrow 0.31C_5H_7O_2N + 1.64C_2H_6O + 1.17CO_2 + 0.464H_2O$$

Kinetic parameters: $\mu_{max} = 5.3$ d^{-1} and $K_S = 3$ g/L.

SOLUTION

The profiles of glucose, microorganisms and ethanol are given by Equations (3.18), (3.20) and (3.21), respectively. In this case:

$$Y_{X/S} = \frac{0.31 \cdot 113}{180} = 0.195\frac{kg}{kg}$$

$$Y_{P/X} = \frac{1.64 \cdot 46}{0.31 \cdot 113} = 2.15\frac{kg}{kg}$$

Therefore, Equations (3.18), (3.20) and (3.21) become:

$$S = \frac{DK_S}{\mu_{max} - D} = \frac{D \cdot 3}{5.3 - D}$$

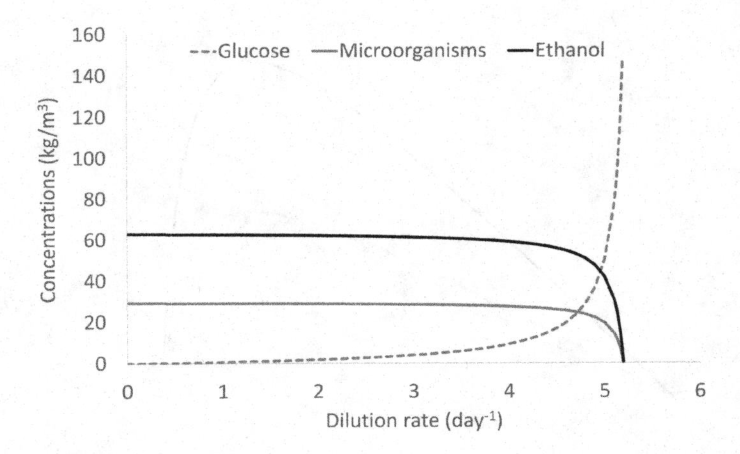

FIGURE 3.7 Results of Example 3.4.

$$X = 0.195(150 - S)$$

$$P = X \cdot 6.68$$

where S, X and P are in kg/m³ and D is in d⁻¹.

In this case, the maximum dilution rate, corresponding to the wash-out of the microorganisms, is:

$$D_{max} = \frac{5.3 \cdot 150}{3 + 150} = 5.196\ d^{-1}$$

The plot of glucose, microorganisms and ethanol concentrations as a function of the dilution rate is shown in Figure 3.7.

The ethanol productivity (=$D \cdot P$) can be calculated using Equation (3.24), and is plotted in Figure 3.8.

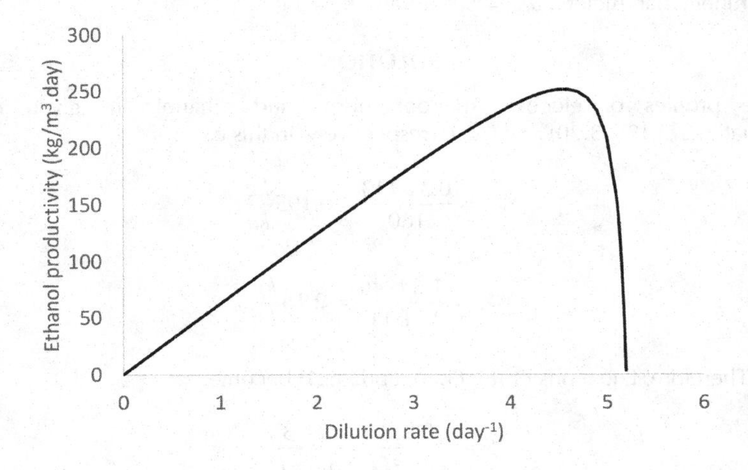

FIGURE 3.8 Ethanol productivity for Example 3.4.

From the graph, the highest ethanol volumetric productivity is obtained for $D = 4.6\ d^{-1}$. For the given feed flow rate, the reactor volume is:

$$V = \frac{Q}{D} = \frac{1{,}000\frac{m^3}{d}}{4.6d^{-1}} = 217.39m^3$$

For $D = 4.6\ d^{-1}$, the ethanol concentration is 54.6 kg/m³, therefore, the ethanol productivity is:

$$Q \cdot P = 1{,}000\frac{m^3}{d} \cdot 54.6\frac{kg}{m^3} = 54{,}600\frac{kg}{d}$$

3.4 MASS BALANCES FOR FED-BATCH REACTORS

The fed-batch reactor sits in the middle ground between the batch and the continuous reactor. In the fed-batch reactor, a volume of feed containing the substrate is fed to the reactor over a period of time. When the volume of the liquid (or liquid-solid mixture) reaches the maximum volume, a fraction of the volume is withdrawn bringing the volume back to the initial value, and a new cycle is started. The operational sequence of the fed-batch reactors is shown in Figure 3.9.

Let us consider what happens during the start-up of the fed-batch reactor. Let us assume that the reactor is inoculated with a low concentration of microorganisms when feeding starts. Initially, substrate concentration increases in the reactor, because the substrate feeding rate exceeds the substrate consumption rate, due to the low biomass concentration. During the cycle, biomass concentration increases, however, and, after the effluent withdrawal which marks the end of the first cycle, at the next cycle the substrate concentration will rise more slowly. After a few cycles, a "quasi-steady state" will be reached when the substrate concentration will be very low and the biomass (and products, if present) concentration will be approximately constant during the cycle.

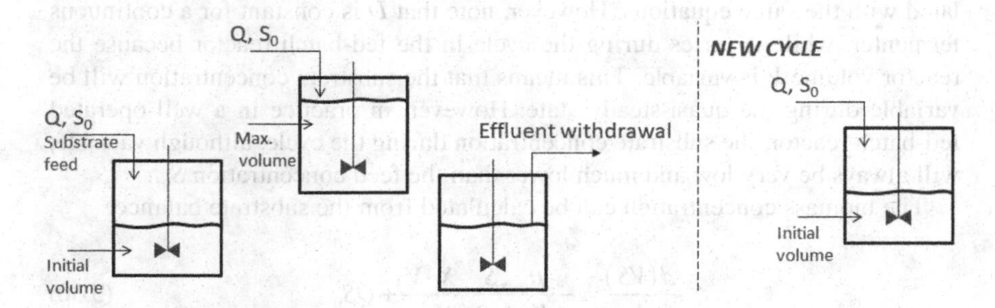

FIGURE 3.9 Operational sequence of the fed-batch reactor.

The calculation of substrate, biomass and product at the quasi-steady state of the fed-batch reactor can be done using mass balances. The mass balance for biomass in the fed-batch reactor is:

$$\frac{d(XV)}{dt} = \frac{\mu_{max} S}{K_S + S} XV \tag{3.25}$$

Equation (3.25) can be re-arranged as:

$$X\frac{dV}{dt} + V\frac{dX}{dt} = \frac{\mu_{max} S}{K_S + S} XV \tag{3.26}$$

Considering that

$$\frac{dV}{dt} = Q \tag{3.27}$$

Equation (3.26) becomes:

$$XQ + V\frac{dX}{dt} = \frac{\mu_{max} S}{K_S + S} XV \tag{3.28}$$

At the "quasi-steady state" it will be that: $\frac{dX}{dt} = 0$ and, therefore, Equation (3.28) will become:

$$\frac{Q}{V} = D = \frac{\mu_{max} S}{K_S + S} \tag{3.29}$$

Equation (3.29) is the same as Equation (3.17) for the continuous fermenter. From Equation (3.29), the concentration of the substrate at the quasi-steady state can be calculated using Equation (3.18):

$$S = \frac{DK_S}{\mu_{max} - D} \tag{3.18}$$

Equations (3.29) and (3.18) show that the substrate concentration in the fed-batch reactor at the quasi-steady state and in the continuous reactor can be calculated with the same equations. However, note that D is constant for a continuous fermenter, while it varies during the cycle in the fed-batch reactor because the reactor volume V is variable. This means that the substrate concentration will be variable during the quasi-steady state. However, in practice in a well-operated fed-batch reactor, the substrate concentration during the cycle, although variable, will always be very low and much lower than the feed concentration S_0.

The biomass concentration can be calculated from the substrate balance:

$$\frac{d(VS)}{dt} = -\frac{\mu_{max} S}{K_S + S} \cdot \frac{X \cdot V}{Y_{X/S}} + QS_0 \tag{3.30}$$

which becomes:

$$V\frac{dS}{dt}+S\frac{dV}{dt}=-\frac{\mu_{max}S}{K_S+S}\frac{X\cdot V}{Y_{X/S}}+QS_0 \tag{3.31}$$

Using Equation (3.27) and observing that at the quasi-steady state $\frac{dS}{dt}=0$, we obtain:

$$Q\left(S_0-S\right)=\frac{\mu_{max}S}{K_S+S}\frac{X\cdot V}{Y_{X/S}} \tag{3.32}$$

Note that Equation (3.32) is the same as Equation (3.15) seen for the continuous fermenter. Equation (3.32) can be re-arranged to give:

$$X=Y_{X/S}(S_0-S)\cong Y_{X/S}S_0 \tag{3.33}$$

Equation (3.33), which is essentially the same as Equation (3.20) already seen for the continuous fermenter, gives the biomass concentration at the quasi-steady state of the fed-batch reactor.

The product concentration at the quasi-steady state of the fed-batch reactor can be calculated with a similar procedure by starting with the mass balance for the product:

$$\frac{d(PV)}{dt}=\frac{\mu_{max}S}{K_S+S}\frac{X\cdot V}{Y_{X/S}}Y_{P/S} \tag{3.34}$$

With similar steps and assumptions as done for the biomass and substrate balance, the product balance Equation (3.34) can be re-arranged at the quasi-steady state as:

$$P=S_0\cdot Y_{P/S} \tag{3.35}$$

Equation (3.35) gives the product concentration at the quasi-steady state in the fed-batch reactor.

Note that Equations (3.18), (3.33) and (3.35) allow to calculate the concentrations of substrate, biomass and product in the fed-batch reactor at quasi-steady state only under the hypothesis that the influent flow rate Q is low enough to allow the microorganisms to maintain a very low substrate concentration ($S \ll S_0$) throughout the feeding cycle.

The biomass and product productivities can also be calculated for the fed-batch reactor. To calculate the product productivity, we need to consider that the total mass of microorganisms in the reactor, X_{tot}, during a cycle at the quasi-steady state is given by:

$$X_{TOT}\left(kg\right)=X\cdot V=Y_{X/S}S_0\left(V_0+Qt\right)=Y_{X/S}S_0V_0+Y_{X/S}S_0Qt \tag{3.36}$$

Obviously, X_{tot} increases during the cycle. The biomass productivity is calculated by taking the derivative of X_{tot} with respect to time, i.e.:

$$\frac{dX_{TOT}}{dt}\left(\frac{kgbiomass}{d}\right) = Y_{X/S}S_0Q \tag{3.37}$$

Similarly, for the product we can define the total mass of product in the reactor, P_{TOT}:

$$P_{TOT} = P \cdot V = Y_{P/S}S_0\left(V_0 + Qt\right) = Y_{P/S}S_0V_0 + Y_{P/S}S_0Qt \tag{3.38}$$

From Equation (4.33), the product productivity can be calculated as:

$$\frac{dP_{TOT}}{dt}\left(\frac{kgproduct}{d}\right) = Y_{P/S}S_0Q \tag{3.39}$$

Note that Equations (3.37) and (3.39) are valid under the same assumptions used to derive the quasi-steady state concentrations, i.e. that the feed flow rate is low enough to ensure that the substrate concentration in the reactor is always much lower than the feed concentration. The biomass and product productivities can also be expressed per unit of reactor volume as done in the case of the continuous fermenter.

Example 3.5

It is planned to produce ethanol by fermentation of glucose by the yeast *Saccharomyces cerevisiae* in a fed-batch reactor. The initial volume of the fermentation broth is V_0 and the final volume is V_{final}. The feed time is t_{feed} and the number of cycles that can be carried out in a year is n_{cycles}. The amount of ethanol which is required to be produced per year is $F_{ethanol}$.

a. Calculate the feed flow rate to the reactor and the required concentration of glucose in the feed.
b. The nitrogen source for microorganisms' growth is $(NH_4)_2SO_4$, which is purchased externally and added to the feed. Calculate the annual cost of $(NH_4)_2SO_4$ in order to supply the minimum amount of nitrogen necessary for the microorganism growth.

Data:

$V_0 = 10$ m^3 (initial volume in the fed-batch reactor)
$V_{final} = 50$ m^3 (final volume in the fed-batch reactor)
$t_{feed} = 2$ h (feed time in the fed-batch reactor)
$n_{cycles} = 2{,}000$ (cycles per year in the fed-batch reactor)
$F_{ethanol} = 1{,}840$ tonnes/year (required ethanol production rate in the fed-batch reactor)
Cost of $(NH_4)_2SO_4$ = £2/kg

Stoichiometry of microorganism growth and ethanol (C_2H_5OH) production:

$$C_6H_{12}O_6 + 0.12NH_3 \rightarrow 0.12C_5H_7O_2N + 1.8C_2H_5OH + 1.8CO_2 + 0.36H_2O$$

SOLUTION

From the fermentation stoichiometry, we get $Y_{P/S} = 0.46$ kg/kg.

a. The feed flow rate to the reactor is:

$$Q = \frac{V_{final} - V_0}{t_{feed}} = \frac{(50-10)m^3}{2h} = 20\frac{m^3}{h}$$

Since there are 2,000 cycles per year, this means that the feed flow rate per year is:

$$Q = 20\frac{m^3}{h}\cdot\frac{2h}{cycle}\cdot\frac{2{,}000\ cycle}{year} = 80{,}000\frac{m^3}{year}$$

The required glucose concentration in the feed is:

$$S_0 = \frac{F_{eth}}{Y_{P/S}Q} = \frac{1.84\cdot 10^6\frac{kg}{year}}{0.46\frac{kg}{kg}\cdot 80{,}000\frac{m^3}{year}} = 50\frac{kg}{m^3}$$

b. The amount of nitrogen required by microbial growth is:

$$Nitrogen\ consumed = 1.84\cdot 10^6\frac{0.12\cdot 14}{1.8\cdot 46} = 37{,}333\frac{kgN}{year}$$

The amount of ammonium sulphate required is equal to 37,333 × 132/28 = 176 × 10^3 kg $(NH_4)_2SO_4$/year.

At a cost of £2/kg, this corresponds to £352,000/year.

3.5 MODIFICATIONS OF MASS BALANCES FOR SECONDARY METABOLITES, MAINTENANCE AND ENDOGENOUS METABOLISM

The equations in previous sections refer to products which are primary metabolites, i.e. produced simultaneously to biomass growth. When the product is a secondary metabolite, there is no production during biomass growth, but there is production after growth has stopped. Secondary metabolites are usually produced in batch fermentation or, in some cases, in fed-batch processes, so in this section we will only see the mass balances for batch fermentation. Using Equations (2.31) and (2.32), the production rate of a secondary metabolite is:

$$\frac{dP}{dt} = \alpha X \tag{3.40}$$

or

$$\frac{dP}{dt} = Y_{P/X} r_X + \alpha X \tag{3.41}$$

Equation (3.41) applies when product production occurs both during biomass growth and when growth has stopped.

Another modification of the mass balances is due to the occurrence of maintenance metabolism. If we assume that biomass metabolism follows the maintenance model according to Equation (2.34) for the rate of substrate removal, then the substrate balance for batch fermentation will be:

$$\frac{dS}{dt} = -\frac{\mu \cdot X}{Y_{X/S}} - m_S X \tag{3.42}$$

The mass balances for biomass and product in batch fermentation will still be given by Equations (3.3) and (3.5).

In a continuous reactor, the maintenance metabolism modifies Equation (3.15) as follows:

$$D(S_0 - S) = \frac{\mu \cdot X}{Y_{X/S}} + m_S X \tag{3.43}$$

Equation (3.43) can be manipulated to obtain the biomass concentration at a steady state:

$$X = \frac{Y_{X/S}(S_0 - S)}{1 + \frac{m_S Y_{X/S}}{D}} \tag{3.44}$$

The comparison of Equations (3.20) and (3.44) indicates that, in a model with the maintenance metabolism, the steady-state biomass concentration in a continuous fermenter is lower than in a model without maintenance. The reason for this is that part of the substrate is used for the maintenance metabolism and not for biomass growth. For a continuous fermenter with maintenance metabolism, the steady-state substrate and product concentrations are still given by Equations (3.18) and (3.21). The substrate concentration will be the same as in the absence of maintenance, while the product concentration will be lower due to the lower biomass concentration.

If we assume the endogenous metabolism model, in a batch reactor the mass balances for the substrate and product will be given by Equations (3.4) and (3.5), while the mass balance Equation (3.3) modifies to:

$$\frac{dX}{dt} = \frac{\mu_{max} S}{K_S + S} X - bX \tag{3.45}$$

In a continuous reactor with endogenous metabolism, the mass balances for substrate and product will remain the same as Equations (3.15) and (3.16), while the biomass balance Equations (3.14) modifies into:

$$\mu \cdot X - bX = DX \tag{3.46}$$

Equation (3.45) can be re-arranged to give:

$$\mu = D + b \tag{3.47}$$

and

$$S = \frac{K_S(D+b)}{\mu_{max} - (D+b)} \tag{3.48}$$

Substituting Equation (3.47) into Equations (3.15) and (3.16) and re-arranging, we obtain the biomass and product concentrations in a continuous reactor for the model with endogenous metabolism:

$$X = \frac{DY_{X/S}(S_0 - S)}{(D+b)} \tag{3.49}$$

$$P = Y_{P/S}(S_0 - S) \tag{3.50}$$

The comparison of Equations (3.18) and (3.48) indicates that, in a model with endogenous metabolism, the steady-state substrate concentration in a continuous fermenter is higher than in a model without endogenous metabolism. The reason is that endogenous metabolism reduces the biomass concentration, therefore, increasing the substrate concentration.

Example 3.6

Compare the batch fermentation profiles in the absence of maintenance metabolism of Example 3.1, with the profiles in the same process with maintenance metabolism. Assume a maintenance parameter $m_S = 2.50$ d^{-1}, and the same parameter values as Example 3.1.

SOLUTION

In this case, the mass balances for substrate, biomass and product become:

$$\frac{dX}{dt} = \frac{5.2S}{0.05+S}X$$

$$\frac{dS}{dt} = -\frac{5.2S}{0.05+S}\frac{X}{0.15} - 2.5X$$

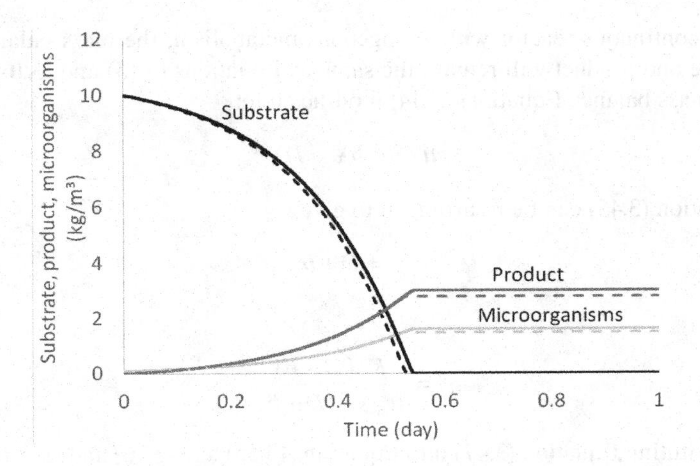

FIGURE 3.10 Solution of Example 3.6, comparison of batch fermentation profiles with (dashed lines) and without (solid lines) maintenance metabolism.

$$\frac{dP}{dt} = \frac{5.2S}{0.05+S} X \cdot 2.70$$

The solution of the mass balances is shown in Figure 3.10. The maintenance metabolism gives higher substrate removal rate with the same biomass and product concentration. The final concentration of biomass and product will be lower than in the absence of maintenance because part of the substrate is used for maintenance rather than for biomass and product formation. With typical values of the maintenance metabolism parameter, the effect of maintenance is quite modest, as shown in Figure 3.10.

Example 3.7

Compare the glucose, ethanol and biomass profiles vs D obtained for the fermenter in Example 3.4, with the profiles in a model with a maintenance coefficient equal to 2.5 d^{-1}. Assume the same stoichiometry as in Example 3.4.

SOLUTION

In a model with maintenance, the mass balances for glucose, ethanol and biomass become:

$$S = \frac{DK_S}{\mu_{max} - D} = \frac{D \cdot 3}{5.3 - D}$$

$$X = \frac{0.195(150 - S)}{1 + \frac{2.5 \cdot 0.195}{D}}$$

$$P = X \cdot 6.68$$

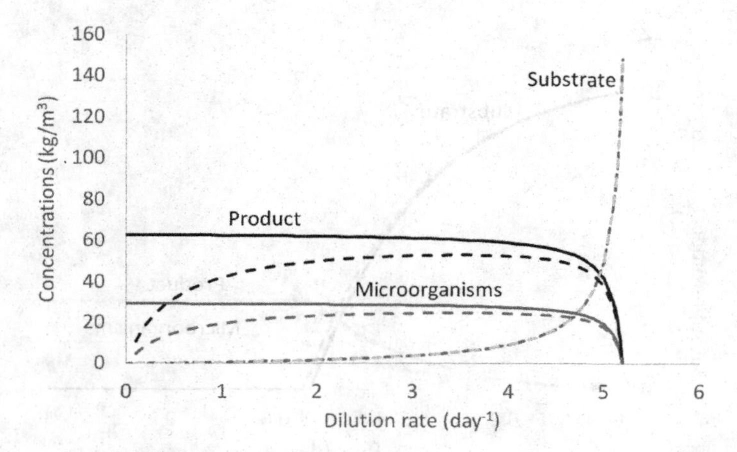

FIGURE 3.11 Solution of Example 3.7: comparison of continuous fermenter profiles with (dashed lines) and without (solid lines) maintenance metabolism.

The solution is shown in Figure 3.11. Maintenance metabolism causes a decrease in product and substrate concentrations, especially at the lowest dilution rate, where the growth is slowest.

Example 3.8

Compare the batch fermentation profiles of Example 3.1, with the profiles in the same process with endogenous metabolism. Assume an endogenous metabolism parameter $b = 0.2\ d^{-1}$, and the same parameter values as Example 3.1.

SOLUTION

In this case, the mass balances for substrate, biomass and product become:

$$\frac{dX}{dt} = \frac{5.2S}{0.05+S}X - 0.2X$$

$$\frac{dS}{dt} = -\frac{5.2S}{0.05+S}\frac{X}{0.15}$$

$$\frac{dP}{dt} = \frac{5.2S}{0.05+S}X \cdot 2.70$$

The solution of these differential equations is reported in Figure 3.12.

Example 3.9

Compare the glucose, ethanol and biomass profiles vs D obtained for the fermenter in Example 3.4, with the profiles in a model with endogenous

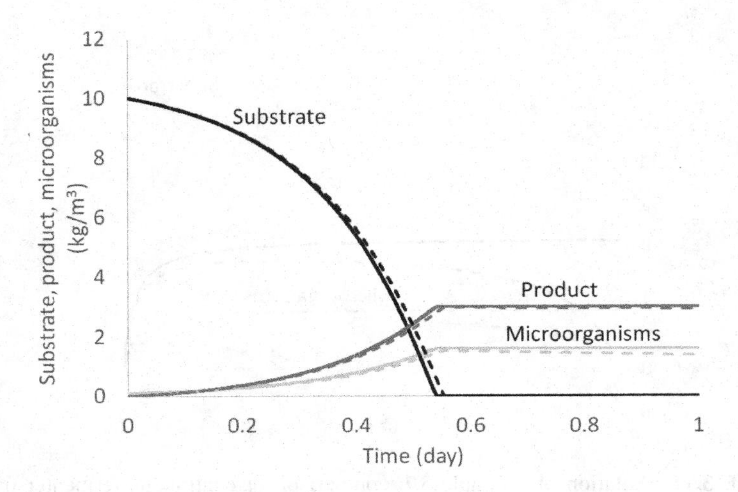

FIGURE 3.12 Solution of Example 3.8, comparison of batch profiles with (dashed lines) and without (solid lines) endogenous metabolism.

metabolism coefficient equal to 0.2 d^{-1}. Assume the same stoichiometry as in Example 3.4.

SOLUTION

The mass balances for glucose, ethanol and biomass give in this case:

$$S = \frac{3(D+0.2)}{5.3-(D+0.2)}$$

$$X = \frac{D \cdot 0.195(150-S)}{(D+0.2)}$$

$$P = 0.42(150-S)$$

The solution of these equations is shown in Figure 3.13.

3.6 USE OF CONTINUOUS REACTORS FOR THE ESTIMATION OF KINETIC PARAMETERS

In Chapter 2, we have seen that kinetic parameters can be estimated from batch tests. Another method is to use continuous cultures. For example, for a microorganism growing in a continuous reactor, the substrate concentration at steady state is obtained by solving Equation (3.17):

$$\frac{\mu_{\max} S}{K_S + S} = D \qquad (3.17)$$

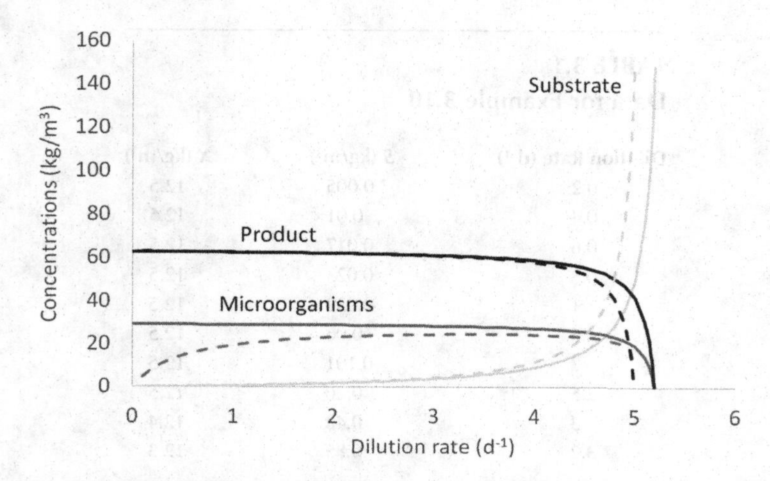

FIGURE 3.13 Solution of Example 3.9, comparison of fermenter profiles with (dashed lines) and without (solid lines) endogenous metabolism.

This equation can be re-arranged as follows:

$$\frac{1}{S} = \frac{\mu_{max}}{K_S}\frac{1}{D} - \frac{1}{K_S} \tag{3.51}$$

Equation (3.51) shows that collecting data of substrate concentration in a fermenter at different dilution rates allows for the calculation of the kinetic parameters μ_{max} and K_S. In particular, a plot of $\frac{1}{S}$ vs $\frac{1}{D}$ should give a straight line with $\frac{\mu_{max}}{K_S}$ as the slope and $-\frac{1}{K_S}$ as the intercept.

The parameter $Y_{X/S}$ can be calculated from Equation (3.52) at any dilution rate:

$$Y_{X/S} = \frac{X}{(S_0 - S)} \tag{3.52}$$

Assuming the microbial metabolism is described by the maintenance model, the parameters μ_{max} and K_S can still be determined using Equation (3.51), while the parameters $Y_{X/S}$ and m_S can be determined by re-arranging Equation (3.44) in a linearised form:

$$\frac{(S_0 - S)}{X} = \frac{1}{Y_{X/S}} + \frac{m_S}{D} \tag{3.53}$$

Equation (3.53) shows that a plot of $\frac{(S_0-S)}{X}$ vs $\frac{1}{D}$ should give a straight line with the slope equal to m_S and the intercept equal to $1/Y_{X/S}$. Note that if $m_S = 0$, Equation (3.53) coincides with Equation (3.52).

TABLE 3.1
Data for Example 3.10

Dilution Rate (d^{-1})	S (kg/m^3)	X (kg/m^3)
0.2	0.005	12.5
0.4	0.01	12.5
0.6	0.017	12.5
0.8	0.023	12.5
1	0.032	12.5
1.5	0.06	12.5
2	0.101	12.5
2.5	0.20	12.5
3	0.48	12.4
3.2	0.85	12.3

Example 3.10

The data in Table 3.1 have been obtained in a continuous fermenter at various dilution rates. Calculate the parameters μ_{max}, K_S and $Y_{X/S}$ (ignore maintenance or endogenous metabolism). The substrate concentration in the feed is 50 kg/m^3.

SOLUTION

μ_{max} and K_S can be calculated from the data in Table 3.1 using Equation (3.51). The plot is reported in Figure 3.14.

From the slope and intercept of Figure 3.13, we calculate $K_S = 0.097$ kg/m^3 and $\mu_{max} = 4.12$ d^{-1}. The parameter $Y_{X/S}$ can be calculated from Equation (3.52) using values at any of the dilution rates, or the average of these values.

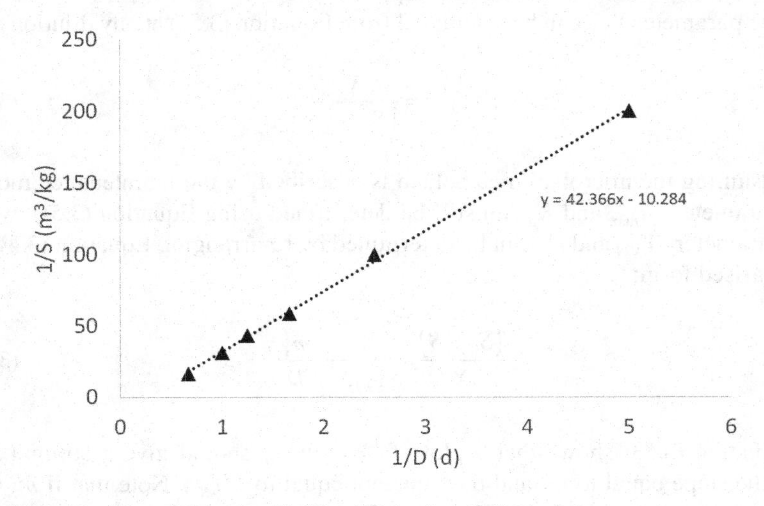

FIGURE 3.14 Plot of Equation (3.51) with the data in Table 3.1.

3.7 COMPARISON OF BATCH, CONTINUOUS AND FED-BATCH REACTORS

Fermentation reactions have some distinctive features in terms of reactor design when compared to non-fermentation reactions. For most non-fermentation reactions, the time required for a batch reaction to go to completion is lower than the residence time required for the same reaction in a continuous-flow reactor. This is because for most non-fermentation reactions, the reaction rate is proportional to the concentration of the reagents, and in batch reactors the reagents' concentration is higher than in continuous-flow reactors. However, for fermentation reactions, the reaction rate depends not only on the substrate concentration, but also on the microorganisms' concentration, and the microorganisms' concentration at the start of batch processes is typically low. For a given kinetics of the fermentation reaction, i.e. for given parameters μ_{max} and K_S, the rate of a batch fermentation process depends on the initial concentration of microorganisms. This is not the case for continuous and fed-batch fermentations at steady state (the "quasi-steady state" for the fed batch), where the concentration of microorganisms is constant with time and depends only on the substrate concentration in the feed but not on the microorganisms' concentration in the inoculum.

Figure 3.15 compares the average reaction rate for a batch fermentation (defined as the initial substrate concentration divided by the time required to

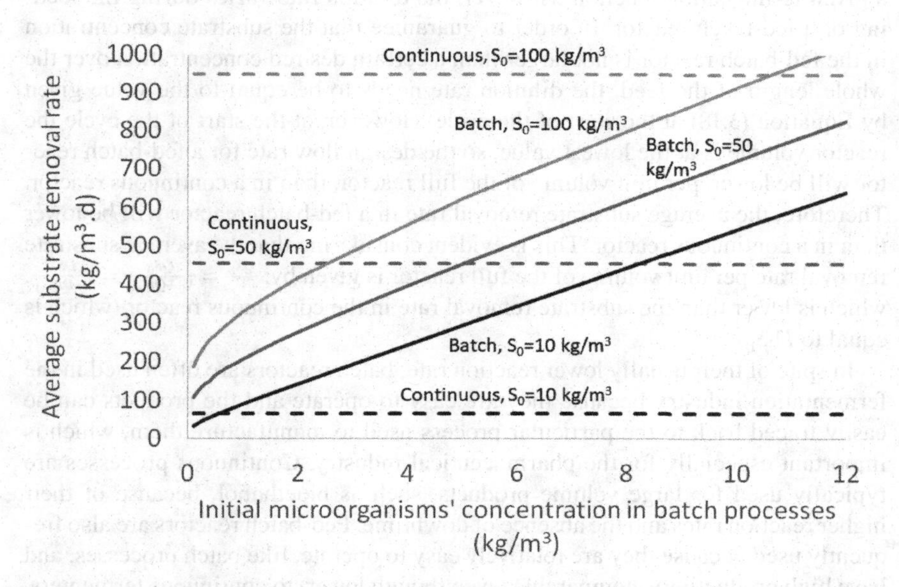

FIGURE 3.15 Comparison of the average substrate removal rate in batch and continuous reactors. The assumed kinetic parameters were: μ_{max}= 10 d^{-1}, K_S = 0.05 kg/m^3 and $Y_{X/S}$ = 0.2 kg/kg. Substrate was assumed to be completely removed when the concentration was 1% of the initial value. The rates for the continuous reactor were calculated for the dilution rate that gave an effluent substrate concentration equal to 1% of the feed concentration (these dilution rates were equal to 6.7, 9.1 and 9.5 d^{-1} for S_0 equal to 10, 50 and 100 kg/m^3, respectively).

achieve 99% removal of the initial substrate) at different initial microorganisms' concentrations with the reaction rate for the same fermentation in a continuous reactor. Depending on the initial microorganisms' concentration in the inoculum, the average substrate removal rate in a batch reactor can be higher or lower than in a continuous reactor, where the rate is independent of the initial inoculum concentration. The higher the initial feed substrate concentration, the more kinetically favoured the continuous reactor becomes compared to the batch, as shown by the higher initial inoculum concentration in the batch required to achieve the same rate as in the continuous fermenter. The reason is that in a continuous process the substrate removal rate is approximately linearly proportional to the substrate concentration in the feed (apart from the small effect of the substrate concentration on the optimum dilution rate), while in a batch reactor the initial rate only depends on the inoculum concentration and not on the substrate concentration (assuming that $S_0 \gg K_S$). In a batch fermenter, the substrate removal rate increases with time and this increase is higher at higher initial substrate concentration (because of the higher biomass concentration that will be obtained); however, the average substrate removal rate during the batch fermentation is less proportional to the initial substrate concentration than for a continuous reactor.

As far as the fed-batch reactor is concerned, the dilution rate that gives a certain effluent substrate concentration is the same as for the continuous reactor, for a given fermentation reaction. However, the dilution rate varies during the feeding of a fed-batch reactor. In order to guarantee that the substrate concentration in the fed-batch reactor is not larger than a certain desired concentration over the whole length of the feed, the dilution rate needs to be equal to the value given by Equation (3.18) at the start of the cycle. However, at the start of the cycle the reactor volume is at the lowest value, so the design flow rate for a fed-batch reactor will be lower, per unit volume of the full reactor, than in a continuous reactor. Therefore, the average substrate removal rate in a fed-batch reactor will be lower than in a continuous reactor. This is evident considering that the average substrate removal rate per unit volume of the full reactor is given by: $\frac{QS_0}{V} = \frac{QS_0}{V_0 + Qt_{feed}} = \frac{DS_0}{1 + Dt_{feed}}$, which is lower than the substrate removal rate in the continuous reactor (which is equal to $D \cdot S_0$).

In spite of their usually lower reaction rate, batch reactors are often used in the fermentation industry because they are easy to operate and the products can be easily traced back to the particular process used to manufacture them, which is important especially for the pharmaceutical industry. Continuous processes are typically used for large volume products, such as bioethanol, because of their higher reaction rates and the absence of downtime. Fed-batch reactors are also frequently used because they are relatively easy to operate, like batch processes, and have high productivity, comparable (even though lower) to continuous fermenters. Fed-batch reactors are particularly convenient when the process conditions have to change during the fermentation, for example, for secondary metabolites, which are only produced at the end of the growth phase when some nutrients become limiting or when there is need to add certain chemicals to trigger the excretion of the metabolites. In these cases, fed-batch reactors are suitable because the feed composition can change during the cycle, which is not easily possible in batch and continuous processes.

Questions and Problems

3.1 Calculate the profiles of substrate, microorganisms and product for the batch fermentation reaction below, where the product is citric acid. The microorganisms have a doubling time of 2.2 h and a K_S of 0.07 kg/m³. The batch is inoculated with a concentration of microorganisms of 400 mg/L and the initial concentration of glucose is 90 kg/m³.

$$C_6H_{12}O_6 + 2.53O_2 + 0.1NH_3 \rightarrow 0.1C_5H_7O_2N$$
$$+ 0.66C_6H_8O_7 + 1.54CO_2 + 3.16H_2O$$

3.2 A microorganism grows in a continuous fermenter. At the dilution rate of 1.0 d⁻¹, the effluent substrate concentration is 0.05 kg/m³. At the dilution rate of 1.3 d⁻¹, the effluent substrate concentration is 0.15 kg/m³. Calculate the values of μ_{max} and K_S for this microorganism.

3.3 A fermentation process is carried out in a fed-batch reactor, with an initial volume of 1 m³. The kinetic parameters of the microorganisms are μ_{max} = 9.6 d⁻¹ and K_S = 0.08 kg/m³. Calculate the maximum feed flow rate that allows for the substrate concentration, during feeding, to remain at any time below 0.1 kg/m³.

3.4 Consider the fermentation reaction for the production of glutamic acid represented by the stoichiometry below.

$$C_6H_{12}O_6 + 2.215O_2 + 0.83NH_3 \rightarrow 0.10C_5H_7O_2N$$
$$+ 0.73C_5H_9O_4N + 3.61H_2O + 1.85CO_2$$

The fermentation is carried out in a continuous reactor. Plot the biomass, substrate, glutamic acid concentrations and oxygen consumption per unit volume of reactor vs the dilution rate. Assume that the kinetic parameters for the microorganisms are μ_{max} = 8.5 d⁻¹ and Ks = 0.06 kg/m³. Assume that the influent feed concentration of the substrate is 100 g/L.

3.5 It is desired to produce 20,000 t/year of ethanol. The stoichiometry of the fermentation is reported below.

$$C_6H_{12}O_6 + 0.12NH_3 \rightarrow 0.12C_5H_7O_2N + 1.8C_2H_5OH + 1.8CO_2 + 0.36H_2O$$

The kinetic parameters of the microorganisms are μ_{max} = 12.1 d⁻¹ and K_S = 0.1 kg/m³. Glucose is available at a concentration of 120 kg/m³ and it is desired to convert at least 99% of it. Compare the volume of reactor required in the three cases:

a. Batch fermentation, starting with an inoculum concentration of 500 mg/L. The downtime between two consecutive batches is 2 h.

b. Continuous fermentation.
c. Fed-batch reactor. Assume that each feed is operated for 11 h, with 1 h downtime between batches.

Assume that the reactors operate all year round (365 d) without interruptions.

3.6 Assume the stoichiometry of citric acid production reported in Problem 3.1. The microorganisms have $\mu_{max} = 10.0\ d^{-1}$ and $K_S = 0.15\ kg/m^3$. Glucose is available at a concentration of 100 kg/m^3 and the desired effluent concentration is 0.5 kg/m^3. Calculate the annual production rate of citric acid per unit volume of reactor under the following process configurations:
a. Batch reactor inoculated with 200 mg/L of microorganisms and with a downtime of 3 h.
b. Continuous reactor.
c. Fed-batch reactor, with a feed length of 11 h and a downtime of 1 h.

3.7 A microorganism is grown in a continuous reactor. The following data (Table 3.2) are obtained as a function of the dilution rate. Assuming a model with maintenance metabolism, calculate the parameters $Y_{X/S}$ and m_S for this microorganism.

TABLE 3.2
Data for Problem 3.7

Dilution Rate (d^{-1})	S (kg/m^3)	X (kg/m^3)
0.1	0.001	8.3
0.5	0.006	23.5
1	0.013	31.2
2	0.03	36.5
3	0.054	39.2
4	0.09	40.1
5	0.15	41.3

3.8 For a certain microorganism, the effect of temperature on the maximum growth rate is described by Equation (2.19), with $Ea = 90{,}000$ J/mol. If the doubling time is 2 h at $T = 30°C$, what will be the doubling time at $T = 35°C$.

3.9 The data in Table 3.3 are obtained in continuous cultures of a microorganism at three different temperatures. Assuming the μ_{max} of the microorganisms depend on temperature according to Equation (2.19), calculate the parameters Ea and A for this microorganism.

TABLE 3.3
Data for Problem 3.9

Dilution Rate (d^{-1})	S, T = 20°C (kg/m^3)	S, T = 25°C (kg/m^3)	S, T = 30°C (kg/m^3)
0.2	0.004	0.002	0.001
0.4	0.008	0.004	0.002
1	0.023	0.01	0.005
2	0.064	0.023	0.01
3	0.16	0.04	0.017
4	0.64	0.064	0.024
5		0.1	0.032
6		0.16	0.042
7		0.28	0.053
8		0.64	0.068

3.10 Consider the anaerobic fermentation of sucrose ($C_{12}H_{22}O_{11}$) to produce lactic acid ($C_3H_6O_3$) by the microorganism *Lactobacillus delbrueckii*:

$$C_{12}H_{22}O_{11} + 0.576NH_3 \rightarrow 0.576C_5H_7O_2N + 3.04C_3H_6O_3 + 0.728H_2O$$

The microorganism grows on sucrose according to the Monod kinetics, with μ_{max} = 1.8 d^{-1} and K_S = 0.4 kg/m^3. The fermentation is carried out in a continuous fermenter. The sucrose concentration in the feed is 100 g/L and it is desired to achieve a sucrose concentration in the effluent of 3 g/L.

Calculate the volume of the reactor required to produce 15 t/d of lactic acid.

3.11 A fermentation process is carried out in two continuous reactors in series. The volumetric flow rate of the feed is 10 m^3/d and the volumes of the reactors are 5 m^3 (first reactor) and 7.5 m^3. The feed to the first reactor is sterile with no product and contains the substrate at a concentration of 100 g/L. The effluent of the first reactor has the following concentrations: Biomass = 5 g/L; Substrate = 3 g/L and Product = 35 g/L. It is known that the microorganisms follow Monod kinetics with μ_{max} = 8.0 d^{-1}.

Calculate:

a. The composition (biomass, substrate and product) at the outlet of the second reactor.
b. The volume which would be required to carry out the process in one single reactor, achieving the same final substrate concentration as in the final effluent of the two-stage process.

3.14 A fermentation process is carried out in a fed-batch reactor. The volume at the start of the cycle is 2 m^3 and the volume at the end of the feed is 10 m^3. The feed (substrate concentration 60 g/L) takes place over 5 h. The concentrations in the effluent are: Biomass 6 g/L and Product 20 g/L. It can be assumed that the substrate concentration is much lower than the feed concentration throughout the feed time. It is now desired to run the process in a chemostat with a feed of 80 g/L aiming to obtain the same effluent concentrations as obtained for the fed-batch reactor. Calculate the required residence time in the continuous reactor and the effluent concentrations of biomass and product.

3.15 A secondary metabolite is produced in a fed-batch reactor in two phases. In the first phase, the reactor is operated as a fed-batch with an initial volume of 1 m^3 and a final volume of 4 m^3. The feed takes place over a time of 3 h. The feed has a substrate concentration of 50 g/L and we assume the substrate concentration in the reactor is negligible at any time. The biomass yield is $Y_{X/S} = 0.15$ g/g. At the end of the feed phase, the pH is changed and the microorganisms start producing the desired product with a rate $r_P = 8.9X$, where r_P is in g/L.d and X is in g/L. The length of the product generation phase is 2 h. At the end of the product production, the effluent is discharged, and the cycle is started again after 1 h of downtime. Calculate the product concentration at the end of the cycle and the product productivity (kg/d). Assume that the process runs for 8,000 h per year.

3.16 A fermentation process is studied at lab-scale in continuous (working volume of 1 L) conditions. The first continuous experiment uses a flow rate of 3 L/d and a feed concentration of 30 g/L. This experiment obtains an effluent substrate concentration of 0.5 g/L, biomass concentration of 4.5 g/L and product concentration of 8.2 g/L. The second experiment uses a flow rate of 5 L/d and a feed concentration of 40 g/L, obtaining an effluent substrate concentration of 1.1 g/L.

Calculate the optimum product productivity (kg/m^3·d) and the product production rate (kg/d) under the conditions of optimum productivity for a full-scale reactor having a volume of 2 m^3 with a feed concentration of 50 g/L. For these conditions, also calculate the biomass concentration.

3.17 Assume that the growth of a microorganism on a substrate is described by the Monod model with maintenance metabolism. Two continuous experiments are carried out, with a substrate concentration in the feed of 5 g/L and with a reactor of 2 L working volume. In the first experiment, with flow rate 0.3 L/h, the biomass and substrate concentration in the reactor are 2.2 and 0.05 g/L, respectively. The second experiment is run with a flow rate of 0.5 L/h, obtaining a substrate concentration of 0.10 g/L and a biomass concentration of 2.4 g/L. Determine the parameters μ_{max}, K_S and m_S for this system.

3.18 A process is carried out in a fed-batch reactor. The initial volume before the feed start is 200 L and the feed concentration is 20 g/L. The kinetic parameters of the microorganisms on the substrate are $\mu_{max} = 7.3$ d^{-1} and $Ks = 0.1$ g/L. Calculate the maximum influent rate that can be used to ensure that, at the quasi-steady state, the substrate concentration is never higher than 1% of the feed concentration.

3.19 Derive the quasi-steady equations for the substrate, biomass and product concentration for a fed-batch reactor assuming the Monod model with (a) maintenance metabolism and (b) endogenous metabolism.

3.20 A process is carried out in a fed-batch reactor. The metabolism of the microorganisms can be described by the Monod model with maintenance with the following kinetic parameters: μ_{max} = 15.0 d^{-1}, K_S = 0.08 kg/m^3, m_S = 0.8 d^{-1}, $Y_{X/S}$ = 0.2 g/g and $Y_{P/S}$ = 0.3 g/g. It is desired that the substrate concentration never exceeds 1% of the feed concentration, which is 30 g/L. Calculate the maximum product and biomass productivity (in kg/d) that is possible to achieve in the fed-batch process, assuming that the initial volume is 1 m^3, the final volume is 10 m^3 and the downtime between end of the cycle and start of a new cycle is 0.5 h. Assume that the flow rate is constant throughout the feed period and that the process operates for 8,000 h/year.

4 Oxygen Transfer

The design of the aeration system is a critical step for aerobic fermentation processes. As examples of oxygen consumption and related energy consumption, cells of *Escherichia coli* growing on glucose can consume up to 0.5 kg·O_2/kg cell dry weight/h and the power requirements for oxygen delivery in fermenters can be higher than 2 kWh/kg·O_2. The aeration system must be designed so that it is able to supply the oxygen required by the microorganisms while minimising energy consumption and capital and operating costs.

4.1 THEORY

We have seen that aerobic fermentations consume oxygen and that the oxygen consumption rate by the microorganisms can be expressed by the following equation:

$$r_{O2biom}\left(\frac{kgO_2}{m^3d}\right) = -\frac{r_X}{Y_{X/O_2}} \tag{2.13}$$

Oxygen needs to be provided to the microorganisms by the aeration system. Typically, air or pure oxygen is pumped into the vessel using spargers (Figure 4.1).

Oxygen transfers from the bubbles to the liquid phase where it is used by the microorganisms for their metabolism. This section describes a theory to express the rate of mass transfer of a species from the gas to the liquid phase, or vice versa.

The mass transfer theory used in this book is based on the "two-film" model, which is schematised in Figure 4.2. In Figure 4.2, oxygen is used as the substance being transferred, but the model can be applied to any substance.

According to the "two-film" model, all the resistance to mass transfer is located in two boundary layers (films), located in the gas and in the liquid phases (gas and liquid films). Oxygen (or in general the species being transferred) is present in the bulk gas phase at a partial pressure p_{O2} (typically expressed in atm) and in the bulk liquid phase at a concentration C_{O2} (expressed in mass or mol units, e.g. mg/L, kg/m^3 or mol/L). In Figure 4.2, oxygen concentration in the bulk gas and bulk liquid phases is not in equilibrium, i.e. p_{O2} and C_{O2} are not in equilibrium with each other. If they were at equilibrium the rate of mass transfer would be 0. The two-film model assumes that at the interface between the liquid and gas phase oxygen concentration on the two sides of the interface is in equilibrium, i.e. $p_{O2,i}$ and $C_{O2,i}$ are in equilibrium with each other. We assume here that the equilibrium relationship is linear at all concentrations, i.e.

$$C_{O2,i} = k_{eq} \cdot p_{O2,i} \tag{4.1}$$

DOI: 10.1201/9781003217275-4

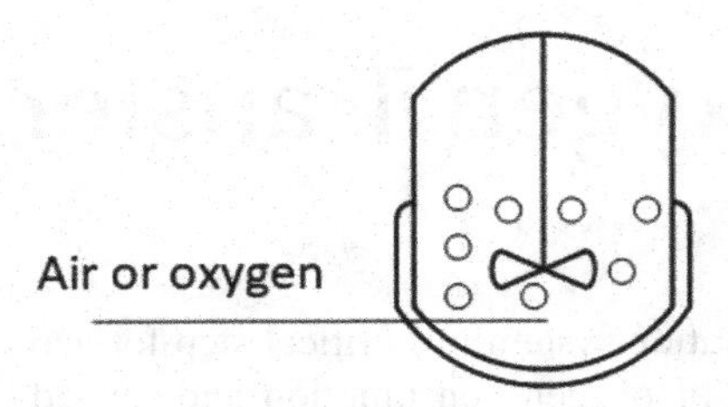

FIGURE 4.1 Conceptual scheme of an aerated fermenter.

Since, according to the model, the mass transfer resistance is only present in the gas and liquid films, the rate of mass transfer per unit of transfer area (or mass transfer flux), J_{O2}, can be expressed as:

$$J_{O2}\left(\frac{kg_{O_2}}{m^2 \cdot d}\right) = k_g \cdot \left(p_{O2} - p_{O2,i}\right) = k_l \cdot \left(C_{O2,i} - C_{O2}\right) \tag{4.2}$$

where k_g and k_l are the local mass transfer coefficients for the gas and liquid phases, respectively. Equation (4.2) defines the coefficients k_g and k_l. We assume that in the gas and liquid films, the movement of oxygen only occurs via molecular diffusion, i.e. oxygen transfer only occurs because of the concentration gradient. With this assumption, an expression for k_g and k_l can be obtained by comparing Equation (4.2) with the corresponding form of Fick's law of diffusion. The general form of Fick's law of diffusion is:

$$J_A = -D_{AB}\frac{dC_A}{dx} \tag{4.3}$$

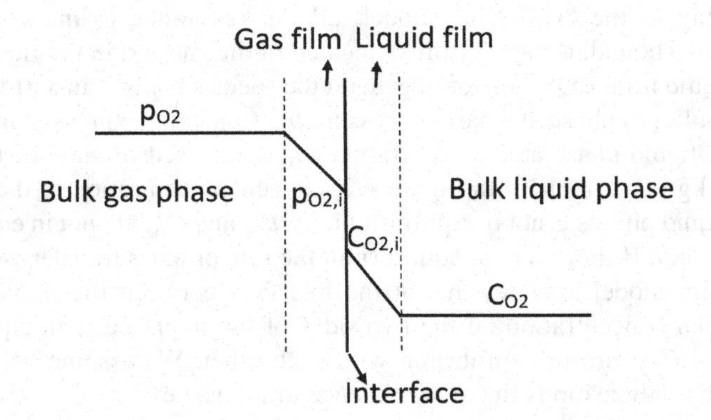

FIGURE 4.2 Scheme of the two-film model (oxygen is used as an example of the substance being transferred).

In Equation (4.3), J_A is the flux (kg/m²·d) of species A in species B, D_{AB} is the diffusivity of species A in species B and dC_A/dx is the concentration gradient of species A in the boundary layer. Applying Equation (4.3) to the flow of oxygen in the gas and liquid films:

$$J_{O2} = -D_{O2,g}\frac{dC_{O2,g}}{dx_g} = -D_{O2,l}\frac{dC_{O2,l}}{dx_l} \tag{4.4}$$

In Equation (4.4), $C_{O2,g}$ and $C_{O2,l}$ are the concentrations of oxygen (kg/m³) in the gas and liquid phases, respectively, $D_{O2,g}$ and $D_{O2,l}$ are the diffusivities of oxygen in the gas and the liquid phase, respectively, and x_l and x_g are the thickness of the gas and liquid films, respectively. It is more usual to represent the oxygen concentration in the gas phase as partial pressure rather than concentration. Using the ideal gas law, the relationship between partial pressure and concentration is:

$$C_{O2,g} = \frac{p_{O2} \cdot MW_{O2}}{RT} \tag{4.5}$$

Using Equation (4.5), Equation (4.4) becomes:

$$J_{O2} = -D_{O2,g} \cdot \frac{MW_{O2}}{RT}\frac{dp_{O2,g}}{dx_g} = -D_{O2,l}\frac{dC_{O2,l}}{dx_l} \tag{4.6}$$

Comparing Equation (4.6) with Equation (4.2), and assuming that the profile of oxygen in the gas and liquid films is linear, we obtain equations for k_g and k_l:

$$k_g = \frac{D_{O2,g}}{x_g} \cdot \frac{MW_{O2}}{RT} \tag{4.7}$$

$$k_l = \frac{D_{O2,l}}{x_l} \tag{4.8}$$

Equations (4.7) and (4.8) show that the local mass transfer coefficients are directly proportional to the oxygen diffusivity in the respective phases, and inversely proportional to the thickness of the films.

Equation (4.2) still does not allow for an easy calculation of the mass transfer rate because, even assuming that the mass transfer coefficients are known from the literature or from experiments, the concentrations at the interface, $p_{O2,i}$ and $C_{O2,i}$, are not known. It is desirable to express the mass transfer rate as a function of the known and easily measurable concentrations p_{O2} and C_{O2}. In order to do this, the concentration C^*_{O2} is introduced (Figure 4.3). C^*_{O2} is the oxygen concentration in the liquid phase that would be in equilibrium with the bulk gas phase, i.e.

$$C^*_{O2} = k_{eq} \cdot p_{O2} \tag{4.9}$$

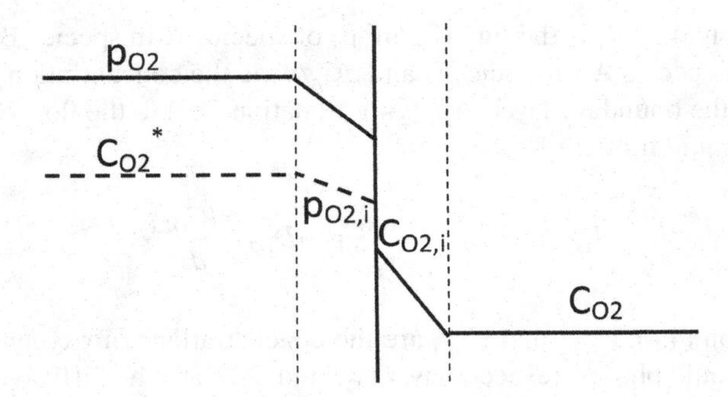

FIGURE 4.3 Two-film model with the introduction of the concentration C^*_{O2}.

While the concentrations p_{O2}, $p_{O2,i}$, $C_{O2,i}$ and C_{O2} exist physically in the system, the concentration C^*_{O2} is fictitious and does not exist physically. However, the advantage of introducing C^*_{O2} is that it can be easily calculated from p_{O2} if the equilibrium relationship is known. Introducing C^*_{O2}, the mass transfer flux can also be expressed as:

$$J_{O2}\left(\frac{kg_{O_2}}{m^2 \cdot day}\right) = k_L \cdot \left(C^*_{O2} - C_{O2}\right) = k_L \cdot \left(k_{eq} p_{O2} - C_{O2}\right) \tag{4.10}$$

Equation (4.10) represents the definition of k_L, the "global" mass transfer coefficient. Global coefficient means that it refers to both phases, while the local coefficients k_g and k_l refer to only one phase. The advantage of using Equation (4.10) over Equation (4.2) for expressing the mass transfer rates lies in the fact that in Equation (4.10) the variables p_{O2} and C_{O2} represent values in the bulk values which are readily measurable, while Equation (4.2) includes the interface values $p_{O2,i}$ and $C_{O2,i}$ which are not measurable.

The value of the global coefficient k_L can be related to the values of the local coefficients k_g and k_l. From Equations (4.1) and (4.2), we have:

$$k_g \cdot \left(p_{O2} - p_{O2,i}\right) = k_l \cdot \left(k_{eq} p_{O2,i} - C_{O2}\right) \tag{4.11}$$

From Equation (4.11), the value of $p_{O2,i}$ can be expressed as a function of the other variables:

$$p_{O2,i} = \frac{k_g p_{O2} + k_l C_{O2}}{k_g + k_{eq} k_l} \tag{4.12}$$

Introducing this value of $p_{O2,i}$ into $k_g\left(p_{O2} - p_{O2,i}\right)$, we obtain:

$$k_g \cdot \left(p_{O2} - p_{O2,i}\right) = k_g \cdot \left(p_{O2} - \frac{k_g p_{O2} + k_l c_{O2}}{k_g + k_{eq} k_l}\right) = \frac{k_g k_l}{k_g + k_{eq} k_l}\left(k_{eq} p_{O2} - C_{O2}\right) \tag{4.13}$$

By equalling Equation (4.13) with Equation (4.10), we obtain an expression for k_L as a function of k_g, k_l and k_{eq}:

$$k_L = \frac{k_g k_l}{k_g + k_{eq} k_l} \tag{4.14}$$

Equation (4.14) shows that the global coefficient k_L depends on the local coefficients k_g and k_l and on the equilibrium constant k_{eq}. Equation (4.14) shows that if the gas has a high solubility (high value of k_{eq}), then $k_L \approx k_g/k_{eq}$, i.e. the controlling resistance is on the gas side. On the other hand, if the gas has a low solubility (low value of k_{eq}), then $k_L \approx k_l$, i.e. the controlling resistance is on the liquid side. Oxygen is a gas with low solubility; therefore, in this case the controlling resistance is on the liquid side.

So far, we have expressed the mass transfer flux J, i.e. the rate of mass transfer per unit mass transfer area. In bioprocesses, we are interested in the volumetric mass transfer rate, i.e. mass transfer rate per unit volume of the biological reactor. We call the volumetric mass transfer rate (again using oxygen as an example) $r_{O2transfer}\left(\frac{kgO_2}{m^3 \cdot d}\right)$. $r_{O2transfer}$ and J_{O2} are related as follows:

$$r_{O2tranfer}\left(\frac{kg_{O_2}}{m^3 \cdot d}\right) = J_{O2}\left(\frac{kg_{O_2}}{m^2 \cdot d}\right)\frac{A}{V} \tag{4.15}$$

In Equation (4.15), A is the total area of the bubbles present in the reactor and V is the liquid volume in the reactor. We call the ratio $\frac{A}{V} = a$, where a represents the specific area of mass transfer, i.e. the mass transfer area per unit volume. Therefore, the desired volumetric mass transfer rate can be expressed as:

$$r_{O2}\left(\frac{kg_{O2}}{m^3 \cdot day}\right) = k_L \cdot a \cdot \left(C^*_{O2} - C_{O2}\right) = k_L a \cdot \left(k_{eq} p_{O2} - C_{O2}\right) \tag{4.16}$$

The two parameters k_L and a have a totally different physical meaning. k_L is related to the resistance to mass transfer, while a accounts for the number and size of the bubbles in the reactor. However, it is difficult with simple experimental means to determine the values of the parameters k_L and a independently. In general, the combined values of $k_L a$ are determined and, therefore, the term $k_L a$ is usually considered one single parameter, rather than two individual parameters. In Equation (4.16), the term $k_L a$ is the mass transfer coefficient and the term $\left(C^*_{O2} - C_{O2}\right)$ is the driving force.

Note that Equation (4.16) can be used to describe the mass transfer of any substance, not just oxygen, from the gas to the liquid phase (or vice versa). For example, if the substance being transferred is carbon dioxide, Equation (4.16) can be written as $r_{CO2transfer}\left(\frac{kg_{CO2}}{m^3 \cdot d}\right) = k_L a_{CO2}\left(k_{eqCO2} p_{CO2} - C_{CO2}\right)$ and similar equations describe the mass transfer for any other substances.

Example 4.1

Calculate the rate of mass transfer for oxygen in a fermenter when the concentration of dissolved oxygen is 2.0 mg/L. Repeat the calculation for a concentration of dissolved oxygen of 6.0 mg/L. Assume that the fermented is sparged with air with C^*_{O2} equal to 9.0 mg/L. The mass transfer coefficient is $k_La = 0.3\ min^{-1}$.

SOLUTION

When $C_{O2} = 2.0$ mg/L

$$r_{O2transfer} = k_La \cdot (C^*_{O2} - C_{O2}) = 0.3 min^{-1} \cdot (9.0 - 2.0)\frac{mg}{L} = 2.1\frac{mg}{L \cdot min}$$

When $C_{O2} = 6.0$ mg/L

$$r_{O2tranfer} = k_La \cdot (C^*_{O2} - C_{O2}) = 0.3 min^{-1} \cdot (9.0 - 6.0)\frac{mg}{L} = 0.9\frac{mg}{L \cdot min}$$

This example shows that the mass transfer rate is higher when the dissolved oxygen concentration is lower, because the driving force is higher.

Example 4.2

When a fermenter is sparged with air, the oxygen mass transfer rate for a dissolved oxygen concentration of 2.0 mg/L is 5,000 kg/m³·d. Calculate the oxygen transfer rate when the same fermenter is sparged with pure oxygen rather than air. Assume that in air the partial pressure of oxygen is 0.21 atm. Assume that the k_La is the same with oxygen and air. Assume that the equilibrium relationship between oxygen in the gas phase and in the liquid phase is given by:

$$C^*_{O2} = 42.9 \cdot p_{O2} \text{ where } C^*_{O2} \text{ is in mg/L and } p_{O2} \text{ is in atm.}$$

SOLUTION

$$r_{O2transferair} = k_La \cdot \left(k_{eq}p_{O2air} - C_{O2}\right)$$

$$r_{O2transferoxygen} = k_La \cdot \left(k_{eq}p_{O2oxygen} - C_{O2}\right)$$

Therefore:

$$r_{O2transferoxygen} = r_{O2transferair}\frac{k_{eq}p_{O2oxygen} - C_{O2}}{k_{eq}p_{O2air} - C_{O2}} = 5{,}000\frac{kg}{m^3d} \cdot \frac{42.9\frac{mg}{L \cdot atm} \cdot 1atm - 2.0}{42.9\frac{mg}{L \cdot atm} \cdot 0.21atm - 2.0}$$

$$= 29{,}200\frac{kg}{m^3d}$$

This example shows that the mass transfer rate is higher when pure oxygen is used, because the driving force is larger.

4.2 MEASUREMENT OF MASS TRANSFER COEFFICIENTS

Many procedures have been reported in the literature to measure $k_L a$. The experiment described here is one of the easiest to measure the $k_L a$ for the oxygen-air system.

The procedure requires an agitated vessel, where air is provided by diffusers. The experimental procedure is very simple:

1. Water is added to the vessel to the desired level and a sensor for dissolved oxygen is inserted into the liquid.
2. Nitrogen is sparged to the vessel in order to bring down the concentration of dissolved oxygen in water. Nitrogen sparging is stopped when oxygen concentration decreases below approximately 2 mg/L.
3. Air (or pure oxygen) at known flow rate is sparged into the vessel. Oxygen concentration starts to rise and oxygen concentration is recorded as a function of time.
4. The experiment is terminated when oxygen concentration reaches a value which is close to the saturation value with the inlet gas.

The $k_L a$ for the system can be calculated by applying the mass balance for oxygen. Indeed:

$$\frac{dC_{O2}}{dt} = k_L a \cdot \left(C^*_{O2} - C_{O2}\right) \tag{4.17}$$

Integrating this equation between time 0, where the oxygen concentration is $C_{O2,in}$, and time t, with generic oxygen concentration C_{O2}, we obtain:

$$ln\left(C^*_{O2} - C_{O2}\right) = ln\left(C^*_{O2} - C_{O2,in}\right) - k_L a \cdot t \tag{4.18}$$

Therefore, from a plot of $\ln\left(C^*_{O2} - C_{O2}\right)$ vs t, the value of $k_L a$ can be calculated as the slope of the line.

It is important to observe that calculation of $k_L a$ based on this procedure is based on several assumptions:

- Both the gas and the liquid phase are perfectly mixed.
- The dynamics of the oxygen probe are negligible compared to the dynamics of oxygen transfer.
- If air is used as gas, the amount of oxygen that transfers to the liquid phase is negligible compared to the amount of oxygen that is fed to the vessel. This ensures that oxygen partial pressure in the bubbles in the vessel is the same as in the atmosphere, and so C^*_{O2} is known and constant during the experiment.

This procedure can be modified to measure the $k_L a$ in the presence of microorganisms. In this case, the fermenter is filled with the suspension of water and active microorganisms at the desired concentration. When the dissolved oxygen concentration is high and close to the maximum value, aeration is stopped and the decrease in oxygen concentration due to the microbial metabolism is recorded. When the oxygen concentration has dropped sufficiently, aeration is started again and the increase in oxygen concentration is recorded. From the mathematical analysis of the data, the k_La can be calculated.

Another method to measure k_La is the sulphite oxidation method. In this method, a concentrated solution of sodium sulphite (Na_2SO_3) is added into the fermenter. In the presence of dissolved oxygen, sulphite is converted into sulphate (Na_2SO_4) with a very fast reaction. The rate at which oxygen transfers into the liquid phase can be determined by measuring the rate of sulphite oxidation into sulphate. From the rate of oxygen transfer, the k_La can be calculated easily.

In the methods for k_La measurement based on the measurement of the dissolved oxygen, oxygen sensors of various types (e.g. polarographic, optical) are used. In some methods, oxygen concentration in the gas phase is measured using mass spectroscopic analysis.

Example 4.3

An experiment is carried out to measure the mass transfer coefficient for oxygen in water. The experimental data are reported in Table 4.1. Calculate the mass transfer coefficient. It is known that $C^*_{O2} = 9.10 \frac{mg}{L}$.

TABLE 4.1
Experimental Data for Example 4.3

Time (min)	Dissolved Oxygen (mg/L)
0	1.69
1	3.36
2	4.67
3	5.90
4	6.68
5	7.26
6	7.72
7	8.09
8	8.36
9	8.57
10	8.73
11	8.90
12	8.96
13	9.01

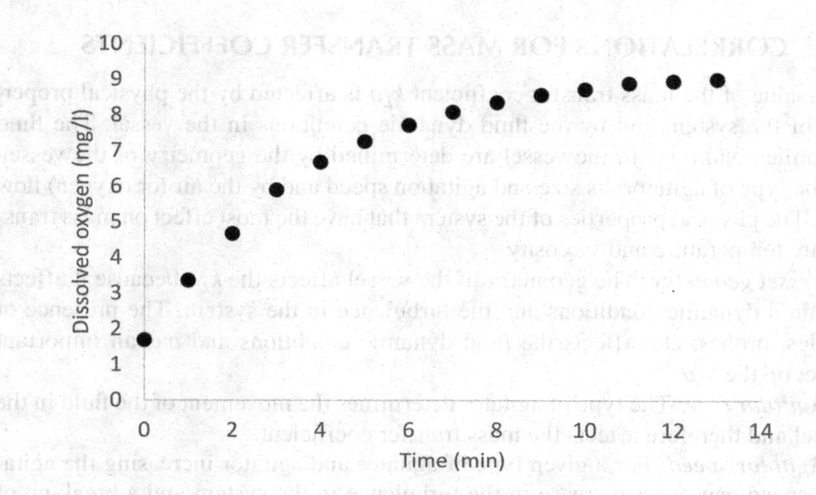

FIGURE 4.4 Plot of the experimental data in Example 4.3.

SOLUTION

The plot of the experimental data in Table 4.1 is shown in Figure 4.4.

To calculate the k_La, the data need to be correlated using Equation (4.18):

$$ln\left(C^*_{O2} - C_{O2}\right) = ln\left(C^*_{O2} - C_{O2,in}\right) - k_La \cdot t$$

The plot of $\ln\left(C^*_{O2} - C_{O2}\right)$ vs time is shown in Figure 4.4. From Figure 4.5, k_La can be obtained from the slope of the curve. In this case $k_La = 0.34$ min^{-1}.

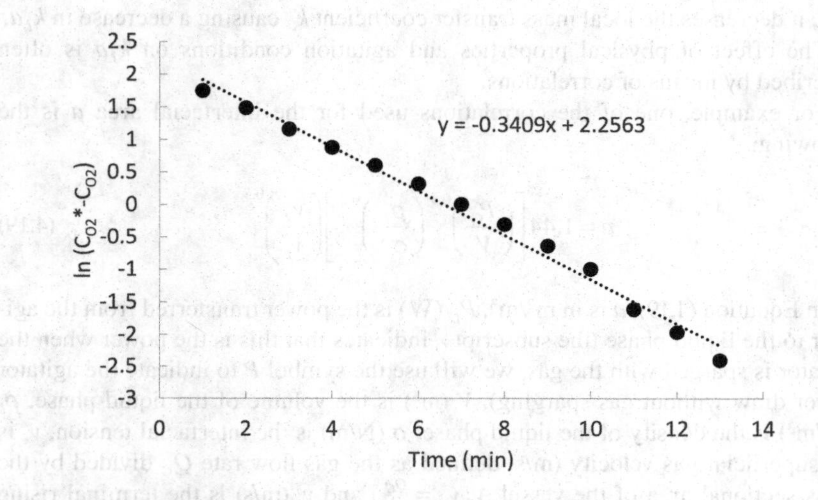

FIGURE 4.5 Plot of $\ln\left(C^*_{O2} - C_{O2}\right)$ vs time for Example 4.3.

4.3 CORRELATIONS FOR MASS TRANSFER COEFFICIENTS

The value of the mass transfer coefficient k_La is affected by the physical properties of the system and by the fluid dynamic conditions in the vessel. The fluid dynamic conditions in the vessel are determined by the geometry of the vessel, by the type of agitator, its size and agitation speed and by the air (or oxygen) flow rate. The physical properties of the system that have the most effect on mass transfer are temperature and viscosity.

Vessel geometry: The geometry of the vessel affects the k_La because it affects the fluid dynamic conditions and the turbulence in the system. The presence of baffles, probes, etc. affects the fluid dynamic conditions and has an important effect on the k_La.

Agitator type: The type of agitator determines the movement of the fluid in the vessel and therefore affects the mass transfer coefficient.

Agitator speed: For a given type of agitator and agitator increasing the agitation speed causes an increase in the turbulence in the system and a break-up of larger gas bubbles into smaller ones. Therefore, increasing the agitator speed will increase both the k_L and the a terms of the coefficient k_La; however, the effect is more important for a.

Air (or oxygen) flow rate: Increasing the gas flow rate increases the number and/or size of bubbles in the system and it therefore increases the a term of the k_La coefficient.

Temperature: Increasing the temperature increases the mobility of the molecules in the liquid phase, which corresponds to an increase in the diffusivity $D_{O2,l}$. In addition, higher temperatures correspond to larger gas bubbles, for the same gas flow rate on a mass basis, and this increases the a term. Therefore, higher temperature corresponds to an increase in k_La.

Viscosity: Higher viscosity increases the size of the liquid and gas films, therefore, it decreases the local mass transfer coefficient k_l, causing a decrease in k_La.

The effect of physical properties and agitation conditions on k_La is often described by means of correlations.

For example, one of the correlations used for the interfacial area a is the following:

$$a = 1.44\left[\left(\frac{P_G}{V}\right)^{0.4}\left(\frac{\rho_L}{\sigma^3}\right)^{0.2}\right]\left(\frac{v_g}{v_t}\right)^{0.5} \tag{4.19}$$

In Equation (4.19), a is in m^2/m^3, P_G (W) is the power transferred from the agitator to the liquid phase (the subscript $_G$ indicates that this is the power when the agitator is sparged with the gas, we will use the symbol P to indicate the agitator power draw without gas sparging), V (m^3) is the volume of the liquid phase, ρ_L (kg/m^3) is the density of the liquid phase, σ (N/m) is the interfacial tension, v_g is the superficial gas velocity (m/s, defined as the gas flow rate Q_G divided by the cross-sectional area of the vessel A, $v_g = \frac{Q_G}{A}$) and v_t (m/s) is the terminal rising velocity of a single bubble. v_t depends on the bubble diameter, which also in turn depends on the agitator power per unit volume, on the gas flow rate and on the physical properties. Correlations for v_t are available in the literature, however, for approximate calculations v_t can be assumed equal to 0.21 m/s.

Equation (4.19) shows that the main parameters that affect the interfacial area a are the agitator power P_G and the gas flow rate (included in the term v_g). As expected, higher values of the power draw per unit volume and of the gas flow rate give higher values of a.

There are also correlations for the mass transfer coefficient k_l. For gas liquid systems in most cases, the gas has low solubility in the liquid phase, so $k_l \approx k_L$. A correlation for k_l is given by Equation (4.20):

$$Sh_L = 2.0 + 0.31\left(\frac{d^3_{\text{bubbles}}\rho_L g}{\mu_L D_{g,L}}\right)^{\frac{1}{3}} \tag{4.20}$$

where Sh_L is the Sherwood number for the liquid phase, defined as: $Sh_L = \frac{k_l \cdot d_{bubbles}}{D_{g,L}}$. In Equation (4.20), k_l is obtained in m/s, d_{bubbles} (m) is the diameter of the gas bubbles, g (m/s^2) is the gravity acceleration, μ_L (Pa·s) is the viscosity of the liquid phase, $D_{g,L}$ (m^2/s) is the diffusivity of the gas in the liquid phase and ρ_L (kg/m^3) is the density of the liquid phase.

Equation (4.20) shows that the main physical property that affects k_l is the diffusivity $D_{g,L}$. k_l increases as the diffusivity increases. This is expected based on the mass transfer theory presented in Section 4.1 (Equation 4.8). k_l is lower for higher values of the liquid viscosity, because higher viscosity corresponds to a larger thickness of the boundary layer (term x_l in Equation 4.8). k_l is essentially independent of the diameter of the bubbles, this means that k_l is independent of the power draw according to this correlation.

For scale-up considerations, it is often useful to measure the mass transfer coefficient in a small vessel, and then use correlations to calculate the coefficient for the large vessel to be used in the commercial process. In these scale-up studies, the physical properties of the gas and liquid phase need to be the same between the small and large vessels. Furthermore, the small and large vessels need to be geometrically similar. Agitated vessels are geometrically similar if they have the same type of agitator and the ratio between all the geometrical dimensions is the same. For example, the ratio between the diameter of the agitator (D_{ag}) and of the vessel (D) should be the same in the large and small vessels:

$$\left(\frac{D_{ag}}{D}\right)_{\text{small}} = \left(\frac{D_{ag}}{D}\right)_{\text{large}} \tag{4.21}$$

If the conditions of same physical properties and geometrically similar vessels are verified, a correlation frequently used in scale-up (or scale-down) studies is the following:

$$k_L a = k\left(\frac{P_G}{V}\right)^a v_g^b \tag{4.22}$$

In Equation (4.22), the variables have the same meaning as in Equation (4.19), i.e. P_G is the agitator power transferred to the liquid phase, V is the volume of the liquid phase and v_g is the superficial gas velocity. Correlations like Equation (4.22) can be used to calculate the value of the mass transfer coefficient for a large vessel once the value of k_La for a smaller vessel is known. The parameter a in Equation (4.22) can be determined by measuring k_La for a range of $\frac{P_G}{V}$ values, keeping v_g constant, and plotting $\ln(k_La)$ vs $\ln\left(\frac{P_G}{V}\right)$. A similar procedure can be used to determine b, by measuring k_La for a range of v_g values, keeping $\frac{P_G}{V}$ constant, and plotting $\ln(k_La)$ vs $\ln(v_g)$. Once a and b are determined, the parameter k in Equation (4.22) can also be calculated from the intercepts of the $\ln(k_La)$ vs $\ln\left(\frac{P_G}{V}\right)$ or $\ln(k_La)$ vs $\ln(v_g)$ plots. For each experiment, the k_La can be determined using the method described in Section 4.2. The $\frac{P_G}{V}$ value can be varied by changing the agitation speed (see Section 4.3.1) and the v_g can be changed by manipulating the gas flow rate.

Correlation (4.22) is valid in the most common case that the fluid dynamic behaviour in the fermenter is, with good approximation, Newtonian, i.e. with the viscosity independent of the shear rate. However, in some cases, due to type of microorganisms present, to their high concentration or to the production of extracellular polymeric substances, the fluid dynamic behaviour in the fermenter is better approximated by the pseudoplastic model, where the viscosity decreases as the shear rate increases. Examples of processes with pseudoplastic behaviour are the production of polyglutamic acid by *Bacillus subtilis*, some fermentation processes with filamentous fungi for the production of pharmaceutically important products, production of alginate and xanthan by various microorganisms. In some cases, the pseudoplastic behaviour is due to the fermentation product (e.g. xanthan or alginate), in other cases is due to the microorganisms. For examples, filamentous fungi at high concentrations tend to become entangled, increasing the viscosity of the medium at low shear rates. When fermentations with filamentous fungi are carried out in batch, initially the rheology of the medium is approximately Newtonian because of the low biomass concentration, however, when the fungi grow and their concentration increases, the viscosity increases and the rheology tends to become pseudoplastic. In these cases, generally, the mass transfer coefficient is lower than when the fluid is Newtonian because of the bubble coalescence in zones with lower shear rate (e.g. in zones far from the agitator). When the fluid behaviour is well described by the pseudoplastic model, Equation (4.22) can be adapted with an additional term which includes the average viscosity in the reactor, calculated empirically, and which reduces the k_La.

Equation (4.22) is no longer valid if the gas flow rate is too high and the agitator becomes unable to disperse the gas in the vessel. This condition is called flooding and is obtained for high gas flow rates combined with low agitator speed. The gas flow rate and the agitator speed in a bioreactor need to be chosen so that flooding is avoided, because flooding leads to low value of the mass transfer coefficient. There are empirical correlations which allow to estimate the gas flow rate that corresponds to the onset of flooding, one example is Equation (4.23). The maximum gas flow rate that we can have in the vessel should be lower than the value of Q_{gF} given by Equation (4.23):

$$Q_{gF} = K\frac{N^3 D_{ag}^{7.5}}{g D^{3.5}} \tag{4.23}$$

In Equation (4.23), Q_{gF} is the gas flow rate at the onset of flooding (m^3/s), N is the agitation speed (s^{-1}), g is the gravitational acceleration (9.8 m/s^2) and D_{ag} and D are the diameter (m) of the agitator and of the vessel, respectively. K is a constant that depends on the type of agitator, it can be assumed equal to 30 for Rushton turbines and equal to 170 for Chemineer BT-6 agitators.

So far, we have discussed the effect of these parameters on the mass transfer coefficient k_La. However, the rate of mass transfer is given by the product of the mass transfer coefficient and of the driving force (Equation 4.16). The driving force is mainly affected by the temperature and pressure of the system. Higher temperature decreases the solubility of gases in liquids, so C^*_{O2} decreases as the temperature increases. Higher pressure increases the solubility of gases in liquids and, therefore, increases the driving force. The composition of the inlet gas used for aeration also affects the driving force. The saturation concentration of oxygen in water, C^*_{O2}, depends on the gas composition and it is the higher, the higher the oxygen content in the gas. If we use atmospheric air, where the oxygen content is approximately 21% in volume, C^*_{O2} is in the approximate range 7–9 mg/L (depending on the process temperature), while if we use pure oxygen, C^*_{O2} is in the approximate range 30–45 mg/L. Therefore, clearly using pure oxygen increases the driving force and the overall mass transfer rate compared to the use of atmospheric air.

4.3.1 Correlations for Power Draw in Agitated Vessels

In order to use Equations (4.19) and (4.22), we need to know the power that the agitator transfers to the liquid phase. This power depends on the agitator type, size and stirring speed and on the physical properties of the fluids. The physical properties that mostly affect the power draw of agitators are density and viscosity of the liquid phase. The agitator properties that affect the power draw are the type of agitator, its size and its rotational speed. For each type of agitator, the power draw is obtained by experimental correlations, available in the literature that relates the power number (P_0) to the Reynolds number (Re).

$$P_0 = \frac{P}{\rho_L N^3 D_{ag}^5} \tag{4.24}$$

$$Re = \frac{\rho_L N D_{ag}^2}{\mu_L} \tag{4.25}$$

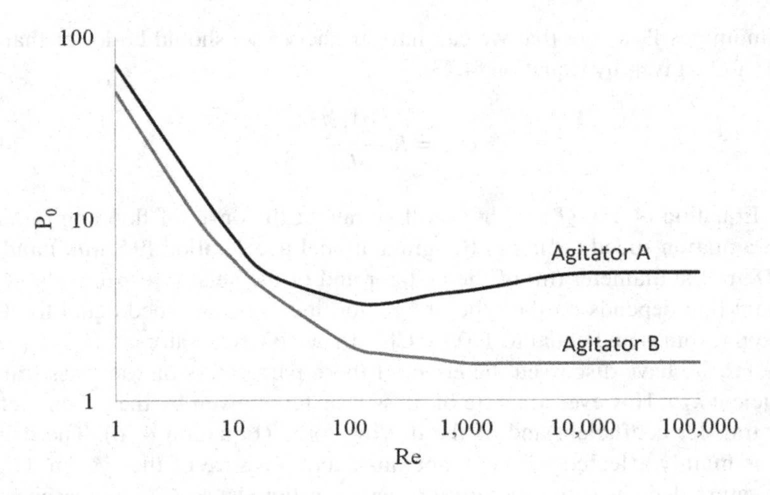

FIGURE 4.6 Typical profiles P_0 vs Re of power curves.

In Equations (4.24) and (4.25), in addition to the symbols already defined, N is the agitator speed. Experimental data indicate that for each type of agitator there is a certain power curve. A power curve (Figure 4.6) is the relationship between P_0 and Re.

If the power curve for the considered agitator is known, then the power draw of the agitator can be calculated immediately as a function of the properties of the fluid (density and viscosity) and as a function of the agitator speed and size.

Once the power draw with no gas flowing has been calculated, the power draw with gas flowing can be corrected using correlations such as Equations (4.26):

$$\frac{P_G}{P} = 1 - 12.2 \frac{Q_G}{ND_{ag}^3} \left(\frac{Q_G}{ND_{ag}^3} < 0.037 \right) \quad (4.26a)$$

$$\frac{P_G}{P} = 0.62 - 1.85 \frac{Q_G}{ND_{ag}^3} \left(\frac{Q_G}{ND_{ag}^3} > 0.037 \right) \quad (4.26b)$$

Equations (4.26a) and (4.26b) were obtained experimentally using disc flat-blade turbines. For this type of agitator, if $\frac{Q_G}{ND_{ag}^3} > 0.11$, the ratio P_G/P can be assumed to be constant and equal to 0.43.

The procedure below shows the calculation of the power draw for a given agitator if the agitator type, its geometry and agitation speed are known:

a. Obtain from the literature the power curve for the agitator you are interested in.
b. From the geometry of the agitator and its speed and from the physical properties of the liquid medium, calculate the Reynolds number (Equation 4.25).

c. With the Reynolds number and the power curve, calculate the Power Number P_0.
d. From the power number, calculate the power transferred from the agitator to the medium (Equation 4.24).

Equation (4.22) can also be used to calculate the power draw that is required for a certain agitator to achieve the desired value of the k_La. Once the power draw is known, power curves can be used to calculate the required agitation speed. In this case, we need an iterative procedure (for given values of the physical properties and of the agitator diameter): we set a value for N, calculate Re, P_G/P, P, then calculate P_0 and N, and repeat the procedure until convergence. In many cases, but not always, the agitation conditions in fermentations fall in the region where P_0 is approx. constant with Re (i.e. independent of Re). In addition, in some cases the ratio P_G/P can be assumed constant and independent of N (e.g. when the ratio $\frac{Q_G}{ND_{ag}^3}$ is higher than 0.11 for flat disc turbines). If P_0 and P_G/P are constant and known values, the procedure to calculate the agitation speed can be greatly simplified: from the required value of P_G, P is calculated. Then, the constant value of P_0 is used to calculate N.

Example 4.4

A fermenter has a diameter of 1.5 m and agitator with a diameter of 0.5 m. The fermenter is sparged with air with a flow rate of 4,000 m³/d. The level of the liquid phase in the reactor is 1.5 m. The power number of the agitator under ungassed conditions is 4.5. The agitator speed is 120 rpm.

Calculate the mass transfer coefficient k_La for oxygen in the fermenter. Assume $v_t = 0.21$ m/s and $d_{bubbles} = 5.0 \times 10^{-4}$ m.
Physical properties:

$$\mu_L = 8.9 \cdot 10^{-4} \frac{kg}{m \cdot s}$$

$$\sigma = 0.072 \frac{N}{m}$$

$$\rho_L = 1{,}000 \frac{kg}{m^3}$$

$$D_{O2,L} = 2.1 \cdot 10^{-9} \frac{m^2}{s}$$

SOLUTION

Equation (4.19) gives the specific surface area of the bubbles. From the power number, we calculate the ungassed power draw using Equation (4.24):

$$P = P_0 \rho N^3 D_{ag}^5 = 4.5 \cdot 1{,}000 \frac{kg}{m^3} \cdot \left(\frac{120 min^{-1}}{60\ s \cdot min^{-1}} \right)^3 (0.5m)^5$$

The gassed power draw is calculated from Equation (4.26). In this case:

$$\frac{Q_G}{ND_{ag}^3} = \frac{\left(\frac{1{,}000 \frac{m^3}{d}}{86{,}400 \frac{s}{d}}\right)}{2s^{-1}(0.5m)^3} = 0.046$$

so Equation (4.26b) will be used:

$$\frac{P_G}{P} = 0.62 - 1.85 \cdot 0.046 = 0.53$$

so:

$$P_G = 0.53 \cdot 1125W = 596W$$

The superficial gas velocity is:

$$v_g = \frac{Q_G}{A} = \frac{\left(\frac{1{,}000}{86{,}400}\right)\frac{m^3}{s}}{\frac{\pi 1.5^2}{4} m^2} = 0.0066 \frac{m}{s}$$

Therefore, the specific surface area of the bubbles is (Equation 4.19):

$$a = 1.44 \left[\left(\frac{596 \frac{kg \cdot m^2}{s^3}}{2.7m^3} \right)^{0.4} \left(\frac{1000 \frac{kg}{m^3}}{0.072 \frac{kg}{s^2}} \right)^{0.2} \right] \left(\frac{0.0066}{0.21} \right)^{0.5} = 14.90 \frac{m^2}{m^3}$$

The Sherwood number for the liquid phase coefficient k_l is given by Equation (4.20):

$$Sh_L = 2.0 + 0.31 \left(\frac{(5 \cdot 10^{-4} m)^3 \, 1{,}000 \frac{kg}{m^3} \cdot 9.81 \frac{m}{s^2}}{8.9 \cdot 10^{-4} \frac{kg}{m \cdot s} 2.1 \cdot 10^{-9} \frac{m^2}{s}} \right)^{\frac{1}{3}} = 28.94$$

From the definition of the Sherwood number:

$$k_l = \frac{Sh_L D_{O2,L}}{d_{bubbles}} = \frac{28.94 \cdot 2.1 \cdot 10^{-9} \frac{m^2}{s}}{5 \cdot 10^{-4} m} = 1.21 \cdot 10^{-3} \frac{m}{s}$$

Since $k_l \approx k_L$, the overall mass transfer coefficient k_La is:

$$k_La = 14.90 \frac{m^2}{m^3} \cdot 1.21 \cdot 10^{-3} \frac{m}{s} = 0.018 s^{-1}$$

Example 4.5

A fermenter has a diameter of 0.9 m and a volume of 1 m³. The fermenter is aerated with an air flow rate of 500 m³/d. The diameter of the agitator is 0.3 m and the agitator speed is 150 rpm. What agitation speed is required to have the same k_La in a geometrically similar vessel, having a volume of 10 m³? The large fermenter is aerated with an air flow rate of 5,000 m³/d. Assume that the power number of the agitator is 4.0 for both vessels.

For these vessels, the mass transfer coefficient is described by the correlation:

$$k_La(s^{-1}) = 0.0015\left(\frac{P_G}{V}\right)^{0.6}\left(\frac{Q_G}{A}\right)^{0.3} \text{ with } P_G/V \text{ in W/m}^3 \text{ and } Q_G/A \text{ in m/s.}$$

SOLUTION

The liquid height for the small fermenter is:

$$H = \frac{4V}{\pi D^2} = \frac{4 \cdot 1m^3}{\pi 0.9^2 m^2} = 1.6m$$

Therefore, the ratio between the liquid height and the diameter of the fermenter is, for the small fermenter:

$$\frac{H}{D} = \frac{1.6m}{0.9m} = 1.8$$

To be geometrically similar, the large fermenter will have the same ratio between the liquid height and the diameter of the fermenter:

$$\left(\frac{H}{D}\right)_{small} = \left(\frac{H}{D}\right)_{large}$$

Therefore, the diameter of the large fermenter will be:

$$D = \sqrt[3]{\frac{4V}{\pi\, 1.8}} = \sqrt[3]{\frac{4 \cdot 10m^3}{\pi\, 1.8}} = 1.9\ m$$

The superficial gas velocities for the small and large vessels are:

$$v_{g,small} = \left(\frac{Q_G}{A}\right)_{small} = \frac{250\frac{m^3}{d}}{\frac{\pi 0.9^2 m^2}{4}} = 393.1m/d = 0.0045m/s$$

$$v_{g,large} = \left(\frac{Q_G}{A}\right)_{large} = \frac{2{,}500\frac{m^3}{d}}{\frac{\pi 1.9^2 m^2}{4}} = 882.2m/d = 0.0102m/s$$

For the small vessel, the ungassed power draw is:

$$P = P_0 \rho N^3 D_{ag}^5 = 4.0 \cdot 1{,}000\frac{kg}{m^3} \cdot \left(\frac{150min^{-1}}{60\frac{s}{min}}\right)^3 (0.3m)^5 = 151.9W$$

The factor $\frac{Q_G}{ND_{ag}^3}$ for the small vessel is:

$$\frac{Q_G}{ND_{ag}^3} = \frac{\left(\frac{250\frac{m^3}{d}}{86,400\frac{s}{d}}\right)}{2.5\ s^{-1}(0.3m)^3} = 0.043$$

So, according to Equation (4.26b), for the small vessel

$$\frac{P_G}{P} = 0.62 - 1.85\frac{Q_G}{ND_{ag}^3} = 0.54$$

For the small vessel, the term gassed power draw is, therefore, equal to:

$$P_G = 0.54 \cdot 151.9W = 82.0W$$

And:

$$\left(\frac{P_G}{V}\right)_{small} = \frac{82.0W}{1m^3} = 82.0\frac{W}{m^3}$$

The power draw per unit volume required for the large vessel is obtained from the mass transfer correlation for the two vessels:

$$\frac{k_L a_{small}}{k_L a_{large}} = \frac{0.0015\left(\frac{P_G}{V}\right)_{small}^{0.6}\left(\frac{Q_G}{A}\right)_{small}^{0.3}}{0.0015\left(\frac{P_G}{V}\right)_{large}^{0.6}\left(\frac{Q_G}{A}\right)_{large}^{0.3}}$$

Therefore, the same $k_L a$ is obtained in the two vessels if:

$$\left(\frac{P_G}{V}\right)_{large} = \left(\frac{\left(\frac{P_G}{V}\right)_{small}^{0.6}\left(\frac{Q_G}{A}\right)_{small}^{0.3}}{\left(\frac{Q_G}{A}\right)_{large}^{0.3}}\right)^{\frac{1}{0.6}} = \left(\frac{82.0^{0.6}0.0045^{0.3}}{0.0102^{0.3}}\right)^{\frac{1}{0.6}} = 54.5\frac{W}{m^3}$$

Note that the power per unit volume required in the large vessel is lower than in the small vessel, because the superficial gas velocity is higher in the large vessel.

The gassed power draw for the large vessel is:

$$P_G = 54.5\frac{W}{m^3} \cdot 10m^3 = 545W$$

We need to calculate the ungassed power draw and the corresponding agitation speed. P and N can be obtained by the simultaneous solutions of Equations (4.26b) and (4.24):

$$\frac{545}{P} = 0.62 - 1.85\frac{\frac{2,500\ m^3}{86,400s}}{N \cdot 0.57^3 m^3} = 0.62 - \frac{0.29}{N}$$

$$P = P_o \rho N^3 D_{ag}^5 = 4.0 \cdot 1{,}000 \frac{kg}{m^3} \cdot N^3 \cdot (0.57m)^5 = 241N^3$$

The simultaneous solution of the two equations (either using Solver in Excel, or by iteration) gives:

$$P = 1141W$$

$$N = 1.35s^{-1}$$

For the large vessel, an agitator speed of 1.35 s^{-1} is required to obtain the same k_La as in the small vessel.

We verify that the ratio $\frac{Q_G}{ND_{ag}^3}$ for the large vessel is larger than 0.037 and lower than 0.11 to justify the use of Equation (4.26b):

$$\frac{Q_G}{ND_{ag}^3} = \frac{\left(\frac{2{,}500 \frac{m^3}{d}}{86{,}400 \frac{s}{d}}\right)}{1.73\ s^{-1}(0.57m)^3} = 0.090$$

The use of Equation (4.26b) is justified.

Example 4.6

The power curve of the desired agitator for a fermentation process is shown in Figure 4.7. The fermenter has a volume of 2 m^3 and a diameter of 1.5 m. The diameter of the agitator is 0.5 m and the agitation speed is 210 rpm. The air flow rate is 300 m^3/d. Calculate the k_La for this vessel.

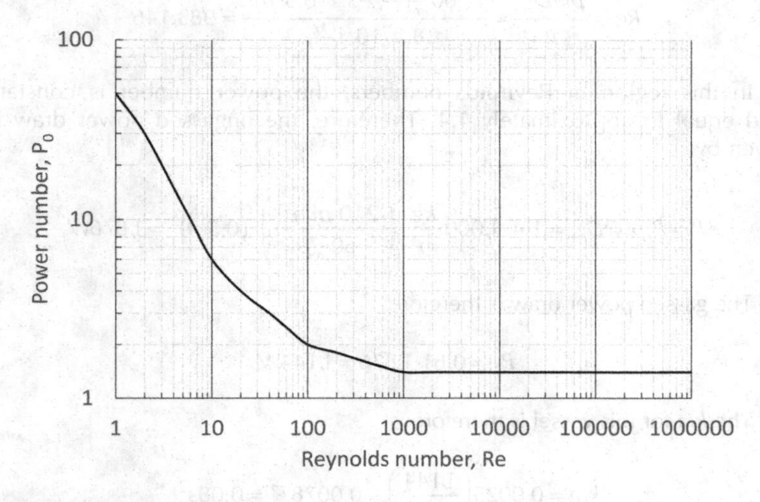

FIGURE 4.7 Power curve for Example 4.6.

For this vessel, the mass transfer coefficient is described by the correlation:

$$k_L a(s^{-1}) = 0.0025\left(\frac{P_G}{V}\right)^{0.7} v_g^{0.2} \text{ (with } P_G \text{ in W, } V \text{ in m}^3\text{, } v_g \text{ in m/s)}$$

SOLUTION

First of all, we calculate the ratio between the gassed and ungassed power draw for this vessel.

$$\frac{Q_G}{ND_{ag}^3} = \frac{\left(\frac{300\frac{m^3}{d}}{86{,}400\frac{s}{d}}\right)}{3.5\ s^{-1}(0.5\ m)^3} = 0.032$$

Therefore, the ratio between P_G and P will be given by Equation (4.26a):

$$\frac{P_G}{P} = 1 - 12.2\frac{Q_G}{ND_{ag}^3} = 1 - 12.2 \cdot 0.032 = 0.61$$

The superficial gas velocity v_g is:

$$v_g = \frac{Q_G}{A} = \frac{\frac{300\ m^3}{86{,}400\ s}}{\frac{\pi 1.5^2 m^2}{4}} = 0.0078 m/s$$

To calculate the ungassed power draw P, we need to use the power curve in Figure 4.5. The Reynolds number is:

$$Re = \frac{\rho N D_{ag}^2}{\mu} = \frac{1{,}000\frac{kg}{m^3} \cdot \frac{210}{60} s^{-1} \cdot 0.5^2 m^2}{8.9 \cdot 10^{-4}\frac{kg}{m \cdot s}} = 983{,}146$$

In this region of Reynolds numbers, the power number is constant and equal to approximately 1.4. Therefore, the ungassed power draw is given by:

$$P = P_0 \rho N^3 D_{ag}^5 = 1.4 \cdot 1{,}000\frac{kg}{m^3} \cdot \left(\frac{210\ min^{-1}}{60\frac{s}{min}}\right)^3 (0.5m)^5 = 1{,}876W$$

The gassed power draw is therefore:

$$P_G = 0.61 \cdot 1{,}876 = 1{,}144W$$

The $k_L a$ for this vessel is therefore:

$$k_L a = 0.0025\left(\frac{1{,}144}{2}\right)^{0.7} 0.0078^{0.2} = 0.08 s^{-1}$$

4.4 OXYGEN BALANCES IN BATCH REACTORS

If the rate of oxygen transfer from the gas to the liquid phase and the rate of oxygen consumption by the microorganisms are known, the mass balance for oxygen in the liquid phase can be written and solved.

For a batch fermentation, the mass balance for oxygen can be written as:

$$\frac{d(VC_{O2})}{dt} = k_L a (C^*_{O2} - C_{O2})V + r_{O2biom}V \tag{4.27}$$

In Equation (4.27), r_{O2biom} is the rate of oxygen consumption by the microorganisms defined in Chapter 2.

Equation (4.27) can be re-arranged by eliminating the constant volume of the liquid phase and by introducing the definition of r_{O2biom} seen in Chapter 2:

$$\frac{dC_{O2}}{dt} = k_L a (C^*_{O2} - C_{O2}) - \frac{\mu_{max} S}{K_S + S} \frac{X}{Y_{X/O2}} \tag{4.28}$$

Equation (4.26) gives the profile of the oxygen concentration in the liquid phase as a function of time during a batch fermentation. Solution of Equation (4.28) requires knowledge of the values of the kinetic parameters μ_{max}, K_S and $Y_{X/O2}$ and of the mass transfer coefficient $k_L a$. It also requires knowledge of the oxygen solubility C^*_{O2} and of an initial condition for the oxygen concentration in the liquid phase. Note that Equation (4.28) cannot be solved in isolation, but it needs to be coupled with the mass balances for substrate and biomass in a batch fermentation seen in Chapter 2.

In summary, the profiles of substrate, microorganisms and dissolved oxygen in batch fermentation are given by the solution of the system of Equations (4.29).

$$\begin{cases} \dfrac{dX}{dt}\left(\dfrac{kgbiomass}{m^3 \cdot d}\right) = \dfrac{\mu_{max} S}{K_S + S} X \\ \dfrac{dS}{dt}\left(\dfrac{kgsubstrate}{m^3 \cdot d}\right) = -\dfrac{\mu_{max} S}{K_S + S} \dfrac{X}{Y_{X/S}} \\ \dfrac{dP}{dt}\left(\dfrac{kgproduct}{m^3 \cdot d}\right) = \dfrac{\mu_{max} S}{K_S + S} X Y_{P/X} \\ \dfrac{dC_{O2}}{dt}\left(\dfrac{kgoxygen}{m^3 \cdot d}\right) = k_L a (C^*_{O2} - C_{O2}) - \dfrac{\mu_{max} S}{K_S + S} \dfrac{X}{Y_{X/O2}} \end{cases} \tag{4.29}$$

Note that, as discussed in Chapter 2, the use of the Monod equation with only the organic substrate as limiting substrate assumes that the concentration of other nutrients and oxygen are not limiting. Therefore, the design of the fermentation

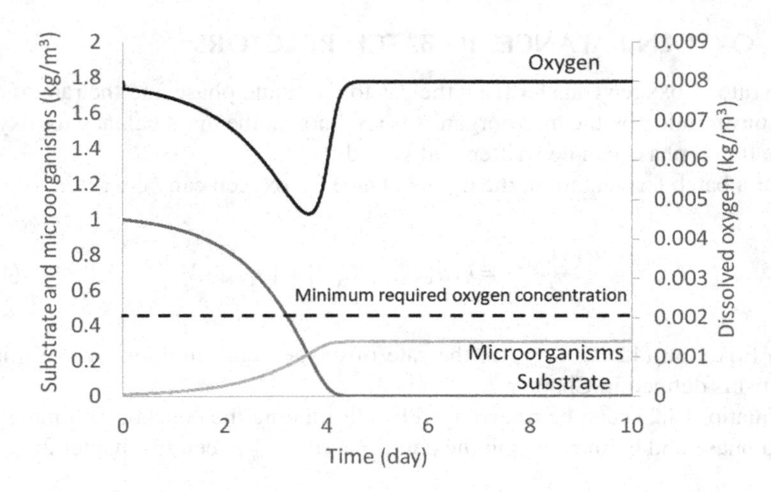

FIGURE 4.8 Typical profiles of oxygen, microorganisms and substrates during a batch fermentation.

process requires making sure that the oxygen concentration always stays above a certain minimum level, below which oxygen would become rate limiting. This minimum oxygen concentration is typically in the range 1–2 mg/L. The typical profiles of X, S, and C_{O2} obtained from the solution of Equation (4.29) are shown in Figure 4.8.

From Figure 4.8 it is evident that the profile of dissolved oxygen follows the profile of the substrate. As microorganisms' concentration increases, the rate of substrate removal increases, and the oxygen concentration starts to decrease more rapidly, because the rate of oxygen consumption is higher than the rate of oxygen transfer. When the substrate is completely removed, oxygen concentration starts to increase because oxygen consumption ceases.

In many aerobic fermentation processes, the value of the mass transfer coefficient k_La determines the maximum value of the initial substrate concentration that can be used in the process. Indeed, as shown in Figure 4.9, if the initial substrate concentration increases, the oxygen concentration decreases to a lower level. The maximum initial substrate concentration that we can have in a given fermentation is the one for which the dissolved oxygen concentration reaches the minimum allowable level (i.e. the concentration corresponding to the rate limiting condition).

The maximum initial substrate concentration can be calculated by trial and error by solving the system of Equations (4.29) with different values of the initial substrate concentration until the oxygen profile reaches the minimum allowable concentration.

In the particular case that the parameter $K_S \rightarrow 0$, the maximum initial substrate concentration that is possible to have in a batch fermentation can be calculated

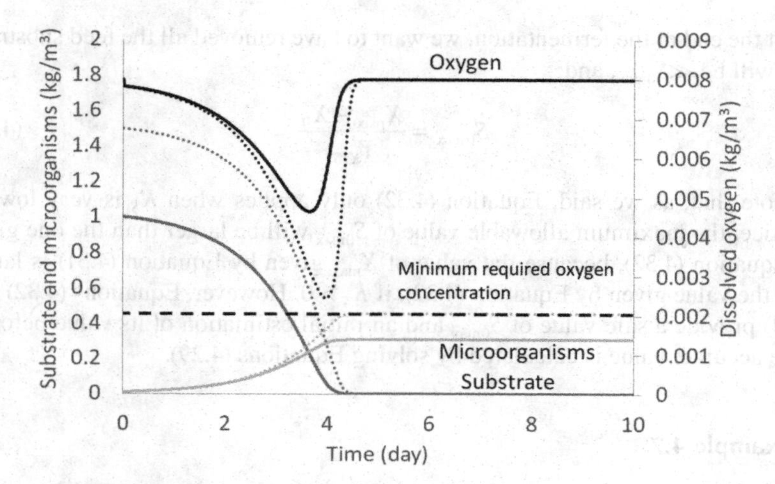

FIGURE 4.9 Effect of the initial substrate concentration on the dissolved oxygen profile in batch fermentation. Solid lines: profiles for initial substrate 1.0 kg/m^3; dotted lines: profiles for initial substrate 1.5 kg/m^3.

analytically without having to solve the system of Equations (4.29). Indeed, the condition of maximum initial substrate concentration is obtained when the oxygen profile reaches its minimum $\left(\frac{dC_{O2}}{dt}\right) = 0$ for $C_{O2} = C_{O2,min}$. The condition that $\frac{dC_{O2}}{dt} = 0$ for $C_{O2} = C_{O2,min}$ corresponds to:

$$\frac{dC_{O2}}{dt} = k_L a\left(C_{O2}^* - C_{O2\,min}\right) - \frac{\mu_{max} S}{K_S + S} \frac{X}{Y_{X/O2}} = 0 \tag{4.30}$$

From Equation (4.30), the maximum biomass concentration that we can have in the fermenter without C_{O2} going below $C_{O2,min}$ is:

$$X_{max} = k_L a\left(C_{O2}^* - C_{O2\,min}\right) \frac{\left(K_S + S\right)}{\mu_{max} S} Y_{X/O2} \tag{4.31}$$

If $K_S \rightarrow 0$, Equation (4.31) allows the calculation of X_{max}:

$$X_{max} = \frac{k_L a\left(C_{O2}^* - C_{O2\,min}\right)}{\mu_{max}} Y_{X/O2} \tag{4.32}$$

Once X_{max} is calculated, the maximum initial substrate concentration can be calculated from:

$$X_{max} - X_0 = Y_{X/S}\left(S_{0\,max} - S\right) \tag{4.33}$$

At the end of the fermentation, we want to have removed all the feed substrate, so S will be $\ll S_{0max}$, and:

$$S_{0\max} = \frac{X_{\max} - X_0}{Y_{X/S}} \tag{4.34}$$

Note that, as we said, Equation (4.32) only applies when K_S is very low. In practice, the maximum allowable value of S_{0max} will be larger than the one given by Equation (4.32), because the value of X_{max} given by Equation (4.31) is larger than the value given by Equation (4.32), if $K_S > 0$. However, Equations (4.32) and (4.34) provide a safe value of S_{0max} and an initial estimation of its value before a more accurate value is calculated by solving Equations (4.29).

Example 4.7

Acetic acid is produced by the aerobic fermentation of ethanol by the microorganism *Acetobacter*. The stoichiometry of the fermentation can be written as follows:

$$C_2H_5OH + O_2 + 0.10NH_3 \rightarrow 0.10C_5H_7O_2N + 0.75CH_3COOH + 0.3H_2O$$

The fermentation is started with an ethanol concentration of 10 g/L and a concentration of microorganisms of 1 g/L. The initial concentration of dissolved oxygen is 7.0 mg/L. The fermenter is sparged with air with a k_La of 0.1 s^{-1} and the saturation concentration of oxygen under the fermentation conditions is 9.0 mg/L. The kinetic parameters are:

$$\mu_{max} = 7d^{-1}; K_S = 0.02\frac{kg}{m^3}$$

Calculate the time profiles of ethanol, biomass, acetic and oxygen during the fermentation.

SOLUTION

The profiles are obtained from the solution of the system of differential Equations (4.29). From the given stoichiometry, we can calculate the parameters $Y_{X/S}$, $Y_{X/O2}$ and $Y_{P/X}$:

$$Y_{X/S} = \frac{0.1 \cdot 113}{46} = 0.25\frac{kg}{kg}$$

$$Y_{X/O2} = \frac{0.1 \cdot 113}{32} = 0.35\frac{kg}{kg}$$

$$Y_{P/X} = \frac{0.75 \cdot 60}{0.1 \cdot 113} = 3.98\frac{kg}{kg}$$

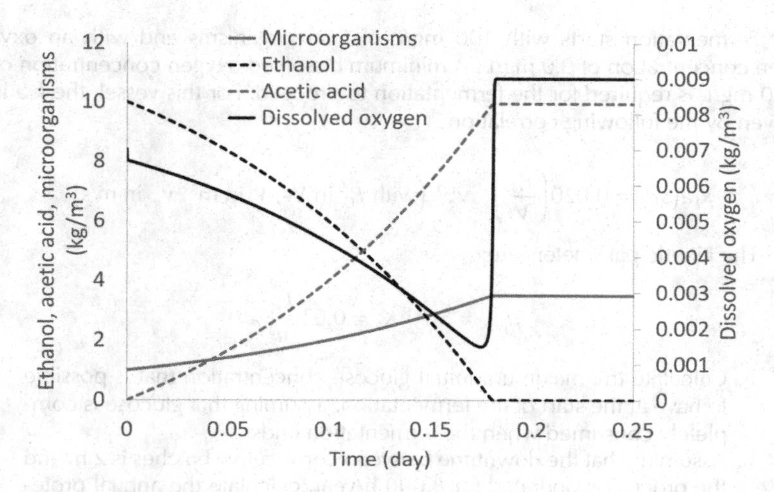

FIGURE 4.10 Calculated profiles for Example 4.7.

The system of Equations (4.29) in this case is:

$$\begin{cases} \dfrac{dX}{dt}\left(\dfrac{kgbiomass}{m^3\cdot d}\right)=\dfrac{7S}{0.02+S}X \\ \dfrac{dS}{dt}\left(\dfrac{kgsubstrate}{m^3\cdot d}\right)=-\dfrac{7S}{0.02+S}\dfrac{X}{0.35} \\ \dfrac{dP}{dt}\left(\dfrac{kgproduct}{m^3\cdot d}\right)=\dfrac{7S}{0.02+S}X\cdot 3.98 \\ \dfrac{dC_{O2}}{dt}\left(\dfrac{kgoxygen}{m^3\cdot d}\right)=0.1\cdot 86,400\left(0.009-C_{O2}\right)-\dfrac{7S}{0.02+S}\dfrac{X}{0.35} \end{cases}$$

The resulting profiles are shown in Figure 4.10.

Example 4.8

The microorganism *Bacillus subtilis* is used for the production of the enzyme protease ($C_3H_5O_2N$). The fermentation reaction is aerobic and uses glucose as substrate and ammonia as nitrogen source. The products are microorganisms, the enzyme, water and carbon dioxide. It is known that 60 g of carbon dioxide and 20 g of microorganisms are produced per 100 g of glucose consumed.

Fermentation is carried out in an agitated vessel having a volume of 20 m³. The vessel has a diameter of 2.1 m. The diameter of the agitator is 0.7 m and the maximum practical agitation speed is 180 rpm. Under the fermentation conditions, the power number of the agitator is 5.0. The vessel is sparged with air with a flow rate of 10,000 m³/d. Under these conditions, we can assume that the ratio between the gassed and ungassed power draw is equal to 0.43.

Fermentation starts with 100 mg/L of microorganisms and with an oxygen concentration of 9.0 mg/L. A minimum dissolved oxygen concentration of 2.0 mg/L is required for the fermentation to proceed. For this vessel, the k_La is given by the following correlation:

$$k_La(s^{-1}) = 0.020\left(\frac{P_G}{V}\right)^{0.5} v_g^{0.3} \left(\text{with } P_G \text{ in W, } V \text{ in m}^3,\ v_g \text{ in m/s}\right)$$

The kinetic parameters are:

$$\mu_{max} = 5d^{-1}; K_S = 0.03\frac{kg}{m^3}$$

a. Calculate the maximum initial glucose concentration that is possible to have at the start of the fermentation, assuming that glucose is completely consumed when the fermentation ends.
b. Assuming that the downtime between consecutive batches is 2 h, and the process is operated for 8,000 h/year, calculate the annual protease productivity (in kg) with this process.

SOLUTION

a. The first step is the calculation of the reaction stoichiometry, which will allow the calculation of the coefficients $Y_{X/S}$, $Y_{P/X}$ and $Y_{X/O2}$. The unbalanced stoichiometry for this reaction is:

$$C_6H_{12}O_6 + aO_2 + bNH_3 \rightarrow cC_5H_7O_2N + dC_3H_5O_2N + eCO_2 + fH_2O$$

We know that:

$$\frac{e \cdot MW_{CO_2}}{MW_{C_6H_{12}O_6}} = 0.60 \quad \Rightarrow \quad e = \frac{0.60 \cdot 180}{44} = 2.45$$

$$\frac{c \cdot MW_{C_5H_7O_2N}}{MW_{C_6H_{12}O_6}} = 0.20 \quad \Rightarrow \quad c = \frac{0.20 \cdot 180}{113} = 0.32$$

The other coefficients *a*, *b*, *d* and *f* can be obtained from the elemental balances of C, H, N and O:

$$\begin{cases} C-balance\text{: } 5c + 3d + e = 6 \\ O-balance\text{: } 6 + 2a = 2c + 2d + 2e + f \\ H-balance\text{: } 12 + 3b = 7c + 5d + 2f \\ N-balancec\text{: } b = c + d \end{cases}$$

From the C balance, we obtain: $d = 0.65$; from the N balance: $b = 0.97$; from the H balance: $f = 4.71$ and from the O balance: $a = 2.78$.

Hence, the balanced stoichiometry is:

$$C_6H_{12}O_6 + 2.78O_2 + 0.97NH_3 \rightarrow 0.32C_5H_7O_2N + 0.65C_3H_5O_2N + 2.45CO_2 + 4.71H_2O$$

The stoichiometric coefficients are therefore:

$$Y_{X/S} = \frac{0.32 \cdot 113}{180} = 0.20 \frac{kg}{kg}$$

$$Y_{X/O2} = \frac{0.32 \cdot 113}{2.78 \cdot 32} = 0.41 \frac{kg}{kg}$$

$$Y_{P/X} = \frac{0.65 \cdot 87}{0.32 \cdot 113} = 1.56 \frac{kg}{kg}$$

The superficial gas velocity is:

$$v_t = \frac{Q_G}{A} = \frac{\frac{10{,}000\ m^3}{86{,}400\ s}}{\frac{\pi 2.1^2\ m^2}{4}} = 0.033 m/s$$

The ungassed power draw is:

$$P = P_o \rho N^3 D_{ag}^5 = 5.0 \cdot 1{,}000 \frac{kg}{m^3} \cdot \left(\frac{180\ min^{-1}}{60 \frac{s}{min}} \right)^3 (0.7m)^5 = 22{,}689\ W$$

And the gassed power draw:

$$P_G = 0.43\ P = 9{,}756 W$$

The $k_L a$ for this vessel is:

$$k_L a\left(s^{-1}\right) = 0.020 \left(\frac{9{,}756}{20} \right)^{0.5} 0.033^{0.3} = 0.159 s^{-1}$$

The mass balances that describe this process are therefore:

$$\begin{cases} \dfrac{dX}{dt}\left(\dfrac{kgbiomass}{m^3 d}\right) = \dfrac{5S}{0.03 + S} X \\ \dfrac{dS}{dt}\left(\dfrac{kgsubstrate}{m^3 d}\right) = -\dfrac{5S}{0.03 + S} \dfrac{X}{0.20} \\ \dfrac{dP}{dt}\left(\dfrac{kgproduct}{m^3 d}\right) = \dfrac{5S}{0.03 + S} X \cdot 1.56 \\ \dfrac{dC_{O2}}{dt}\left(\dfrac{kgoxygen}{m^3 d}\right) = 0.033 \cdot 86{,}400 \left(0.009 - C_{O2}\right) - \dfrac{5S}{0.03 + S} \dfrac{X}{0.20} \end{cases}$$

These mass balances can be solved for various values of the initial substrate concentration. The maximum allowable initial substrate

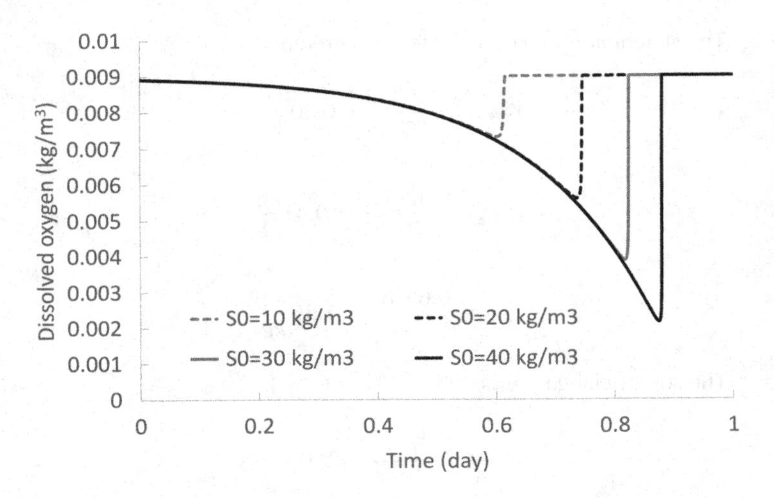

FIGURE 4.11 Solution of Example 4.8.

concentration is the value for which the oxygen concentration drops to 2.0 mg/L without going below this value. Figure 4.11 shows the oxygen profiles for various values of the initial substrate concentration in the batch. The maximum allowable initial substrate concentration is 40 kg/m³.

b. When the substrate concentration is 40 kg/m³, the product concentration at the end of each batch is 12.57 kg/m³ (from the reaction stoichiometry). Since the volume of the reactor is 20 m³, in each batch 251.4 kg of enzyme are produced. The substrate is completely removed in 0.90 d = 21.6 h. Adding the 2 h of downtime, the total time of a batch is 23.6 h. The process is operated for 8,000 h per year, so the number of batches in a year is equal to 339. So, in 1 year, this fermenter will produce 339 × 251.4 = 85,225 kg of enzyme.

4.5 OXYGEN BALANCE IN CONTINUOUS REACTORS

In addition to calculating the required residence time in continuous reactor design for aerobic fermentations, it is also important to design the aeration system so that it provides the oxygen required by the microorganisms. Design of the aeration system means essentially selecting an aeration system that is able to transfer oxygen with the required mass transfer coefficient k_La. The k_La that is required for a certain fermentation can be calculated from a mass balance of oxygen on the continuous fermenter (Figure 4.12).

In Figure 4.12, in addition to the symbols already defined, C_{O20} and C_{O2} are the concentrations of dissolved oxygen in the feed and in the reactor, respectively. For a continuous fermenter at a steady state, the mass balance of oxygen can be

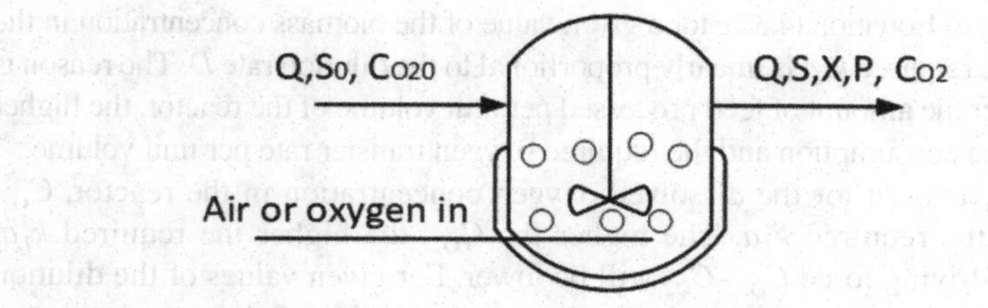

FIGURE 4.12 Scheme of a continuous fermenter including oxygen transfer for oxygen mass balance.

done using the general equation for mass balances Equation (3.1). In this case, Equation (3.1) becomes:

$$Q \cdot C_{O20} + r_{O2transf} V = Q \cdot C_{O2} + \left(-r_{O2biom}\right) \cdot V \tag{4.35}$$

In Equation (4.35), the terms $r_{O2transf}$ and r_{O2biom} are expressed as seen earlier:

$$r_{O2transf}\left(\frac{kg_{O_2}}{m^3 d}\right) = k_L a \cdot \left(C^*_{O2} - C_{O2}\right)$$

$$r_{O2biom}\left(\frac{kg_{O2}}{m^3 d}\right) = -\frac{\mu_{max} S}{K_S + S} \frac{X}{Y_{X/O2}}$$

with $C^*_{O2} = k_{eq} p_{O2}$, where p_{O2} is taken as the partial pressure of oxygen in the influent gas, ignoring the decrease in oxygen concentration in the gas bubbles due to oxygen transfer from the gas phase.

Once the equations for $r_{O2transf}$ and r_{O2biom} are substituted, Equation (4.35) becomes:

$$Q \cdot C_{O20} + k_L a \cdot \left(C^*_{O2} - C_{O2}\right) V = Q \cdot C_{O2} + \frac{\mu_{max} S}{K_S + S} \frac{X}{Y_{X/O2}} \tag{4.36}$$

Equation (4.36) can be re-arranged into Equation (4.37):

$$k_L a = \frac{D \cdot \left[C_{O2} - C_{O20} + \frac{X}{Y_{X/O2}}\right]}{\left(C^*_{O2} - C_{O2}\right)} \tag{4.37}$$

Equation (4.37) can be used to calculate the value of $k_L a$ that the aeration system needs to provide in order to maintain the desired concentration of dissolved oxygen in the reactor. Note that at the numerator of Equation (4.37), the term representing the oxygen consumption by the microorganisms $\left(\frac{X}{Y_{X/O2}}\right)$ is usually much larger than the term representing the transport of oxygen in and out of the reactor (C_{O2}-C_{O20}).

According to Equation (4.37), for a given value of the biomass concentration in the reactor, X, the required k_La is linearly proportional to the dilution rate D. The reason is that the higher the amount of feed processed per unit volume of the reactor, the higher are the oxygen consumption and the required oxygen transfer rate per unit volume.

The value chosen for the dissolved oxygen concentration in the reactor, C_{O2}, also affects the required k_La. The higher the C_{O2}, the higher the required k_La, because the driving force C_{O2}^*-C_{O2} will be lower. For given values of the dilution rate and of the biomass concentration in the reactor, it is beneficial to maintain the required k_La at the lowest possible level, because higher value of the mass transfer coefficient will require higher values of the agitator power draw and/or higher values of the inlet gas flow rate with consequently higher energy costs. Therefore, the oxygen concentration in a bioreactor needs to be maintained at the lowest possible level that ensures oxygen doesn't become rate limiting for the growth of the microorganisms. Figure 4.13 shows the general profile of the required k_La vs the dissolved oxygen concentration in the reactor, C_{O2}.

The required k_La also depends on the substrate concentration in the feed, S_0, because the microorganisms' concentration X depends on S_0. Higher values of S_0 will require higher k_La values, because the concentration of the microorganisms and their oxygen consumption rate will be higher. To show this effect, Equation (4.37) can be re-arranged into Equation (4.38) by replacing X as a function of S_0 (Equation 3.20), and by solving for S_0, under the assumption that $S \ll S_0$ (i.e. assuming that the dilution rate D is always low enough to guarantee (almost) complete substrate removal):

$$S_0 = \left[\frac{k_La}{D}(C_{O2}^* - C_{O2}) - (C_{O2} - C_{O20})\right]\frac{Y_{X/O2}}{Y_{X/S}} \tag{4.38}$$

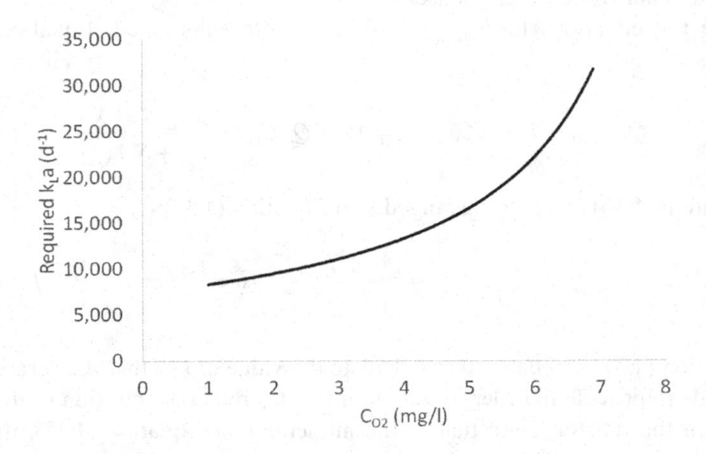

FIGURE 4.13 Plot of Equation (4.37) showing the effect of the oxygen concentration in the continuous fermenter on the required k_La. Plot obtained with $D = 5$ d^{-1}, $C_{O20} = 5$ mg/L, $X = 2$ kg/m^3, $Y_{X/O2} = 0.15$ kg/kg and $C_{O2}^* = 9$ mg/L.

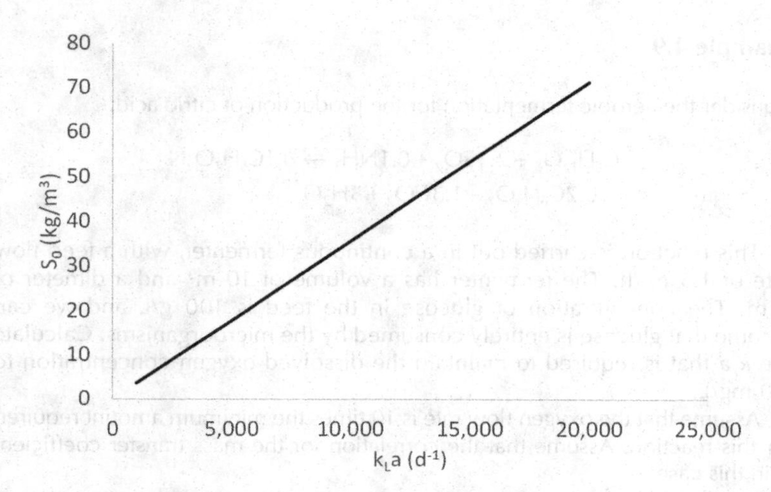

FIGURE 4.14 Plot of Equation (4.38) showing the effect of the mass transfer coefficient on the maximum substrate concentration that we can have in the feed of a continuous fermenter to maintain the oxygen concentration at the desired level. Plot obtained with $C_{O2} = 1.5$ mg/L, $C^*_{O2} = 9$ mg/L, $D = 5$ d^{-1}, $C_{O20} = 5$ mg/L, $X = 2$ kg/m^3, $Y_{X/O2} = 0.15$ kg/kg and $Y_{X/S} = 0.063$ kg/kg.

Equation (4.38), plotted in Figure 4.14, gives the maximum substrate concentration that we can have in the feed of a continuous fermenter to maintain the oxygen concentration at the desired value C_{O2}. The higher the k_La in the fermenter, the higher the substrate concentration that we can have in the feed and therefore the higher the productivity of the process. Therefore, if the higher process productivity outweighs the higher energy costs associated with higher k_La values, it may be beneficial to increase the mass transfer coefficient to increase the productivity of the process. Note that in Equation (4.38), as in (4.37), the term due to oxygen transfer is usually much higher than the contribution of the dissolved oxygen in and out of the reactor (C_{O2}–C_{O20}).

The efficiency of oxygen transfer from the gas to the liquid phase can also be calculated. The efficiency (η) can be defined as the ratio between the oxygen transferred from the gas to the liquid phase and the total oxygen fed to the gas phase. For a continuous fermenter, the equation for the efficiency is:

$$\eta\left(\frac{\text{kgO}_2\text{transferred}}{\text{kgO}_2\text{provided}}\right) = \frac{k_LaV\left(C^*_{O2} - C_{O2}\right)}{Q_{air}\rho_{air} \cdot 0.23} \tag{4.39}$$

In Equation (4.39), the factor 0.23 represents the mass fraction of oxygen in air. The same equation can also be used to calculate the efficiency for a batch reactor, however, in a batch process the oxygen concentration C_{O2} will change during the process and so the efficiency will also be variable with time.

Example 4.9

Consider the aerobic fermentation for the production of citric acid:

$$C_6H_{12}O_6 + 2.35O_2 + 0.1NH_3 \rightarrow 0.1C_5H_7O_2N$$
$$+ 0.7C_6H_8O_7 + 1.3CO_2 + 3H_2O$$

This reaction is carried out in a continuous fermenter, with a feed flow rate of 1.5 m^3/h. The fermenter has a volume of 10 m^3 and a dimeter of 2 m. The concentration of glucose in the feed is 100 g/L and we can assume that glucose is entirely consumed by the microorganisms. Calculate the k_La that is required to maintain the dissolved oxygen concentration to 3.0 mg/L.

Assume that the oxygen flow rate is 10 times the minimum amount required for this reaction. Assume that the correlation for the mass transfer coefficient is in this case:

$$k_La(s^{-1}) = 0.050\left(\frac{P_G}{V}\right)^{0.63} v_g^{0.40} \text{ with } P_G \text{ in W, V in } m^3,\ v_g \text{ in m/s}$$

What is the required ungassed power draw? Assume that under these conditions $P_G/P = 0.43$.

Assume that the saturation concentration of oxygen in water under the fermentation conditions is $C^*_{O2} = 9.0$ mg/L.

SOLUTION

We start from the oxygen balance, which is, with the hypotheses of this problem:

$$k_La\left(C^*_{O2} - C_{O2}\right)V = Q \cdot C_{O2} + \left(-r_{O2biom}\right) \cdot V$$

The glucose consumption rate is equal to $100\frac{kg}{m^3} \cdot 1.5\frac{m^3}{h} = 150\frac{kg}{h}$.

From the reaction stoichiometry, 0.42 kg of oxygen is consumed by the microorganisms per 1 kg of glucose removed. Therefore, the oxygen consumption by the microorganisms is:

$$\left(-r_{O2biom}\right)*V = 150\frac{kg}{h} \cdot 0.42\frac{kgO_2}{kgglucose} = 63\frac{kgO_2}{h}$$

Therefore, the required k_La is:

$$k_La = \frac{20\frac{m^3}{h} \cdot 0.003\frac{kg}{m^3} + 63\frac{kg}{h}}{(0.009 - 0.003)\frac{kg}{m^3} 10\ m^3} = 1{,}051\ h^{-1} = 0.29\ s^{-1}$$

To calculate the superficial gas velocity v_g we need the gas flow rate. Since the gas flow rate is 10 times the minimum requirement, this will be equal to 630 kg_{O2}/h. Using the ideal gas law:

$$V_{O2} = \frac{nRT}{P} = \frac{\frac{630{,}000 g}{32 \frac{g}{mol}} \cdot 0.0821 \frac{L \cdot atm}{mol \cdot K} \cdot 298.15K}{1 atm} = 481{,}912 \frac{L}{h}$$

Since atmospheric air is made of 21% in volume by oxygen, the air flow rate will be:

$$V_{air} = \frac{481{,}912}{0.21} = 2{,}294{,}823 \frac{L}{h} = 0.637 \frac{m^3}{s}$$

The superficial gas velocity is:

$$v_g = \frac{Q_G}{A} = \frac{0.637 \frac{m^3}{s}}{\frac{\pi 2^2 m^2}{4}} = 0.203 m/s$$

Therefore, the gassed power draw required:

$$P_G = V\left(\frac{k_L a}{0.050 \cdot v_t^{0.40}}\right)^{\frac{1}{0.63}} = 10\left(\frac{0.29}{0.050 \cdot 0.203^{0.40}}\right)^{\frac{1}{0.63}} = 448W$$

Therefore, the ungassed power draw will be $P_G/0.43 = 1{,}042$ W.

4.6 CONSIDERATIONS FOR SCALE-UP AND SCALE-DOWN

It is often desirable to reproduce the results of a lab-scale fermentation at larger (commercial) scale (scale-up), or, on the other hand, to replicate a large-scale fermentation at lab-scale (scale-down). Scale-up is made with the obvious aims to obtain more product or more microorganisms or to process more wastewater, and the aims of scale-down are often to investigate at lab-scale how altering the process conditions (e.g. the residence time, temperature or pH) affects the fermentation. We have already seen an example of a scale-up calculation in Example 4.5. In this section, we will cover the general theory of scale-up/scale-down for oxygen transfer in fermentation processes, assuming they are carried out in agitated vessels.

If we want to maintain the same dissolved oxygen concentration in the scaled-up or scaled-down vessel as in the original fermentation, the aeration and agitation system needs to be designed appropriately. We assume that the temperature, pH, residence time, feed composition and all the other fermentation parameters are kept constant on scale-up or scale-down, so here we will only focus on oxygen transfer. We will see in the next chapter how to deal with scale-up/scale-down for temperature control.

The oxygen balance in a continuous fermenter is given by Equation (4.35). This equation shows that, in order to maintain the same oxygen concentration in the scaled-up/scale-down vessel as in the original vessel, the k_La must be maintained to the same value. Indeed, with the same feed composition and residence time, the substrate and biomass concentration in the vessels will be the same. In order to maintain the same k_La in the scaled-up/scaled-down vessels, assuming that, as it is good practice, the scale-up/scale-down is made using geometrically similar vessels, we need to look at the correlation for k_La (4.22). According to this correlation, maintaining the same k_La means maintaining the same value of the term $\left(\frac{P_G}{V}\right)^a v_g^b = \left(\frac{P_G}{V}\right)^a \left(\frac{Q_G}{A}\right)^b$. In this equation, the terms V and A are fixed by the required volumes and scale-up/scale-down ratio, while the values of P_G and Q_G in the scale-up/scale-down vessel need to be calculated or assigned. If we assign a value for Q_G for the scale-up/scale-down vessel, then the value of P_G for this vessel will be calculated, vice versa if the value of P_G for the scale-up/scale-down vessel is assigned, the value of Q_G for this vessel will be calculated. In our example procedure explained below, we assume that the scale-up/scale-down is made maintaining the same gas flow rate per unit volume of reactor, i.e. $\left(\frac{Q_G}{V}\right)_{large} = \left(\frac{Q_G}{V}\right)_{small}$. This is not an absolute requirement but it is often a convenient way to scale-up/scale-down the gas flow rate. We will also assume, for simplicity's sake, that the ratio $\frac{P_G}{P}$ is the same in the large and small vessels and that the power number P_0 is independent of the Reynolds number for both vessels (flat region of the P_0 vs Re curve). These assumptions simplify the analysis but don't alter the significance of the results.

With the assumptions in the paragraph above, the scale-up/scale-down problem means finding the value of the power draw P_G required in the scale-down/scaled-down vessel once the geometry of the vessels and the agitation conditions in the original vessel are known. We assume we know the scale-up/scale-down ratio, i.e. the ratio between the vessel volumes, $\frac{V_{large}}{V_{small}}$. As we have seen, the condition of same k_La for the large and small vessels is:

$$\left(\frac{P_G}{V}\right)_{\text{large}}^a \left(\frac{Q_G}{A}\right)_{\text{large}}^b = \left(\frac{P_G}{V}\right)_{\text{small}}^a \left(\frac{Q_G}{A}\right)_{\text{small}}^b \tag{4.40}$$

Equation (4.40) can be re-arranged as:

$$\left(\frac{P_{\text{G,large}}}{P_{\text{G,small}}}\right)^a = \left(\frac{V_{\text{large}}}{V_{\text{small}}}\right)^a \left(\frac{A_{\text{large}}}{A_{\text{small}}}\right)^b \left(\frac{Q_{\text{G,small}}}{Q_{\text{G,large}}}\right)^b \tag{4.41}$$

With the conditions that the gas flow rate per unit volume remains constant on scale-up/scale-down, Equation (4.41) can be re-arranged as:

$$\left(\frac{P_{\text{G,large}}}{P_{\text{G,small}}}\right)^a = \left(\frac{V_{\text{large}}}{V_{\text{small}}}\right)^{a-b} \left(\frac{A_{\text{large}}}{A_{\text{small}}}\right)^b \tag{4.42}$$

For cylindrical vessels, the volume and the cross-sectional area are given by:

$$V = \frac{\pi D^2}{4} H \tag{4.43}$$

$$A = \frac{\pi D^2}{4} \tag{4.44}$$

In Equations (4.43) and (4.44), D is the vessel diameter and H the vessel height. We define α as the ratio between the height and diameter of the vessel:

$$\alpha = \frac{H}{D} \tag{4.45}$$

For geometrically similar vessels the parameter α will stay constant on scale-up/scale-down. With the introduction of α we can write the vessel volume as:

$$V = \frac{\alpha \pi D^3}{4} \tag{4.46}$$

From Equations (4.44) and (4.46), we obtain:

$$\frac{D_{\text{large}}}{D_{\text{small}}} = \left(\frac{V_{\text{large}}}{V_{\text{small}}}\right)^{\frac{1}{3}} \tag{4.47}$$

$$\frac{A_{\text{large}}}{A_{\text{small}}} = \left(\frac{V_{\text{large}}}{V_{\text{small}}}\right)^{\frac{2}{3}} \tag{4.48}$$

Introducing Equation (4.48) in Equation (4.42), we obtain:

$$\left(\frac{P_{\text{G,large}}}{P_{\text{G,small}}}\right)^{a} = \left(\frac{V_{\text{large}}}{V_{\text{small}}}\right)^{a-\frac{b}{3}} \Rightarrow \frac{P_{\text{G,large}}}{P_{\text{G,small}}} = \left(\frac{V_{\text{large}}}{V_{\text{small}}}\right)^{1-\frac{b}{3a}} \tag{4.49}$$

Equation (4.49) can be rewritten in terms of the power per unit volume:

$$\frac{\frac{P_{\text{G,large}}}{V_{\text{large}}}}{\frac{P_{\text{G,small}}}{V_{\text{small}}}} = \left(\frac{V_{\text{large}}}{V_{\text{small}}}\right)^{-\frac{b}{3a}} \tag{4.50}$$

Equations (4.49) and (4.50) show that in the scale-up/scale-down, the ratio between the power draw to be provided in the vessels is proportional to the scale-up/scale-down ratio; however, the proportionality is less than linear and the power draw per unit volume decreases on scale-up. The reason for the lower than linear increase of $\frac{P_{\text{G,large}}}{P_{\text{G,small}}}$ with $\frac{V_{\text{large}}}{V_{\text{small}}}$ is that, with the assumption of $\left(\frac{Q_G}{V}\right)_{\text{large}} = \left(\frac{Q_G}{V}\right)_{\text{small}}$, the

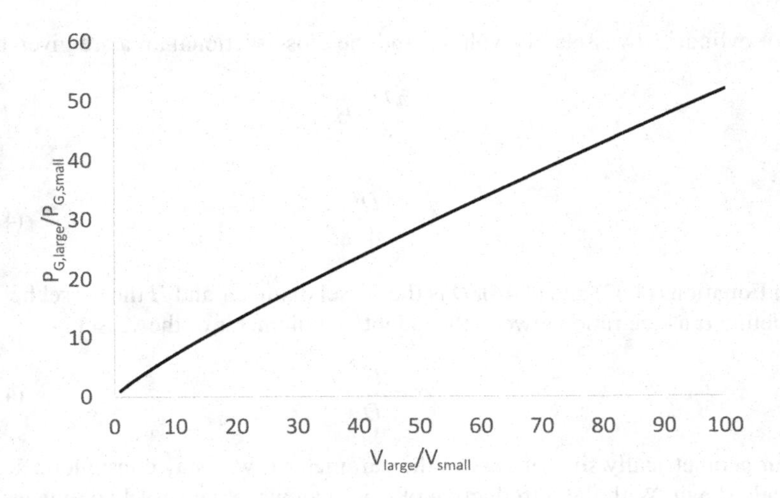

FIGURE 4.15 Plot of Equation (4.49) with $a = 0.7$ and $b = 0.3$.

superficial gas velocity increases as $\frac{V_{large}}{V_{small}}$ increases, giving a higher contribution to the k_La. This is positive since it allows to achieve the same k_La in large reactors with lower power-to-volume ratios than in small reactors, limiting the increase in agitator energy consumption on scale-up. The relative role of the superficial gas velocity and of the power draw depends of course on the values of the coefficients a and b of Equation (4.22). Figures 4.15 and 4.16 show, for arbitrary but realistic values of the parameters a and b, the shape of the curves defined by Equations (4.49) and (4.50).

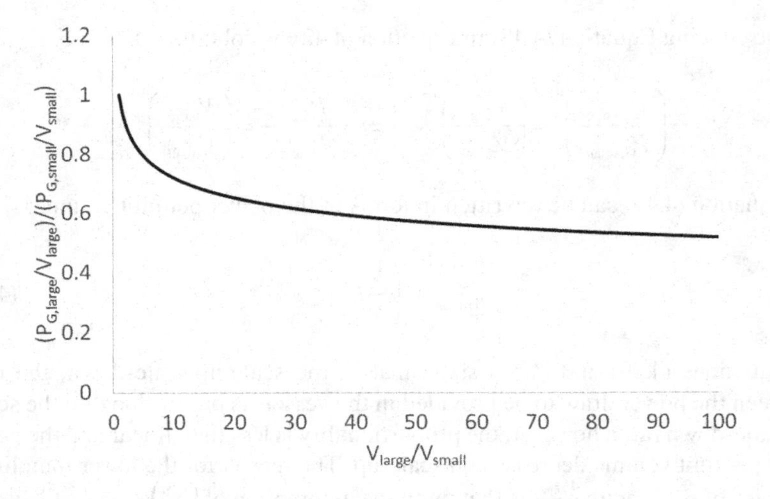

FIGURE 4.16 Plot of Equation (4.50) with $a = 0.7$ and $b = 0.3$.

Once the required power in the scaled-up/scale-down vessel has been calculated with Equations (4.49) or (4.50), the next step is to calculate the agitation speed that is required in the scaled-up/scaled-down vessel if the agitation speed in the original vessel is known.

The relationship between the agitator power draw and the power number is given by Equation (4.24). Since we are assuming that $\frac{P_G}{P}$ and P_0 stay constant with scale-up/scale-down, and the density of the medium is also constant, Equation (4.24) leads to Equation (4.51):

$$\left(\frac{P_{G,\text{large}}}{P_{G,\text{small}}}\right) = \left(\frac{N_{\text{large}}}{N_{\text{small}}}\right)^3 \left(\frac{D_{ag,\text{large}}}{D_{ag,\text{small}}}\right)^5 \tag{4.51}$$

For geometrically similar vessels, we have:

$$\frac{D_{ag,\text{large}}}{D_{ag,\text{small}}} = \frac{D_{\text{large}}}{D_{\text{small}}}\left(\frac{V_{\text{large}}}{V_{\text{small}}}\right)^{\frac{1}{3}} \tag{4.52}$$

Inserting Equation (4.52) into (4.50), combining with Equation (4.48) and rearranging we obtain the equation for the required agitation speed in the scaled-up/scaled-down vessel:

$$\frac{N_{\text{large}}}{N_{\text{small}}} = \left(\frac{V_{\text{large}}}{V_{\text{small}}}\right)^{-\frac{1}{9}\left(2+\frac{b}{a}\right)} \tag{4.53}$$

Equation (4.53) shows that the agitation speed required in the scaled-up/scaled-down vessel is inversely proportional to the ratio $\frac{V_{\text{large}}}{V_{\text{small}}}$. This is because of the large effect of the agitator diameter (fifth power) on the power draw. Figure 4.17 shows the plot of $\frac{N_{\text{large}}}{N_{\text{small}}}$ vs $\frac{V_{\text{large}}}{V_{\text{small}}}$ for the same values of the parameters a and b used in Figure 4.15.

Equations (4.49) and (4.53) allow for the calculation of the power draw and agitation speed in scaled-up/scaled-down vessels. The same methodology shown in this section can be used for the scale-up/scale-down of fermentation vessels under different assumptions, e.g. $\left(\frac{Q_G}{V}\right)_{\text{large}} \neq \left(\frac{Q_G}{V}\right)_{\text{small}}$, $\left(\frac{P_G}{P}\right)_{\text{large}} \neq \left(\frac{P_G}{P}\right)_{\text{small}}$ or $P_{0,\text{large}} \neq P_{0,\text{small}}$.

Example 4.10

A fermentation process is carried out at commercial scale in a 10 m³ vessel. The vessel is operated as a continuous fermenter and stirred with a 0.80 m agitator rotating at 90 rpm, giving a gassed power draw of 5.8 kW. In order to identify strategies to optimise the fermentation, it is desired to scale down the process to a 100 L pilot scale fermenter, geometrically similar to the one at commercial scale. It is desired to maintain the same dissolved oxygen concentration

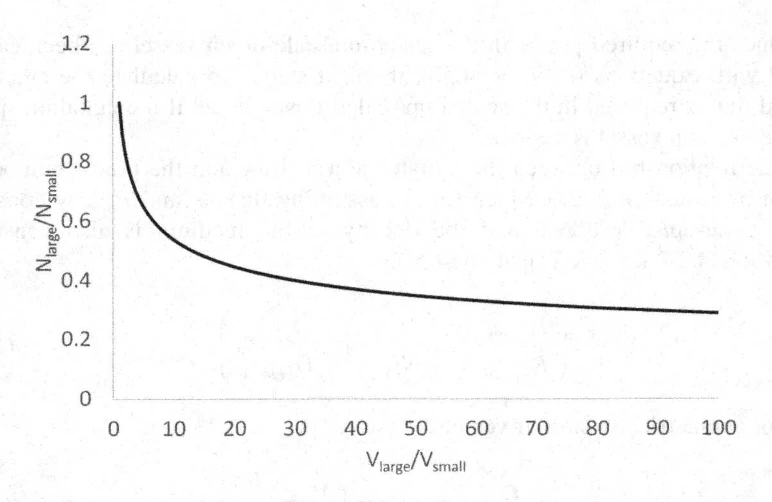

FIGURE 4.17 Plot of Equation (4.52) with $a = 0.7$ and $b = 0.3$.

in the small reactor as in the large reactor. Assume that all the fermentation parameters (temperature, pH and residence time) remain constant with the scale-down. Assume that the gas flow rate per unit volume, the ratio between the gassed and ungassed power draw and the power number remain constant on scale-down. For this vessel, the k_La is described by Equation (4.22) with $a = 0.65$ and $b = 0.28$.

Calculate the required gassed power draw and agitation speed in the small reactor.

SOLUTION

The ratio $\frac{V_{large}}{V_{small}}$ is 100. In this case, Equation (4.46) gives:

$$\frac{P_{G,large}}{P_{G,small}} = (100)^{1-\frac{0.28}{3 \cdot 0.65}} = 51.5$$

The required gassed power draw in the small reactor is therefore:

$$P_{G,small} = \frac{5.8\ kW}{51.5} = 0.11kW$$

The required agitation speed is given by Equation (4.49), which in this case is:

$$\frac{N_{large}}{N_{small}} = 100^{-\frac{1}{9}\left(2+\frac{0.28}{0.65}\right)} = 0.29$$

The required agitation speed in the small reactor is therefore:

$$N_{small} = \frac{N_{large}}{0.29} = 310rpm$$

Questions and Problems

4.1 Explain which factors affect the oxygen transfer rate in agitated vessels.

4.2 An experiment has been carried out in an agitated vessel to measure the mass transfer coefficient k_La, obtaining the graph in Figure 4.18. Calculate the rate of oxygen transfer in this vessel (kg/m³·min), under the conditions of the experiment, if the equilibrium concentration of oxygen in water is 9.2 mg/L and the oxygen concentration in the fermentation medium is 2.1 mg/L.

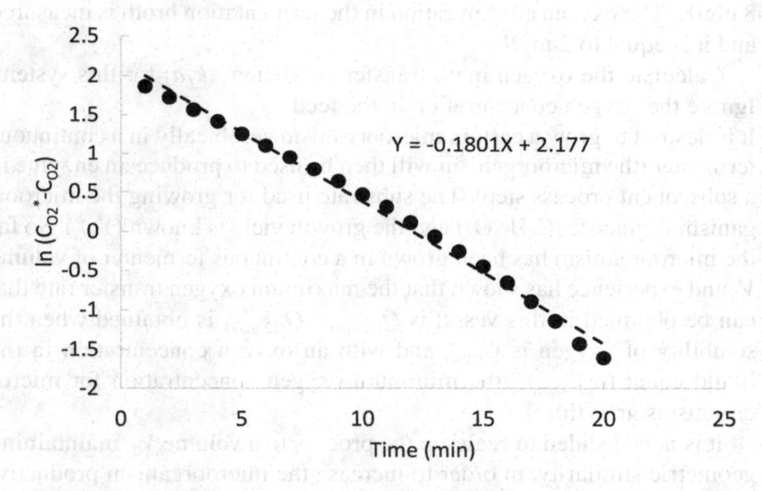

FIGURE 4.18 Plot for Problem 4.2. The dashed line represents the linear regression of the experimental data.

4.3 A fermenter has a diameter of 2 m, an agitator diameter of 0.6 m and a liquid height of 2 m. What are the diameter, agitator diameter and liquid height for a geometrically similar vessel having a working volume of 15 m³?

4.4 A fermenter has a diameter of 0.9 m and a volume of 1 m³. The fermenter is aerated with an air flow rate of 250 m³/d. The diameter of the agitator is 0.3 m and the agitator speed is 150 rpm. What agitation speed is required to have the same k_La in a geometrically similar vessel, having a volume of 10 m³? The large fermenter is aerated with an air flow rate of 2,500 m³/d. Assume that the power number of the agitator is 4.0 for both vessels. Assume that the ratio P_G/P is the same for both vessels in all conditions.

For these vessels, the mass transfer coefficient is described by the correlation:

$$k_La = \alpha\left(\frac{P_G}{V}\right)^{0.62}\left(\frac{Q_G}{A}\right)^{0.34} \quad \text{with } P_G/V \text{ in W/m}^3 \text{ and } Q_G/A \text{ in m/s.}$$

4.5 The stoichiometry of a microorganism growing on glucose as carbon source can be described by the following equation:

$$C_6H_{12}O_6 + 3O_2 + 0.48NH_3 \rightarrow 0.48C_6H_{10}NO_3 + 4.32H_2O + 3.12CO_2$$

The microorganism grows in a continuous fermenter operated at a dilution rate of 0.5 d^{-1}. The glucose concentration is 40 g/L in the influent and 4 g/L in the effluent. The fermenter is aerated with air, and the oxygen concentration in equilibrium with air can be assumed to be 8 mg/L. The oxygen concentration in the fermentation broth is measured and it is equal to 2 mg/L.

Calculate the oxygen mass transfer coefficient (k_La) for this system. Ignore the oxygen concentration in the feed.

4.6 It is desired to grow a certain microorganism aerobically in a continuous fermenter (the microorganism will then be used to produce an enzyme in a subsequent process step). The substrate used for growing the microorganism is glucose ($C_6H_{12}O_6$) and the growth yield is known ($Y_{X/S}$). So far the microorganism has been grown in a continuous fermenter of volume V_1 and experience has shown that the maximum oxygen transfer rate that can be obtained in this vessel is $Q_{O2,max1}$. $Q_{O2,max1}$ is obtained when the solubility of oxygen is C_{O2sat} and with an oxygen concentration in the liquid equal to C_{O2min} (the minimum oxygen concentration for microorganisms growth).

It is now decided to scale-up the process to a volume V_2, maintaining geometric similarity, in order to increase the microorganism productivity. In scaling up the process the superficial gas velocity is kept constant, while the power per unit volume is changed, due to limitations in the maximum power available in the scaled-up reactor: $\left(\frac{P_G}{V}\right)_2 = \gamma\left(\frac{P_G}{V}\right)_1$. Assume that the ratio P_G/P is the same in all conditions. Calculate the maximum microorganism productivity (in kg/d) that can be obtained in the scaled-up reactor.

Assumptions:

a. Assume that the products of glucose fermentation are only microorganisms (C5H7O2N), water and carbon dioxide. Assume that nitrogen is supplied in the feed as ammonia (NH3).
b. Assume that for the considered vessels the mass transfer coefficient for oxygen in water (kLa) is given by: $k_La = \alpha\left(\frac{P_G}{V}\right)^{0.7}\left(\frac{Q_G}{A}\right)^{0.2}$.
c. In carrying out the mass balances for oxygen in the fermenter, ignore the contribution of the dissolved oxygen in the inlet and outlet streams.

Data:

$Q_{O2,max1} = 50$ g/min (maximum oxygen transfer rate in reactor of volume V1)

$C_{O2sat} = 8$ mg/L (oxygen concentration in water/fermentation broth at equilibrium with air)

$C_{O2min} = 1$ mg/L	(minimum oxygen concentration to be maintained in the reactor)
$Y_{X/S} = 0.3$ g/g	(biomass produced per unit of glucose consumed)
$\gamma = 0.5$	(ratio of the power per unit volume in the two reactors)
$V_1 = 1$ m³	(volume of the small reactor)
$V_2 = 10$ m³	(volume of the scaled-up reactor)

4.7 It is desired to produce a certain microorganism at a high rate in a continuous process. The microorganism grows aerobically using acetic acid (CH_3COOH) as substrate and ammonia as nitrogen source and producing new microorganisms (assume an empirical formula $C_5H_7O_2N$), carbon dioxide and water. The fermenter will be operated with a substrate concentration in the feed equal to 20 g/L and at a dilution rate $D = 0.9\,\mu_{max}$, where μ_{max} is the maximum growth rate of the microorganism.

A batch test on the same microorganisms have produced the data in Figure 4.19 below. The batch test was carried out with an initial acetic acid concentration equal to 1 kg/m³ and, at the end of the test when all the substrate was removed, the observed production of microorganisms was 0.3 kg/m³.

In this graph, the variable on the Y-axis is $\frac{X}{(-r_S)}$, where X is the biomass concentration and $(-r_S)$ is the substrate removal rate, and the variable on the X-axis is $\frac{1}{S}$ where S is the substrate concentration.

What minimum value of k_La needs to be provided in the continuous fermenter, in order to maintain the oxygen concentration at least to 2 mg/L?

Assume that air is used to provide oxygen in the fermenter. Assume that the solubility of oxygen in water under the conditions of the fermenter is 8.5 mg/L. Consider negligible the oxygen concentration in the feed.

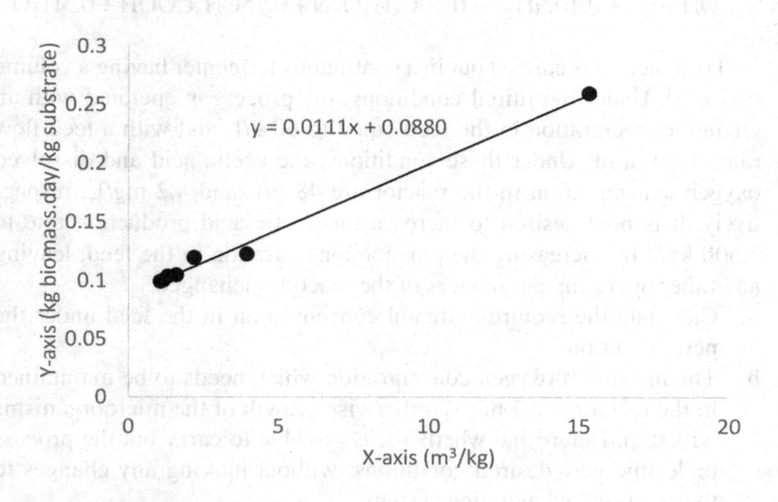

FIGURE 4.19 Plot for Problem 4.7.

4.8 Consider the fermentation reaction for the production of citric acid below:

$$C_6H_{12}O_6 + 2.35O_2 + 0.1NH_3 \rightarrow 0.1C_5H_7O_2N + 0.7C_6H_8O_7 + 1.3CO_2 + 3H_2O$$

Assume that the fermentation is carried out in batch and that each batch is inoculated with 100 mg/L of the microorganism, already acclimated to the substrate. Aeration is carried out by diffusers with a constant air flow rate.

It is desired to maintain a constant oxygen concentration in the liquid phase of 4 mg/L throughout all the fermentation and this will be achieved by adjusting the stirring speed of the agitator.

Calculate the stirring speed that will be required after 4.5 h from the start of the fermentation.

Assumptions and data:

a. For the particular reactor where the fermentation takes place and for the applied air flow rate, the mass transfer coefficient (k_La) is related to the agitation power (P) by the equation $k_La = 0.020 \cdot P^{0.8}$ and the agitation power relates to the stirring speed (N) according to the equation $P = 2.40 \cdot N^3$. In these equations, k_La is in s^{-1}, P is in W and N is in s^{-1}.
b. The microorganism follows Monod growth kinetics with $S \gg K_S$.
c. Under the conditions of the fermentation, the doubling time of the microorganism is 58 min and the oxygen concentration in the liquid phase in equilibrium with the air phase is 8.6 mg/L.

4.9 The reaction by which acetic acid bacteria produce acetic acid (CH_3COOH) from ethanol is the following:

$$C_2H_5OH + O_2 + 0.10NH_3 \rightarrow 0.10C_5H_7O_2N + 0.75CH_3COOH + 0.3H_2O$$

The reaction is carried out in a continuous fermenter having a volume of 10 m^3. Under the initial conditions, the process is operated with an ethanol concentration in the feed equal to 60 g/L and with a feed flow rate of 20 m^3/d. Under these conditions, the acetic acid and dissolved oxygen concentration in the reactor are 48 g/L and 4.2 mg/L, respectively. It is now desired to increase the acetic acid production rate to 1,500 kg/d by increasing the ethanol concentration in the feed, leaving any other operating parameters of the reactor unchanged.

a. Calculate the required ethanol concentration in the feed under the new conditions.
b. The minimum oxygen concentration which needs to be maintained in the reactor is 1.5 mg/L, otherwise growth of the microorganisms will stop. Determine whether it is possible to carry out the process under the new desired conditions, without making any changes to the aeration and agitation system.

Note: Ignore any changes in volume due to the water produced by the reaction and any effects of the microorganisms' concentration on the mass transfer coefficient. Assume that the saturation concentration for oxygen in water is in all conditions 8.0 mg/L.

4.10 An aerobic fermentation process is carried out as a continuous process. The reactor has a volume of 5 m^3 and a cylindrical shape. The agitator rotates at 180 rpm. In this reactor, the oxygen concentration in the fermentation medium is 4.0 mg/L.

It is desired to scale-up the process in a 25 m^3 vessel, geometrically similar to the smaller vessel. In the scale-up, the concentration of the feed and the residence time are kept constant. Furthermore, the air flow rate per unit volume of reactor is also kept constant. To save energy, in the large vessel it is desired to maintain the oxygen concentration at 2.0 mg/L. It is expected that the metabolism of the microorganisms will not be affected by the change in oxygen concentration.

Calculate the agitator speed which is required in the larger vessel.

The power number of the agitator $\left(P_0 = \frac{P}{\rho_L N^3 D_{ag}^5}\right)$ can be assumed to be the same in all fermentation conditions. Assume that the density of the liquid phase is the same in all conditions. For these vessels, the mass transfer coefficient is described by the following correlation:

$$k_L a = k\left(\frac{P_G}{V}\right)^{0.75}\left(\frac{Q_G}{A}\right)^{0.45}$$

Assume that the ratio P_G/P is the same in all conditions and that the equilibrium concentration of oxygen in water is $C^*_{O2} = 8.0$ mg/L in all conditions. In the oxygen balance ignore the contribution of the oxygen in the feed and in the outlet stream.

4.11 A fermentation is carried out in a lab-scale continuous process having a volume of 5 L, with an agitator running at 120 rpm. For vessels of this geometry, the mass transfer coefficient for oxygen in water is given by the following correlation:

$$k_L a = 0.03\left(\frac{P_G}{V}\right)^{0.69}\left(\frac{Q_G}{A}\right)^{0.39}$$

In this correlation, $k_L a$ is in s^{-1}, P_G in W, V in m^3, Q_G in m^3/s and A in m^2. Assume that in all conditions $P_G/P = 0.43$ and $P_0 = 3.6$.

It is now desired to scale-up the process using a 30 L reactor, to be operated as a continuous fermenter with the same feed and residence time as the lab-scale reactor. In the scale-up process, it is desired to maintain the superficial gas velocity at the same value as for the lab-scale reactor. Calculate the agitation rate that is required in the large reactor to maintain the same dissolved oxygen as in the lab-scale reactor.

Assume that the oxygen saturation concentration will stay the same in the large and small reactors.

4.12 Under certain agitation and gas flow conditions, the mass transfer rate for oxygen in a fermenter at 25°C is equal to 0.14 kg/m^3·s, with a dissolved oxygen concentration of 2.5 mg/L. Calculate the mass transfer rate for oxygen in the same vessel when the temperature is 35°C (physical properties in Table 4.2), for the same agitation conditions, mass flow of the inlet air and dissolved oxygen concentration. Assume that the parameters a and k_l are described by Equations (4.19) and (4.20), respectively. Assume that the interfacial tension, the liquid density (1,000 kg/m^3), the superficial gas velocity and the terminal gas velocity do not depend on temperature. Assume that the diameter of gas bubbles is in all conditions equal to 3.0×10^{-3} m.

TABLE 4.2
Physical Properties for Problem 4.12

Property	25°C	35°C
C^*_{O2} (mg/L)	9.3	7.5
Diffusivity of oxygen in water (cm^2/s)	2.2×10^{-5}	3.0×10^{-5}
Viscosity of water (Pa·s)	8.90×10^{-4}	7.20×10^{-4}

4.13 A fermentation reaction is characterised by the following values of the parameters: $Y_{X/S} = 0.10$ kg/kg, $Y_{X/O2} = 0.18$ kg/kg and $Y_{P/X} = 5.20$ kg/kg. It is desired to carry out this fermentation in a 2 m^3 continuous fermenter with a residence time of 8 h, maintaining the oxygen concentration at at least 2.0 mg/L. The maximum k_La that is possible to achieve in this vessel is 0.087 s^{-1} and the saturation concentration for oxygen is 9.0 mg/L. Calculate the maximum product productivity (in kg/d) that is possible to achieve in this process and the maximum substrate concentration that we can have in the feed, assuming that the effluent substrate concentration is negligible compared to the one in the feed.

4.14 A continuous fermentation process is operated with a residence time of 12 h. Under these conditions, the substrate in the feed is converted practically completely. The dissolved oxygen concentration is 3.5 mg/L and the biomass concentration is 4.0 kg/m^3. In order to increase the productivity of the process, it is proposed to decrease the residence time to 10 h, still expecting total conversion of the substrate. Determine whether it is possible to decrease the residence time to 10 h, while maintaining a dissolved oxygen concentration of at least 1.5, without any changes to the agitation rate and to the gas flow rate. Assume for the parameter $Y_{X/O2}$ a value of 0.25 kg/kg and for the oxygen saturation in water a value of 9.2 mg/L.

4.15 Assuming that we have only available 1 fermenter of a given volume, describe a series of experiments aimed at measuring the parameters of k, a and b of Equation (4.22).

4.16 Consider Equation (4.36) for the required k_La in a continuous process. Derive modified equations for a microbial metabolism based on the Monod model with (a) maintenance and (b) endogenous metabolism.

4.17 A fermentation process is carried out in batch in a 3 m³ reactor. The kinetic parameters of the fermentation are $\mu_{max} = 4.8\ d^{-1}$, $Y_{X/S} = 0.20$ kg/kg and $Y_{X/O2} = 0.25$ kg/kg. The batch is started with a substrate concentration of 20 kg/m³ and inoculated with biomass concentration of 0.1 kg/m³. The fermentation is stopped when the substrate concentration reaches 0.2 kg/m³ and it can be assumed that, during all the fermentation, $S \gg K_S$. It is desired to maintain the oxygen concentration to 3.0 mg/L throughout the length of the batch by manipulating the air flow rate while keeping the agitator power draw constant. The oxygen saturation is 9.0 mg/L.

Derive and plot an equation for the time profile of the air flow rate during this batch process.

The k_La for this reactor is given by the following correlation:

$$k_La = 0.045\left(\frac{P_G}{V}\right)^{0.60}\left(\frac{Q_G}{A}\right)^{0.42}$$

In this correlation, k_La is in s^{-1}, P_G in W, V in m³, Q_G in m³/s and A in m². Assume that in all conditions $P_G = 1.0$ kW. Assume $A = 2.8$ m².

4.18 A fermentation process is carried out in a continuous process having an agitator of 1.3 m diameter. Under the standard process conditions, the agitation speed is 180 rpm and the dissolved oxygen concentration in the reactor is 4.0 mg/L. To save energy, it is desired to reduce the oxygen concentration to 2.0 mg/L by reducing the agitation speed, without changing the gas flow rate or any other fermentation parameters. It is believed that the performance of the microorganisms will not be affected by the lower oxygen concentration.

Calculate the new required agitation rate and the energy saved over 1 year (assume the reactor operates for 8,000 h/year). The mass transfer coefficient is described by the following correlation:

$$k_La = 0.025\left(\frac{P_G}{V}\right)^{0.70}\left(\frac{Q_G}{A}\right)^{0.26}$$

In this correlation, k_La is in s^{-1}, P_G in W, V in m³, Q_G in m³/s and A in m². Assume that the ratio $P_G/P = 0.43$ and $P_0 = 4.2$ in all conditions. In the oxygen balance, ignore the contributions of the mass flow of dissolved oxygen with the influent and outlet streams. Assume the saturation concentration for oxygen in water is 8.5 mg/L and the parameter $Y_{X/O2}$ is 0.15 kg/kg.

4.19 A continuous fermenter with a working volume of 30 m^3 is aerated with an air flow rate of 3,000 m^3/h. The inlet air (composed of 21% oxygen by volume) is at 25°C and 1 atm. A fermentation process is carried out characterised by a dilution rate of 1.9 d^{-1} and a biomass concentration of 3.5 kg/m^3. For this process, $Y_{X/O2}$ = 0.29 kg/kg. Calculate the efficiency of oxygen transfer in this process. In the oxygen balance, ignore the contributions of the mass flow of dissolved oxygen with the influent and outlet streams. Assume ideal gas law for the inlet stream with R = 0.0821 L·atm/mol·K.

4.20 A fermentation process is characterised by the following kinetic parameters: μ_{max} = 5.8 d^{-1}; K_S = 0.08 kg/m^3; $Y_{X/S}$ = 0.15 kg/kg; $Y_{X/O2}$ = 0.22 kg/kg and $Y_{P/S}$ = 0.40 kg/kg. It is desired to run the process in a continuous fermenter having a working volume of 10 m^3, with an influent substrate concentration of 45 kg/m^3 and an effluent substrate concentration not higher than 0.1 kg/m^3. Calculate the product productivity (kg/d) and the k_La required for this process, ignoring the contributions of the oxygen mass flow in the inlet and outlet streams. Assume that the oxygen saturation is 9 mg/L and the desired oxygen concentration in the reactor is 2 mg/L.

5 Heat Generation and Heat Balances

Microorganisms only work in a certain temperature range and have an optimum temperature for their metabolism. Most *Saccharomyces cerevisiae* strains used for bioethanol production have an optimum temperature in the range 30–35°C, while the optimum temperature for citric acid production with *Aspergillus niger* and other fungi is in the range 25–30°C. Anaerobic digestion processes, which use mixed microbial culture to convert organic waste and biomass into methane, are usually operated in the temperature range 35–40°C.

If the fermentation temperature goes outside the optimum range, i.e. below the minimum or above the maximum allowable temperature, the performance of the process will be severely compromised. It is, therefore, important to make sure that the temperature of the fermentation remains in the right range.

5.1 ENTHALPY OF REACTION

Like any other chemical reactions, fermentation reactions have heat effects. Fermentation reactions may generate or absorb heat. For a generic chemical reaction:

$$A + \beta B \rightarrow \gamma C + \delta D \tag{5.1}$$

The heat of reaction is defined as:

$$\Delta H_r \left(\frac{J}{mol} \right) = \gamma H_C + \delta H_D - H_A - \beta H_B \tag{5.2}$$

In Equation (5.2), the terms H_i represent the enthalpies of the various chemical species, calculated according to the general Equation (5.3):

$$H_i(T) \left(\frac{J}{mol} \right) = \lambda H_{fi}^0 + \left[\lambda_{phasechange/mixing} (25^\circ\ C) \right] + \int_{25}^{T} c_{Pi} dt \tag{5.3}$$

In Equation (5.3), the term ΔH_{fi}^0 is the enthalpy of formation of species i in its reference state at 25°C, λ is the enthalpy associated with the phase change from the reference state to the final state and/or to the dissolution of the substance and c_P is the specific heat of the substance in its final state.

DOI: 10.1201/9781003217275-5

If at the temperature T the substance is still in its reference state, if we ignore any heat of mixing and if the specific heat is constant in the temperature range 25°C-T, then Equation (5.3) can be simplified into:

$$H_i(T)\left(\frac{J}{mol}\right) = \Delta H^0_{fi} + c_{Pi}(T-25) \tag{5.4}$$

The standard enthalpy of reaction, ΔH^0_r, is the enthalpy of the reaction when all the species are in their reference state at 25°C (Equation 5.5):

$$\Delta H^0_r\left(\frac{J}{mol}\right) = \gamma\Delta H^0_{fC} + \delta\Delta H^0_{fD} - \Delta H^0_{fA} - \beta\Delta H^0_{fB} \tag{5.5}$$

Note that for the generic reaction (Equation 5.1), the enthalpy of reaction, ΔH_r (J/mol), is expressed per mol of reagent A converted. From the enthalpy of reaction, the rate of heat generation per unit volume of reactor can be calculated:

$$r_H\left(\frac{J}{m^3d}\right) = (-r_A)(-\Delta H_r) \tag{5.6}$$

where r_A is the rate of reaction referred to component A, as mol/m³·s. The rate of heat generation can be expressed as a function of the production/consumption rate of other reactants if the stoichiometry of the reaction is known. For example, if the rate of production of component C is known (r_C, mol/m³·s) and the reaction stoichiometry is Equation (5.1), then the rate of heat generation is:

$$r_H\left(\frac{J}{m^3d}\right) = \frac{r_C}{\gamma}(-\Delta H_r) \tag{5.7}$$

Example 5.1

Consider the fermentation stoichiometry for citric acid production (data in Table 5.1):

$$C_6H_{12}O_6 + 2.35O_2 + 0.1NH_3 \rightarrow 0.1C_5H_7O_2N$$
$$+ 0.7C_6H_8O_7 + 1.3CO_2 + 3H_2$$

a. Calculate the enthalpy of the reaction at 25°C and with each substance in its reference state.
b. Calculate the actual enthalpy of reaction under the fermentation conditions (35°C), where glucose, oxygen, ammonia, citric acid and water are in the liquid phase, microorganisms are solid and carbon dioxide is a gas.

TABLE 5.1
Physical Properties and Enthalpies for Example 5.1

Substance	Reference State	ΔH_f^0 (J/kg)	Enthalpy of Phase Change (any T)	Specific Heat (J/kg·K)
Glucose	Solid	-7.074×10^6	(Solid to dissolved in water) 6.11×10^4	(Solid) 1.22×10^3
Oxygen	Gas	0	(Gas to dissolved in water) -3.656×10^5	(Gas) 9.10×10^2
Ammonia	Gas	-2.717×10^6	(Gas to dissolved in water) -1.794×10^6	(Gas) 1.95×10^2
Biomass	Solid	-8.000×10^6	—	(Solid) 3.00×10^3
Citric acid	Solid	-1.021×10^7	(Solid to dissolved in water) 9.484×10^4	(Solid) 1.18×10^3
Carbon dioxide	Gas	-9.000×10^6	—	(Gas) 8.71×10^2
Water	Liquid	-1.600×10^7	—	(Liquid) 4.19×10^3

SOLUTION

a. Standard enthalpy of formation:

$$\Delta H_r^0 = 0.1\Delta H_{f,C5H7O2N}(\text{solid}, 25°C) + 0.7\Delta H_{f,C6H8O7}(\text{solid}, 25°C)$$

$$+1.3\Delta H_{f,CO2}(\text{gas}, 25°C) + 3\Delta H_{f,H2O}(\text{water}, 25°C)$$

$$-\Delta H_{f,C6H12O6}(\text{solid}, 25°C) - 2.35\Delta H_{f,O2}(\text{gas}, 25°C) 0.1\Delta H_{f,NH3}(\text{gas}, 25°C)$$

$$= 0.1\left(-8.000 \cdot 10^6 \frac{J}{kg}\right) 113 \frac{kg}{kmol} + 0.7\left(-1.021 \cdot 10^7 \frac{J}{kg}\right) 192 \frac{kg}{kmol}$$

$$+1.3\left(-9.000 \cdot 10^6 \frac{J}{kg}\right) 44 \frac{kg}{kmol} + 3\left(-1.6 \cdot 10^7 \frac{J}{kg}\right) 18 \frac{kg}{kmol}$$

$$-\left(-7.074 \cdot 10^6 \frac{J}{kg}\right) 180 \frac{kg}{kmol} - 0 - 0.1\left(-2.717 \cdot 10^6 \frac{J}{kg}\right) 17 \frac{kg}{kmol} = -1.563 \cdot 10^9 \frac{J}{kmol}$$

b. Under the actual fermentation conditions:

$$\Delta H_r = 0.1H_{C5H7O2N}(\text{solid}, 35°C) + 0.7H_{C6H8O7}(\text{liquid}, 35°C)$$

$$+H_{CO2}(\text{gas}, 35°C) + 3H_{H2O}(\text{liquid}, 35°C) - H_{C6H12O6}(\text{liquid}, 35°C)$$

$$-2.35H_{O2}(\text{gas}, 35°C) - 0.1H_{NH3}(\text{liquid}, 35°C)$$

$$H_{C5H7O2N}(\text{solid}, 35°C) = -8.000 \cdot 10^6 \frac{J}{kg} \cdot 113 \frac{kg}{kmol} + 3.00 \cdot 10^3 \frac{J}{kg°C} \cdot 113 \frac{kg}{kmol}(35-25)°C = -9.006 \cdot 10^8 \frac{J}{kmol}$$

$$H_{C6H8O7}(\text{liquid}, 35°C) = -1.021 \cdot 10^7 \frac{J}{kg} \cdot 192 \frac{kg}{kmol} + 9.484 \cdot 10^4 \frac{J}{kg} \cdot 192 \frac{kg}{kmol} + 1.18 \cdot 10^3 \frac{J}{kg°C} \cdot 192 \frac{kg}{kmol}(35-25)°C = -1.938 \cdot 10^9 \frac{J}{kmol}$$

$$H_{CO2}(\text{gas}, 35°C) = -9.000 \cdot 10^6 \frac{J}{kg} \cdot 44 \frac{kg}{kmol} + 8.71 \cdot 10^2 \frac{J}{kg°C} \cdot 44 \frac{kg}{kmol}(35-25)°C = -3.956 \cdot 10^8 \frac{J}{kmol}$$

$$H_{H2O}(\text{liquid}, 35°C) = -1.600 \cdot 10^7 \frac{J}{kg} \cdot 18 \frac{kg}{kmol} + 4.19 \cdot 10^3 \frac{J}{kg°C} \cdot 18 \frac{kg}{kmol}(35-25)°C = -2.872 \cdot 10^8 \frac{J}{kmol}$$

$$H_{C6H12O6}(\text{dissolved}, 35°C) = -7.074 \cdot 10^6 \frac{J}{kg} \cdot 180 \frac{kg}{kmol} + 6.11 \cdot 10^4 \frac{J}{kg} \cdot 180 \frac{kg}{kmol} + 1.22 \cdot 10^3 \frac{J}{kg°C} \cdot 180 \frac{kg}{kmol}(35-25)°C = -1.240 \cdot 10^9 \frac{J}{kmol}$$

$$H_{O2}(\text{dissolved}, 35°C) = 0 + 3.656 \cdot 10^5 \frac{J}{kg} \cdot 32 \frac{kg}{kmol} + 9.10 \cdot 10^2 \frac{J}{kg°C} \cdot 32 \frac{kg}{kmol}(35-25)°C = -1.199 \cdot 10^7 \frac{J}{kmol}$$

$$H_{NH3}(\text{dissolved}, 35°C) = -2.717 \cdot 10^6 \frac{J}{kg} \cdot 17 \frac{kg}{kmol} - 1.794 \cdot 10^6 \frac{J}{kg} \cdot 17 \frac{kg}{kmol} + 1.95 \cdot 10^2 \frac{J}{kg°C} \cdot 17 \frac{kg}{kmol}(35-25)°C = -7.665 \cdot 10^7 \frac{J}{kmol}$$

Therefore:

$$\Delta H_r = -1.428 \cdot 10^9 \frac{J}{kmol}$$

Example 5.2

Consider the fermentation reaction of Example 5.1, with $\Delta H_r^0 = -1.563 \cdot 10^6 \frac{J}{mol}$. Assuming that all the reagents and products are in their reference state and at the reference temperature, what is the rate of heat generation when the rate of citric acid ($C_6H_8O_7$) production is 10 kg/m³·d?

SOLUTION

From the reaction stoichiometry, we produce 0.75 kg of citric acid per kg of glucose converted. To produce 10 kg/m³·d of citric acid, we need to convert 10/0.75 = 13.33 kg of glucose/m³·d. In mol terms, this is 74.07 mol of glucose/m³·d and the rate of heat generation is:

$$r_H\left(\frac{J}{m^3d}\right) = 74.07\frac{mol\ \text{glucose}}{m^3d}\left(1.563 \cdot 10^6 \frac{J}{mol\ \text{glucose}}\right) = 1.158 \cdot 10^8 \frac{J}{m^3d}$$

5.2 RATE OF HEAT GENERATION OF FERMENTATION REACTIONS

In fermentation reactions, it is often useful to express the rate of heat generation as a function of the rate of biomass growth. To link the rate of heat generation with the rate of biomass growth, we introduce the coefficient Y_H, defined as:

$$Y_H\left(\frac{kg}{J}\right) = \frac{\text{Biomass produced}}{\text{Heat generated}} \tag{5.8}$$

Using the coefficient Y_H, the rate of heat generation r_H can be expressed as:

$$r_H\left(\frac{J}{m^3d}\right) = \frac{r_X\left(\frac{\text{kgbiomass}}{m^3d}\right)}{Y_H\left(\frac{\text{kgbiomass}}{J}\right)} = \frac{\mu \cdot X}{Y_H} \tag{5.9}$$

Note that according to these definitions, Y_H and r_H are positive for exothermic reactions, which generate heat and have a negative enthalpy of reaction ΔH_r. Most fermentation processes are exothermic. The coefficient Y_H can be calculated from the enthalpy of the fermentation reaction. If the enthalpy of the fermentation reaction is not known, Y_H can still be calculated if the enthalpy of combustion of the substrate and of the microorganisms is known, and if the growth yield $Y_{X/S}$ is also known. The calculation of the coefficient Y_H from the enthalpies of combustion of the substrate and biomass is shown with reference to the fermentation stoichiometry Equation (5.10), as an example, but the equations obtained have general validity:

$$C_6H_{12}O_6 + 2.55O_2 + 0.60NH_3 \rightarrow 0.60C_5H_7O_2N + 3CO_2 + 3.9H_2O \tag{5.10}$$

Using Equation (5.10), the enthalpy of the reactants in their reference state at 25°C is (the enthalpy of formation of molecular oxygen is 0, as it is an element in its reference state):

$$H_{\text{reactants}} = 0.60 H_{\text{NH3}} + H_{\text{Glucose}} \tag{5.11}$$

And the enthalpy of the products is:

$$H_{\text{products}} = 0.60 H_{\text{biom}} + 3H_{\text{CO2}} + 3.9 H_{\text{H2O}} \tag{5.12}$$

Since we are calculating the enthalpy of reaction with the substances in their reference state at 25°C, the enthalpy of each species will be equal to their enthalpy of formation. The enthalpy of the reaction (Equation 5.10) can be written, therefore, as:

$$\Delta H_R = 0.60 \Delta H_{f,\text{biom}} + 3\Delta H_{f,\text{CO2}} + 3.9 \Delta H_{f,\text{H2O}} - 0.60 \Delta H_{f,\text{NH3}} - \Delta H_{f,\text{Glucose}} \tag{5.13}$$

Equation (5.13) can be re-written by introducing the combustion reaction for the substrate (glucose) and for the biomass.

For glucose, the combustion reaction is:

$$C_6H_{12}O_6 + 6O_2 \rightarrow 6CO_2 + 6H_2O \tag{5.14}$$

The enthalpy of reaction (5.14) at 25°C is:

$$\Delta H_{\text{comb,glucose}} \left(\frac{J}{mol \text{ glucose}} \right) = 6\Delta H_{f,\text{CO2}} + 6\Delta H_{f,\text{H2O}} - \Delta H_{f,\text{glucose}} \tag{5.15}$$

Similarly for the biomass, its combustion reaction is:

$$C_5H_7O_2N + 5O_2 \rightarrow 5CO_2 + 2H_2O + NH_3 \tag{5.16}$$

The enthalpy of reaction (5.16) at 25°C is:

$$\Delta H_{\text{comb,biom}} \left(\frac{J}{mol \text{ biom}} \right) = 5\Delta H_{f,\text{CO2}} + 2\Delta H_{f,\text{H2O}} + \Delta H_{f,\text{NH3}} - \Delta H_{f,\text{biom}} \tag{5.17}$$

We can now eliminate the enthalpies of formation of glucose and biomass from Equation (5.13) using Equations (5.15) and (5.17). We obtain:

$$\Delta H_R \left(\frac{J}{mol} \right) = \Delta H_{\text{comb,glucose}} - 0.60 \Delta H_{\text{comb,biom}} \tag{5.18}$$

Equation (5.18) shows that the enthalpy of a fermentation reaction can be calculated from the enthalpy of combustion of the substrate and of the biomass.

For the fermentation reaction (5.10), the coefficient Y_H can be calculated as:

$$Y_H\left(\frac{kg}{J}\right)=\frac{\text{Biomass produced}}{\text{Heat generated}}=\frac{\frac{\text{Biomass produced}}{mol\ glucose}}{\frac{\text{Heat generated}}{mol\ glucose}}=\frac{0.60MW_{\text{biomass}}}{\left(-\Delta H_R\right)} \quad (5.19)$$

The coefficient Y_H can be expressed as a function of the enthalpies of combustion of the substrate and biomass by introducing Equation (5.18) into Equation (5.19):

$$Y_H=\frac{0.60MW_{\text{biom}}}{\left(-\Delta H_{\text{comb,glucose}}\right)-0.60\left(-\Delta H_{\text{comb,biom}}\right)} \quad (5.20)$$

In Equation (5.20), the minus sign has been used before the terms $\Delta H_{\text{comb,glucose}}$ and $\Delta H_{\text{comb,biom}}$ because these enthalpies are negative but Y_H needs to be positive by definition. Equation (5.20) shows the expression for the calculation of the Y_H for the reaction stoichiometry (5.10). However, Equation (5.20) can be made more general by substituting the factor 0.60 with the yield coefficient $Y_{X/S}$. Indeed, $Y_{X/S}$ can be expressed as:

$$Y_{X/S}\left(\frac{kg\text{ biomass}}{kg\text{ substrate}}\right)=\frac{0.60MW_{\text{biom}}}{MW_{\text{glucose}}} \quad (5.21)$$

Combining Equation (5.21) with Equation (5.20), we obtain:

$$Y_H=\frac{Y_{X/S}}{\frac{\left(-\Delta H_{\text{comb.glucose}}\right)}{MW_{\text{glucose}}}-\frac{Y_{X/S}\cdot\left(-\Delta H_{\text{comb,biom}}\right)}{MW_{\text{biom}}}} \quad (5.22)$$

Equation (5.22) gives the general expression for the calculation of Y_H once $Y_{X/S}$ and the enthalpies of combustion of biomass and substrate are known. Note that so far the enthalpies of all the reaction have been expressed in J/mol. However, the terms $\frac{\left(-\Delta H_{\text{comb,biom}}\right)}{MW_{\text{glucose}}}$ and $\frac{\left(-\Delta H_{\text{Comb.biom}}\right)}{MW_{\text{biom}}}$ correspond to the enthalpies of combustion of the substrate and biomass expressed in J/g. Equation (5.22) is, therefore, equivalent to Equation (5.23):

$$Y_H=\frac{Y_{X/S}}{\left(-\Delta H_{\text{comb,glucose}}\right)\left(\frac{J}{g}\right)-Y_{X/S}\cdot\left(-\Delta H_{\text{comb,biom}}\right)\left(\frac{J}{g}\right)} \quad (5.23)$$

In summary, the coefficient Y_H, which allows the calculation of the rate of heat generation per unit of biomass produced, can be calculated from the growth yield and from the enthalpies of combustion of the substrate and biomass. As stated above, Y_H can also be calculated from the enthalpy of the fermentation reaction.

Equation (5.23) can be extended to anaerobic reactions with other products in addition to microorganisms. For example, let us consider the anaerobic fermentation of glucose for the production of ethanol:

$$C_6H_{12}O_6 + 0.31NH_3 \rightarrow 0.31C_5H_7O_2N + 1.64C_2H_6O + 1.17CO_2 + 0.464H_2O \quad (5.24)$$

The enthalpy of this reaction is:

$$\Delta H_R = 0.31\Delta H_{f,\text{biom}} + 1.64\Delta H_{f,\text{C2H6O}} + 1.17\Delta H_{f,\text{CO2}} + 0.464\Delta H_{f,\text{H2O}} - 0.31\Delta H_{f,\text{NH3}} - \Delta H_{f,\text{Glucose}} \quad (5.25)$$

The combustion of ethanol is described by the following reaction:

$$C_2H_6O + 3O_2 \rightarrow 2CO_2 + 3H_2O \quad (5.26)$$

With enthalpy of combustion of:

$$\Delta H_{\text{comb,C2H6O}}\left(\frac{J}{mol\ ethanol}\right) = 2\Delta H_{f,\text{CO2}} + 3\Delta H_{f,\text{H2O}} - \Delta H_{f,\text{C2H6O}} \quad (5.27)$$

Using Equations (5.27), (5.17) and (5.15), the enthalpy of the reaction (5.24) can be written as:

$$\Delta H_R\left(\frac{J}{mol}\right) = \Delta H_{\text{comb,glucose}} - 0.31\Delta H_{\text{comb,biom}} - 1.64\Delta H_{\text{comb,ethanol}} \quad (5.28)$$

The coefficient Y_H in this case is given by:

$$Y_H = \frac{0.31MW_{biom.}}{\left(-\Delta H_{\text{comb,glucose}}\right) - 0.31\left(-\Delta H_{\text{comb,glucose}}\right) - 1.64\left(-\Delta H_{\text{comb,ethanol}}\right)} \quad (5.29)$$

The definitions of $Y_{X/S}$ and $Y_{P/S}$ are, in this case:

$$Y_{X/S}\left(\frac{kg\text{biomass}}{kg\text{substrate}}\right) = \frac{0.31MW_{biom}}{MW_{glucose}} \quad (5.30)$$

$$Y_{P/S}\left(\frac{kg\text{biomass}}{kg\text{substrate}}\right) = \frac{1.64MW_{\text{ethanol}}}{MW_{\text{glucose}GLUCOSE}} \quad (5.31)$$

Using Equations (5.30) and (5.31), the coefficient Y_H can be written as:

$$Y_H = \frac{Y_{X/S}}{\left(-\Delta H_{\text{comb,glucose}}\right)\left(\frac{J}{g}\right) - Y_{X/S}\cdot\left(-\Delta H_{\text{comb,biom}}\right)\left(\frac{J}{g}\right) - Y_{P/S}\cdot\left(-\Delta H_{\text{comb.ethanol}}\right)\left(\frac{J}{g}\right)} \quad (5.32)$$

Equation (5.32) can be written in a more general form:

$$Y_H = \frac{Y_{X/S}}{\left(-\Delta H_{\text{comb,substrate}}\right)\left(\frac{J}{g}\right) - Y_{X/S}\cdot\left(-\Delta H_{\text{comb,biom}}\right)\left(\frac{J}{g}\right) - \Sigma_i Y_{Pi/S}\cdot\left(-\Delta H_{\text{comb,products}}\right)\left(\frac{J}{g}\right)} \quad (5.33)$$

In Equation (5.33), $\Delta H_{\text{comb,products}}$ accounts for the enthalpy of combustion of all the products of the fermentation reaction except biomass (which is included separately in Equation 5.33), carbon dioxide and water (which cannot be combusted). Using the general Equation (5.33), the rate of heat generation in a fermentation reaction can be written as:

$$r_H\left(\frac{J}{m^3 d}\right) = \frac{r_X}{Y_H} = \frac{r_X}{Y_{X/S}}\left[\left(-\Delta H_{\text{comb,substrate}}\right)\left(\frac{J}{g}\right) - Y_{X/S}\cdot\left(-\Delta H_{\text{comb,biom}}\right)\left(\frac{J}{g}\right)\right.$$

$$\left. - \Sigma_i Y_{Pi/S}\cdot\left(-\Delta H_{\text{comb,products}}\right)\left(\frac{J}{g}\right)\right] = \left(-r_S\right)\left[\left(-\Delta H_{\text{comb,substrate}}\right)\left(\frac{J}{g}\right)\right. \quad (5.34)$$

$$\left. - Y_{X/S}\cdot\left(-\Delta H_{\text{comb,biom}}\right)\left(\frac{J}{g}\right) - \Sigma_i Y_{Pi/S}\cdot\left(-\Delta H_{\text{comb,products}}\right)\left(\frac{J}{g}\right)\right]$$

Equation (5.34) shows that fermentation reactions which mainly oxidise the substrate with low generation of microorganisms and products (other than carbon dioxide and water) generate higher heat than reactions with high generation of microorganisms and products. Typically, although not exclusively, reactions with products formation are anaerobic reactions, and for these reactions the heat generated per unit of substrate removed is much lower than for aerobic reactions.

Example 5.3

Calculate the coefficient Y_H (kg/J), the enthalpy of the reaction (J/mol) and the rate of heat generation (J/m³·h) when the growth rate is 1 kg/m³·h for the following reactions:

a. Aerobic conversion of glucose to microorganisms, water and carbon dioxide, with the production of 0.38 kg of microorganisms per 1 kg of glucose removed.
b. Anaerobic conversion of glucose to ethanol, microorganisms, water and carbon dioxide, with the production of 0.19 kg of microorganisms and 0.42 kg of ethanol per 1 kg of glucose removed.

Standard enthalpies of combustion: glucose = -1.556×10^7 J/kg; biomass = -1.503×10^7 J/kg and ethanol = -2.957×10^7 J/kg.

SOLUTION

a. From Equation (5.23):

$$Y_H = \frac{0.38\frac{kg}{kg}}{1.556\cdot 10^7\frac{J}{kg} - 0.38\cdot 1.503\cdot 10^7\frac{J}{kg}} = 0.386\cdot 10^{-7}\frac{kg}{J}$$

If $Y_{X/S} = 0.38$ kg/kg, the stoichiometric coefficient for biomass in the fermentation reaction is 0.60, so, using Equation (5.18):

$$\Delta H_R\left(\frac{J}{mol}\right) = \Delta H_{comb,glucose} - 0.60\Delta H_{comb,biom}$$

$$= -1.556\cdot 10^7\frac{J}{kg}\cdot 0.180\frac{kg}{mol} - 0.60\left(-1.503\cdot 10^7\frac{J}{kg}\cdot 0.113\frac{kg}{mol}\right)$$

$$= -1.78\cdot 10^6\frac{J}{mol}$$

The rate of heat generation is given by Equation (5.9):

$$r_H\left(\frac{J}{m^3\cdot day}\right) = \frac{1\frac{kgbiomass}{m^3\cdot d}}{0.386\cdot 10^{-7}\frac{kg}{J}} = 2.59\cdot 10^7\frac{J}{m^3 d}$$

b. In this case $Y_{X/S} = 0.19$ kg/kg and $Y_{P/S} = 0.42$ kg/kg.
From Equation (5.32):

$$Y_H = \frac{0.19\frac{kg}{kg}}{1.556\cdot 10^7\frac{J}{kg} - 0.19\cdot 1.503\cdot 10^7\frac{J}{kg} - 0.42\cdot 2.957\cdot 10^7\frac{J}{kg}} = 6.67\cdot 10^{-7}\frac{kg}{J}$$

From Equation (5.28):

$$\Delta H_R\left(\frac{J}{mol}\right) = \Delta H_{comb,glucose} - 0.31\Delta H_{comb,biom} - 1.64\Delta H_{comb,ethanol}$$

$$= -1.556\cdot 10^7\frac{J}{kg}\cdot 0.180\frac{kg}{mol} - 0.31\left(-1.503\cdot 10^7\frac{J}{kg}\cdot 0.113\frac{kg}{mol}\right)$$

$$-1.64\left(-2.957\cdot 10^7\frac{J}{kg}\cdot 0.046\frac{kg}{mol}\right) = -4.35\cdot 10^4\frac{J}{mol}$$

The rate of heat generation is given by Equation (5.9):

$$r_H\left(\frac{J}{m^3\cdot d}\right) = \frac{1\frac{kgbiomass}{m^3\cdot d}}{6.67\cdot 10^{-7}\frac{kg}{J}} = 1.50\cdot 10^6\frac{J}{m^3 d}$$

Note that the rate of heat generation is much higher for the aerobic reaction than for the anaerobic one.

5.3 HEAT BALANCES

The general form of heat balances is:

$$\begin{aligned}\text{Enthalpy accumulated} &= \text{Enthalpy in} - \text{Enthalpy out}\\ &+ \text{Enthalpy added / removed}\end{aligned} \tag{5.35}$$

5.3.1 Heat Balances for Continuous Fermentation

Let us consider the continuous fermenter in Figure 5.1 at steady state.

The enthalpy balance Equation (5.35) for the fluid in the reactor becomes:

$$\text{Enthalpy in} - \text{Enthalpy out} + \text{Enthalpy added / removed} = 0 \tag{5.36}$$

The term "Enthalpy in" is given by:

$$\text{Enthalpy in} = \sum_i \dot{m}_i H_i\left(T_{IN}\right) \tag{5.37}$$

where m_i is the mass flow rates and H_i is the enthalpies (per unit of mass) of the components in the feed.

The term "Enthalpy out" is given by:

$$\text{Enthalpy out} = \sum_i \dot{m}_i H_i\left(T_R\right) \tag{5.38}$$

where the enthalpy terms are calculated at the outlet temperature T_R.

The term "Enthalpy added/removed" corresponds to the rate of enthalpy provided/removed to/from the fluid in the reactor by the external system and it is positive if enthalpy is added and negative if enthalpy is removed. For an adiabatic reactor, the Enthalpy added/removed is equal to zero. If heat transfer occurs via a completely mixed jacket, with the fluid in the jacket at a uniform temperature T_J, the term Enthalpy added/removed is given by:

$$\text{Enthalpy added / removed} = UA\left(T_J - T_R\right) \tag{5.39}$$

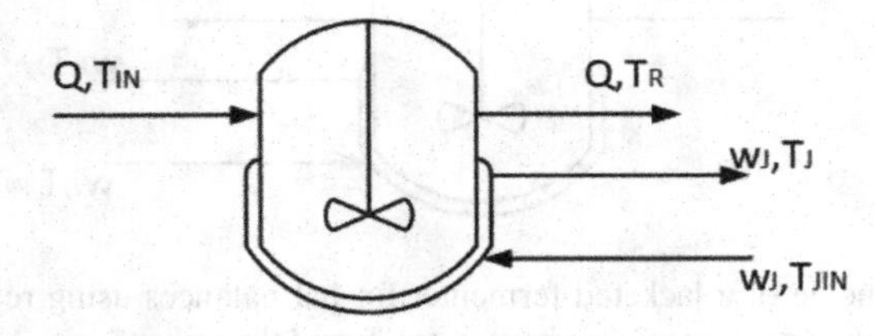

FIGURE 5.1 Generic scheme of a continuous fermenter for heat balances.

In Equation (5.39) if $T_J > T_R$ (positive term), enthalpy is added to the fluid in the reactor, if $T_J < T_R$ enthalpy is removed. Therefore, in this case the enthalpy balance becomes:

$$\sum_i \dot{m}_i H_i(T_{IN}) - \sum_i \dot{m}_i H_i(T_R) + UA(T_J - T_R) = 0 \tag{5.40}$$

We want now to show that the enthalpy balance (5.40) can be written, if we ignore the enthalpy in with any inlet gas stream, as:

$$w_{\text{feed}} C_{P\text{feed}}(T_{IN} - T_R) + r_H V = UA(T_R - T_J) \tag{5.41}$$

In Equation (5.41), w_{feed} represents the mass flow rate of the feed and $C_{P\text{feed}}$ its specific heat.

The equivalence between Equations (5.40) and (5.41) will be shown using, as an example, a typical stoichiometry for the anaerobic fermentation to ethanol:

$$C_6H_{12}O_6 + 0.12NH_3 \rightarrow 0.12C_5H_7O_2N + 1.8C_2H_5OH + 1.8CO_2 + 0.36H_2O \tag{5.42}$$

We will develop the heat balances referring to the scheme in Figure 5.2.

For the reaction stoichiometry (5.42), the various terms in the enthalpy balance (5.36) are shown below.

Enthalpy in:

$$w_{\text{NH3},IN} H_{\text{NH3}}(T_{IN}) + w_{\text{S},IN} H_{\text{S}}(T_{IN}) + w_{\text{H2O},IN} H_{\text{H2O}}(T_{IN}) \tag{5.43}$$

Enthalpy out:

$$\begin{aligned} & w_{\text{X},OUT} H_{\text{X}}(T_R) + w_{\text{ETH},OUT} H_{\text{ETH}}(T_R) + w_{\text{CO2},OUT} H_{\text{CO2}}(T_R) \\ & \quad + w_{\text{H2O},OUT} H_{H2O}(T_R) + w_{\text{S},OUT} H_S(T_R) \\ & \quad + w_{\text{NH3},OUT} H_{\text{NH}_3}(T_R) \end{aligned} \tag{5.44}$$

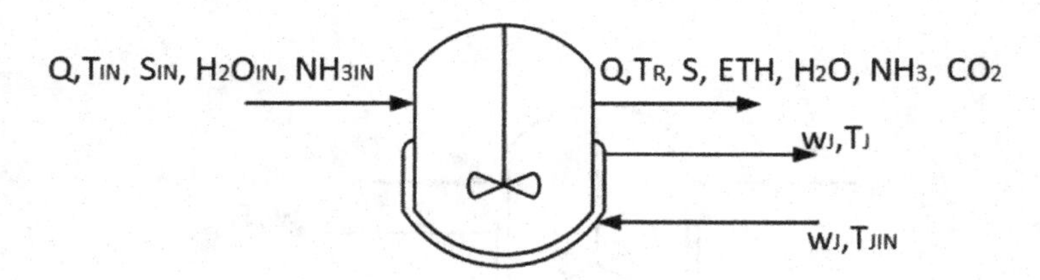

FIGURE 5.2 Scheme of a jacketed fermenter for hat balances using reaction (5.42). Q represents the volumetric flow rate (units, e.g. m^3/d) and the terms S_{IN}, S, H_2O_{IN}, H_2O, etc., represent the concentrations of the components (units, e.g. kg/m^3).

Enthalpy removed:

$$UA(T_R - T_J) \tag{5.39}$$

In Equations (5.43) and (5.44) and in the rest of this section, the term w_i represent the mass flow rate (in units of, e.g. kg/d) of component i at the inlet or outlet of the reactor. Each term w_i can be calculated by the product of the volumetric flow rate Q and the component concentration, e.g. $w_{S,IN} = Q \cdot S_{IN}$, $w_{ETH,OUT} = Q \cdot ETH$.

The overall enthalpy balance is shown by Equation (5.45):

$$\begin{aligned} w_{\mathrm{NH3},IN} H_{\mathrm{NH3}}(T_{IN}) + w_{\mathrm{S},IN} H_{\mathrm{S}}(T_{IN}) + w_{\mathrm{H2O},IN} H_{\mathrm{H2O}}(T_{IN}) &= w_{X,OUT} H_X(T_R) \\ &+ w_{\mathrm{ETH},OUT} H_{\mathrm{ETH}}(T_R) + w_{\mathrm{CO2},OUT} H_{\mathrm{CO2}}(T_R) + w_{\mathrm{H2O},OUT} H_{\mathrm{H2O}}(T_R) \\ &+ w_{\mathrm{S},OUT} H_S(T_R) + w_{\mathrm{NH3},OUT} H_{\mathrm{NH3}}(T_R) + UA(T_R - T_J) \end{aligned} \tag{5.45}$$

Equation (5.45) can be shown to correspond to Equation (5.41) if we do several re-arrangements. We start by writing the enthalpy out term (5.44) as a function of the reaction rate. We express the outlet concentration of each species as a function of the biomass concentration using the equations in (5.46)

$$\begin{cases} \mathrm{ETH}\left(\dfrac{kg}{m^3}\right) = \dfrac{1.8MW_{\mathrm{ETH}}}{0.12MW_{\mathrm{BIOM}}} X \\ \mathrm{CO_2}\left(\dfrac{kg}{m^3}\right) = \dfrac{1.8MW_{\mathrm{CO2}}}{0.12MW_{\mathrm{BIOM}}} X \\ \mathrm{H_2O}\left(\dfrac{kg}{m^3}\right) = H_2O_{IN} + \dfrac{0.36MW_{\mathrm{H2O}}}{0.12MW_{\mathrm{BIOM}}} X \\ S\left(\dfrac{kg}{m^3}\right) = S_{IN} - \dfrac{MW_{\mathrm{S}}}{0.12MW_{\mathrm{BIOM}}} X \\ \mathrm{NH_3}\left(\dfrac{kg}{m^3}\right) = NH_{3IN} - \dfrac{0.12MW_{\mathrm{NH3}}}{0.12MW_{\mathrm{BIOM}}} X \end{cases} \tag{5.46}$$

Substituting Equation (5.46) into Equation (5.44), the enthalpy out term becomes:

$$\begin{aligned} & w_{X,OUT} H_X(T_R) + w_{X,OUT} \frac{1.8MW_{\mathrm{ETH}}}{0.12MW_{\mathrm{BIOM}}} H_{\mathrm{ETH}}(T_R) + w_{X,OUT} \frac{1.8MW_{\mathrm{CO2}}}{0.12MW_{\mathrm{BIOM}}} H_{\mathrm{CO2}}(T_R) \\ & + w_{\mathrm{H2O},IN} H_{\mathrm{H2O}}(T_R) + w_{X,OUT} \frac{0.36MW_{\mathrm{H2O}}}{0.12MW_{\mathrm{BIOM}}} H_{\mathrm{H2O}}(T_R) + w_{S,IN} H_S(T_R) \\ & + w_{X,OUT}\left(-\frac{MW_S}{0.12MW_{\mathrm{BIOM}}}\right) H_S(T_R) + w_{\mathrm{NH3},IN} H_{\mathrm{NH3}}(T_R) \\ & + w_{X,OUT}\left(-\frac{MW_{\mathrm{NH3}}}{MW_{\mathrm{BIOM}}}\right) H_{\mathrm{NH3}}(T_R) \end{aligned} \tag{5.47}$$

Equation (5.47) can be re-written by noting the equivalence between enthalpies expressed on a mass and on a molar basis:

$$H_i\left(\frac{J}{kg}\right)=\frac{H_{i,mol}\left(\frac{J}{kmol}\right)}{MW_i} \tag{5.48}$$

By substituting Equations (5.48) into (5.47), the enthalpy out can be re-written as:

$$\begin{aligned}
& w_{X,OUT}\frac{H_{X,mol}(T_R)}{MW_{BIOM}}+w_{X,OUT}\frac{1.8}{0.12MW_{\mathrm{BIOM}}}H_{\mathrm{ETH},mol}\left(T_R\right)\\
& +w_{X,OUT}\frac{1.8}{0.12MW_{\mathrm{BIOM}}}H_{\mathrm{CO2},mol}\left(T_R\right)\\
& +w_{\mathrm{H2O},IN}H_{\mathrm{H2O}}\left(T_R\right)+w_{X,OUT}\frac{0.36}{0.12MW_{\mathrm{BIOM}}}H_{\mathrm{H2O},mol}\left(T_R\right)\\
& +w_{S,IN}H_S\left(T_R\right)-w_{X,OUT}\frac{1}{0.12MW_{\mathrm{BIOM}}}H_{S,mol}\left(T_R\right)\\
& +w_{\mathrm{NH3},IN}H_{\mathrm{NH3}}\left(T_R\right)-w_{X,OUT}\frac{1}{MW_{\mathrm{BIOM}}}H_{\mathrm{NH3},mol}\left(T_R\right)
\end{aligned} \tag{5.49}$$

The enthalpy of reaction (5.42) can be written as:

$$\begin{aligned}
\Delta H_{R,mol} &= 0.12H_{X,mol}+1.8H_{\mathrm{ETH},mol}+1.8H_{\mathrm{CO2},mol}\\
&\quad +0.36H_{\mathrm{H2O},mol}-H_{S,mol}-0.12H_{\mathrm{NH3},mol}
\end{aligned} \tag{5.50}$$

By substituting Equation (5.50) into Equation (5.49), the enthalpy out term becomes:

$$w_{\mathrm{H2O},IN}H_{\mathrm{H2O}}\left(T_R\right)+w_{S,IN}H_S\left(T_R\right)+w_{\mathrm{NH3},IN}H_{NH3}\left(T_R\right)+w_{X,OUT}\frac{\Delta H_R\left(T_R\right)}{0.12MW_{\mathrm{BIOM}}} \tag{5.51}$$

For reaction (5.42), the coefficient Y_H is defined as:

$$Y_H=\frac{0.12MW_{\mathrm{substrate}}}{\left(-\Delta H_{R,mol}\right)} \tag{5.52}$$

Equation (5.52) becomes:

$$\left(-\Delta H_{R,mol}\right)=\frac{0.12MW_{\mathrm{biomass}}}{Y_H} \tag{5.53}$$

In addition, from the mass balance for the biomass in the fermenter we have:

$$r_X V = QX$$

Introducing these equations in Equation (5.51), we have:

$$w_{\mathrm{H2O},IN} H_{\mathrm{H2O}}(T_R) + w_{S,IN} H_S(T_R) + w_{\mathrm{NH3},IN} H_{NH3}(T_R) - \frac{r_X V}{Y_H} \quad (5.54)$$

Equation (5.54) is equal to:

$$w_{\mathrm{H2O},IN} H_{\mathrm{H2O}}(T_R) + w_{S,IN} H_S(T_R) + w_{\mathrm{NH3},IN} H_{\mathrm{NH3}}(T_R) - r_H V \quad (5.55)$$

Equation (5.55) represents the enthalpy out as a function of the rate of heat generation. Using Equation (5.55), the overall heat balance (5.45) can be re-written as:

$$w_{\mathrm{NH3},IN} H_{\mathrm{NH3}}(T_{IN}) + w_{S,IN} H_S(T_{IN}) + w_{\mathrm{H2O},IN} H_{\mathrm{H2O}}(T_{IN}) = w_{\mathrm{NH3},IN} H_{\mathrm{NH3}}(T_R) + w_{S,IN} H_S(T_R) + w_{\mathrm{H2O},IN} H_{\mathrm{H2O}}(T_R) - r_H V + UA(T_R - T_J) \quad (5.56)$$

The term on the left-hand side of Equation (5.56) is the rate at which enthalpy enters the reactor with the inlet feed:

$$w_{\mathrm{NH3},IN} H_{\mathrm{NH3}}(T_{IN}) + w_{S,IN} H_S(T_{IN}) + w_{\mathrm{H2O},IN} H_{\mathrm{H2O}}(T_{IN}) = w_{\mathrm{feed}} H_{\mathrm{feed}}(T_{IN}) \quad (5.57)$$

On the right-hand side of Equation (5.56), the first term corresponds to the enthalpy flow of a stream having the flow rate and composition of the feed but the temperature of the reactor:

$$w_{\mathrm{NH3},IN} H_{\mathrm{NH3}}(T_R) + w_{S,IN} H_S(T_R) + w_{\mathrm{H2O},IN} H_{\mathrm{H2O}}(T_R) = w_{\mathrm{feed}} H_{\mathrm{feed}}(T_R) \quad (5.58)$$

Using Equations (5.57) and (5.58), the enthalpy balance for the fluid in the reactor, Equation (5.56), becomes:

$$w_{\mathrm{feed}} H_{\mathrm{feed}}(T_{IN}) - w_{\mathrm{feed}} H_{\mathrm{feed}}(T_R) = -r_H V + UA(T_R - T_J) \quad (5.59)$$

Equation (5.59) can be re-arranged into Equations (5.60) and (5.41). Feeds of the fermenters are usually diluted water-based mixtures and in many cases the enthalpy of the feed can be approximated by the enthalpy of water:

$$w_{\mathrm{feed}} \left[H_{\mathrm{feed}}(T_{IN}) - H_{\mathrm{feed}}(T_R) \right] + r_H V = UA(T_R - T_J) \quad (5.60)$$

$$w_{\mathrm{feed}} C_{P\mathrm{feed}}(T_{IN} - T_R) + r_H V = UA(T_R - T_J) \quad (5.41)$$

In summary, the enthalpy balance for the fluid in the reactor (Equation 5.40) can also be expressed by Equation (5.41). Note that $w_{\text{feed}} = Q\rho_{\text{feed}}$ and that the density and specific heat of the feed to fermenters are often similar to the values of water because fermentation feeds are often diluted water solutions. Equation (5.41) is a convenient form to use if the reaction rate and the enthalpy of reaction are known. On the other hand, Equation (5.40) is the equation to use to calculate the heat effects if the inlet and outlet compositions are known but the reaction stoichiometry and enthalpy are not known.

Equation (5.41) can be solved for T_R (expressing r_H according to Equation (5.9) and using the steady-state mass balances for the continuous fermentation process):

$$T_R = \frac{\frac{w_{\text{feed}} C_{p\text{feed}} T_{IN}}{UA} + \frac{w_{\text{feed}} X}{\rho_{\text{feed}} UA Y_H} + T_J}{1 + \frac{w_{\text{feed}} C_{p\text{feed}}}{UA}} \tag{5.61}$$

Equation (5.61) shows that, if the jacket is used for cooling ($T_J < T_R$), higher value of the UA term corresponds to lower value of the temperature in the reactor. This shows that higher values of the heat transfer coefficient and/or of the heat transfer area are a benefit for temperature control in fermenters. Assuming that the rate of reaction and the heat of reaction are known (these are included in the terms X and Y_H in Equation 5.61), Equation (5.61) allows the calculation of the reactor temperature T_R if the jacket temperature T_J is known. To determine T_J, we use the heat balance for the fluid in the jacket, which can be derived using the same general form, Equation (5.35), which in this case becomes:

$$W_J C_{PJ} \left(T_{JIN} - 25\right) + UA\left(T_R - T_J\right) = W_J C_{PJ} \left(T_J - 25\right) \tag{5.62}$$

Equation (5.62) immediately becomes:

$$UA\left(T_R - T_J\right) = W_J C_{PJ} \left(T_J - T_{JIN}\right) \tag{5.63}$$

So, in summary the temperatures of the reactor and of the fluid in the jacket (T_R and T_J) can be calculated by the simultaneous solution of Equations (5.61) (or of its equivalent Equations 5.41 and 5.63). Instead of Equations (5.61) or (5.41), depending on the data available, the general form of the heat balance for the fluid in the reactor Equation (5.40) can be used.

Equations (5.61) and (5.63) can be used to understand the effect of the fermentation parameters and fermenter design on the process temperature T_R and how to control this temperature within the desired range. By solving Equation (5.63) for T_J, we obtain Equation (5.64) and by substituting the value of T_J in Equation (5.61) we obtain Equation (5.65), which shows T_R as a function of the fermentation and fermenter design parameters:

$$T_J = \frac{UAT_R + w_J C_{PJ} T_{JIN} T_{IN}}{UA + w_J C_{PJ}} \tag{5.64}$$

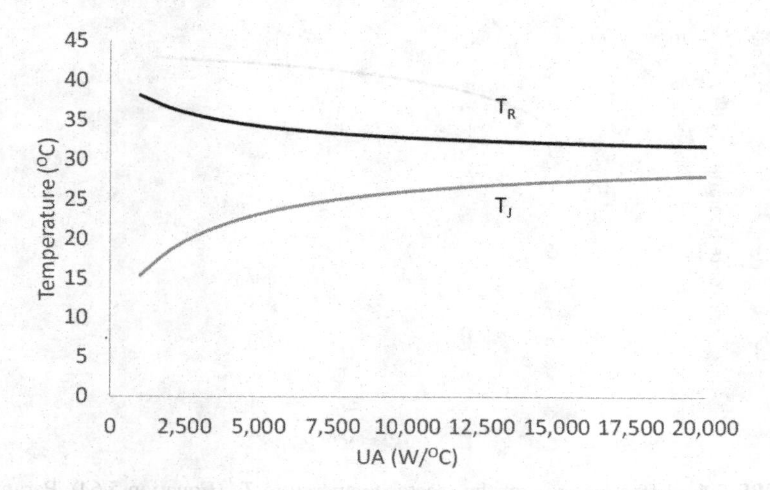

FIGURE 5.3 Effect of *UA* on the reactor temperature T_R (Equation 5.63) and on the jacket temperature T_J (Equation 5.64). Parameter values used: $w_{\text{feed}} = 2$ kg/s, $C_{P\text{feed}} = C_{PJ} =$ 4,186 J/kg·°C, $T_{IN} = 25$°C, $\rho_{\text{feed}} = 1000$ kg/m³, $X = 2$ kg/m³, $Y_H = 3.0 \times 10^{-8}$ kg/J, $w_J = 1$ kg/s and $T_{JIN} = 10$°C.

$$T_R = \frac{\frac{w_{\text{feed}} C_{P\text{feed}} T_{IN}}{UA} + \frac{w_{\text{feed}} X}{\rho_{\text{feed}} U A Y_H} + \frac{w_J C_{PJ} T_{JIN}}{UA + w_J C_{PJ}}}{1 + \frac{w_{\text{feed}} C_{P\text{feed}}}{UA} - \frac{UA}{UA + w_J C_{PJ}}} \tag{5.65}$$

Equation (5.65) shows that, for a cooling jacket with $T_{JI} < T_R$, T_R decreases as *UA* increases. Therefore, increasing *UA* ensures that the process temperature T_R does not exceed the desired value. The effect of *UA* on T_R and T_J is shown in Figure 5.3. With the data used for Figure 5.3, assuming that the maximum allowable temperature in the reactor is 35°C, a *UA* value of at least 4,000 W/°C is required.

Equation (5.64) also shows the effect of the feed flow rate on the reactor temperature T_R (Figure 5.4). Assuming that X remains unaffected by w_{feed}, i.e. that w_{feed} is always low enough to allow the (almost) complete removal of the feed substrate, increasing w_{feed} increases the rate of heat generation because more substrate is reacted in the same reactor volume. Therefore, everything else being constant, the temperature T_R tends to increase as the feed flow rate increases. However, T_R tends to an asymptotic value for high values of w_{feed} because increasing w_{feed} also increases the rate of heat removal with the outlet stream.

It is important to observe that the biomass concentration X that appears in Equation (5.65) depends on the substrate concentration in the feed. Higher substrate concentration in the feed gives higher values of the biomass concentration, assuming that the residence time is large enough to guarantee (almost) complete substrate removal (under this assumption $X = Y_{X/S} S_0$). If we replace X with $Y_{X/S} \cdot S_0$ in Equation (5.65), we are able to show the effect of the substrate concentration

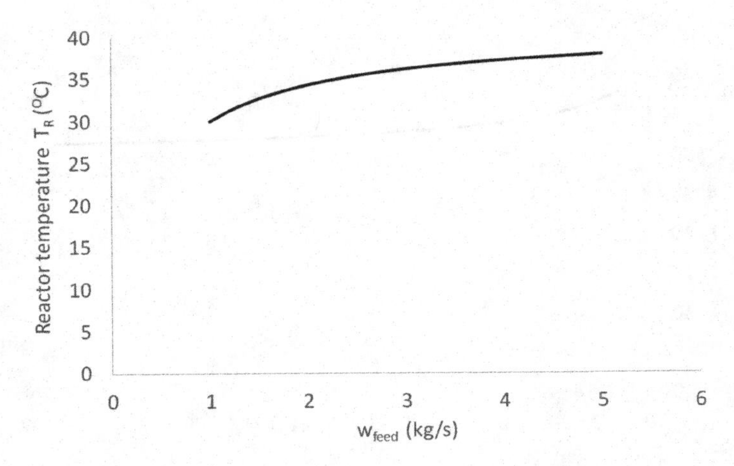

FIGURE 5.4 Effect of w_{feed} on the reactor temperature T_R (Equation 5.64). Parameter values used: UA = 5,000 W/°C, C_{Pfeed} = C_{PJ} = 4,186 J/kg·°C, T_{IN} = 25°C, ρ_{feed} = 1000 kg/m³, X = 2 kg/m³, Y_H = 3.0 × 10⁻⁸ kg/J, w_J = 1 kg/s and T_{JIN} = 10°C.

in the feed on the reaction temperature T_R (Figure 5.5). As expected, the reactor temperature T_R increases as the feed concentration S_0 increases because of the higher rate of heat generation.

As a further analysis of heat transfer in continuous fermenters, Equation (5.65) can be manipulated to show how the maximum substrate concentration in the feed that we can have without exceeding a certain maximum temperature for the

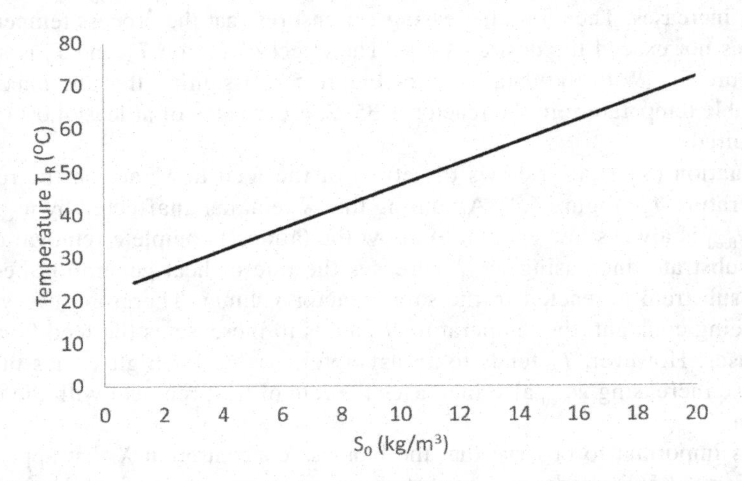

FIGURE 5.5 Effect of substrate concentration in the feed (under the assumption that the effluent substrate is negligible compared to the influent) on the reactor temperature T_R. Parameter values used: w_{feed} = 2 kg/s, UA = 5,000 W/°C, C_{Pfeed} = C_{PJ} = 4,186 J/kg·°C, T_{IN} = 25°C, ρ_{feed} = 1000 kg/m³, $Y_{X/S}$ = 0.4 kg/kg, Y_H = 3.0·10⁻⁸ kg/J, w_J = 1 kg/s and T_{JIN} = 10°C.

microorganisms depends on the heat removal properties of the system. If we rearrange Equation (5.65) solving for S_0, we obtain Equation (5.66):

$$S_0 = \frac{T_R\left(1+\frac{w_{feed}C_{Pfeed}}{UA}-\frac{UA}{UA+w_JC_{PJ}}\right)-\frac{w_{feed}C_{Pfeed}T_{IN}}{UA}-\frac{w_JC_{PJ}T_{JIN}}{UA+w_JC_{PJ}}}{\frac{w_{feed}Y_{X/S}}{\rho_{feed}UAY_H}} \quad (5.66)$$

Equation (5.66) shows that by increasing the value of UA, which is proportional to the heat transfer rate from the vessel, we obtain an increase of the maximum substrate concentration that we can have in the continuous fermentation feed without exceeding a maximum desired reactor temperature. Therefore, increasing the heat removal capacity of the vessel can lead to an increase of the fermentation productivity, i.e. of the amount of substrate that we can process in a given reactor volume. This effect is shown in Figure 5.6.

Overall, the analysis in this section shows that heat transfer, analogously to oxygen mass transfer (Chapter 4), can limit the performance of fermentation process. Therefore, it is important to optimise the heat transfer system according to the requirements of the fermentation.

Note that Equation (5.41) does not consider explicitly the enthalpy of the inlet gas phase, if there is any. In aerobic fermentation reactions, there is an inlet gas composed of oxygen and, if air is used, nitrogen. In many cases, the enthalpy of the inlet gas gives only a minor contribution to the enthalpy balance and can be

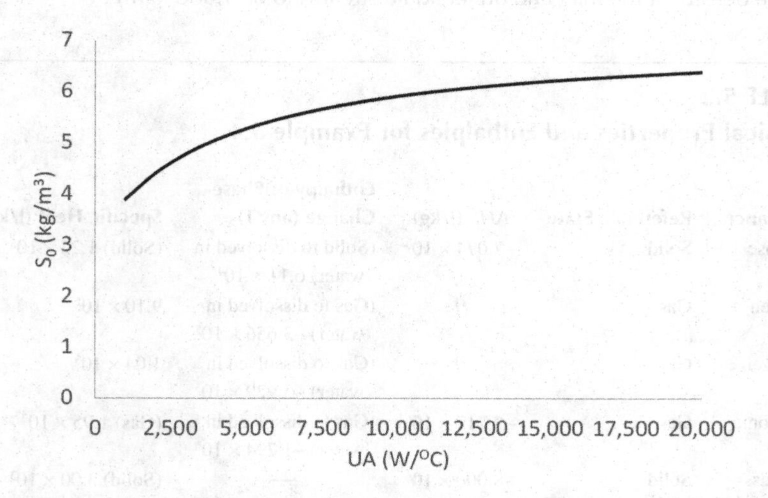

FIGURE 5.6 Effect of *UA* on the maximum substrate concentration that we can have in the feed of a continuous fermentation process (under the assumption that the effluent substrate is negligible compared to the influent) without exceeding the maximum desired reactor temperature T_R. Parameter values used: $T_R = 35°C$, $UA = 5{,}000$ W/°C, $C_{Pfeed} = C_{PJ} =$ 4,186 J/kg·°C, $T_{IN} = 25°C$, $\rho_{feed} = 1000$ kg/m^3, $Y_{X/S} = 0.4$ kg/kg, $Y_H = 3.0 \times 10^{-8}$ kg/J, $w_J =$ 1 kg/s and $T_{JIN} = 10°C$.

ignored, however, if needed it can be accounted for easily with Equation (5.67), which is a modification of Equation (5.41):

$$w_{feed}C_{Pfeed}\left(T_{IN}-T_R\right)+w_{gas}C_{Pgas}\left(T_{IN}-T_R\right)+r_H V=UA\left(T_R-T_J\right) \quad (5.67)$$

In Equation (5.67), w_{gas} is the total mass flow rate of the inlet gas and C_{Pgas} the specific heat of this stream. In Equation (5.67), it is assumed that the gas phase enters and leaves the reactor at the same temperatures T_{IN} and T_R as the liquid phase. For an exothermic reaction, the effect of the enthalpy of the gas phase, accounted for in Equation (5.67), is to reduce the temperature increase (T_{IN}–T_R) of the fermentation mixture caused by the heat of reaction.

Example 5.4

Consider the fermentation reaction (data in Table 5.2):

$$C_6H_{12}O_6+2.35O_2+0.1NH_3 \rightarrow 0.1C_5H_7O_2N+0.7C_6H_8O_7+1.3CO_2+3H_2O$$

This reaction is carried out in a continuous fermenter. The total inlet mass flow rate is 20,000 kg/d. The glucose concentration in the influent is 15 g/L and 95% of it is converted in the fermenter. The temperature of the inlet liquid and gas streams is 18°C. The fermenter is sparged with air (composed of 21% oxygen and 79% nitrogen in volume) having a flow rate of 7,500 m^3/d. Assume the density of the inlet and outlet liquid streams to be 1,000 kg/m^3.

TABLE 5.2
Physical Properties and Enthalpies for Example 5.4

Substance	Reference State	ΔH_f^0 (J/kg)	Enthalpy of Phase Change (any T)	Specific Heat (J/kg·K)
Glucose	Solid	-7.074×10^6	(Solid to dissolved in water) 6.11×10^4	(Solid) 1.22×10^3
Oxygen	Gas	0	(Gas to dissolved in water) -3.656×10^5	9.10×10^2
Nitrogen	Gas	0	(Gas to dissolved in water) -3.929×10^5	1.04×10^3
Ammonia	Gas	-2.717×10^6	(Gas to dissolved in water) -1.794×10^6	(Gas) 1.95×10^2
Biomass	Solid	-8.000×10^6	—	(Solid) 3.00×10^3
Citric acid	Solid	-1.021×10^7	(Solid to dissolved in water) 9.484×10^4	(Solid) 1.18×10^3
Carbon dioxide	Gas	-9.000×10^6	—	8.71×10^2
Water	Liquid	-1.600×10^7	—	4.19×10^3

Assume that the concentration of ammonia in the influent is equal to the stoichiometric amount required for the conversion of 15 g/L of glucose.
Assume the dissolved oxygen in the inlet and outlet streams is negligible and ignore its contribution to the enthalpy balance.

a. Calculate the temperature in the reactor, assuming adiabatic operation. Compare the use of Equations (5.40) and (5.41), assuming, for Equation (5.41) that the physical properties of the feed are the ones of water. In the use of Equation (5.41), assume that the enthalpy of reaction is equal at all temperatures at its standard value.
b. Calculate the flow rate of the cooling fluid that is required to maintain the fermentation temperature at 38°C. Assume the cooling fluid is available at 10°C and assume $UA = 800$ W/°C. The heat capacity of the cooling fluid is 4.19×10^3 J/kg·°C.

SOLUTION

a. The problem can be solved with either Equation (5.40) or (5.41), with $UA = 0$.
Equation (5.40) in this case is:

$$\dot{m}_{glucose,in}H_{glucose}\left(T_{IN}\right)+\dot{m}_{O2,ingas}H_{O2}\left(T_{IN}\right)+\dot{m}_{N2,ingas}H_{N2}\left(T_{IN}\right)$$
$$+\dot{m}_{water,in}H_{water}\left(T_{IN}\right)+\dot{m}_{NH3,in}H_{NH3}\left(T_{IN}\right)=\dot{m}_{glucose,out}H_{glucose}\left(T_{R}\right)$$
$$+\dot{m}_{O2,outgas}H_{O2}\left(T_{R}\right)+\dot{m}_{N2,outgas}H_{N2}\left(T_{R}\right)+\dot{m}_{NH3,out}H_{NH3}\left(T_{R}\right)$$
$$+\dot{m}_{biom,out}H_{biom}\left(T_{R}\right)+\dot{m}_{citric,out}H_{citric}\left(T_{R}\right)$$
$$+\dot{m}_{CO2,outgas}H_{CO2}\left(T_{R}\right)+\dot{m}_{water,out}H_{water}\left(T_{R}\right)$$

We need to calculate the mass flow rates of all the species in and out of the process.

$$\dot{m}_{Glucose,in}=\frac{20{,}000\frac{kg}{day}\cdot 15\frac{kg}{m^3}}{1{,}000\frac{kg}{m^3}}=300\frac{kg}{d}$$

$$\dot{m}_{O2,ingas}=7.5\cdot 10^6\frac{l}{day}\cdot 0.21\frac{l}{l}\cdot\frac{1atm\cdot mol\cdot K}{0.0821 l\cdot atm\cdot 291.15K}\cdot 0.032\frac{kg}{mol}$$
$$=2{,}108\frac{kg}{d}$$

$$\dot{m}_{N2,ingas}=7.5\cdot 10^6\frac{l}{day}\cdot 0.79\frac{l}{l}\cdot\frac{1atm\cdot mol\cdot K}{0.0821 l\cdot atm\cdot 291.15K}\cdot 0.028\frac{kg}{mol}$$
$$=6{,}940\frac{kg}{d}$$

$$\dot{m}_{NH3,in} = \frac{300\frac{kg}{day}}{180\frac{kg}{kmol}} \cdot 0.1\frac{molNH_3}{molGlucose} \cdot 17\frac{kg}{kmol} = 2.8\frac{kg}{d}$$

$$\dot{m}_{water,in} = (20{,}000 - 300 - 2.8)\frac{kg}{d} = 19{,}697.2\frac{kg}{d}$$

$$\dot{m}_{Glucose,out} = 300(1-0.95)\frac{kg}{day} = 15\frac{kg}{d}$$

$$\dot{m}_{O2,out} = 2{,}108\frac{kg}{d} - 300 \cdot 0.95\frac{kg}{d} \cdot 0.42\frac{kgO2}{kgglucose} = 1988\frac{kg}{d}$$

$$\dot{m}_{N2,outgas} = 6{,}940\frac{kg}{d}$$

$$\dot{m}_{NH3,out} = 2.8\frac{kg}{d} - 300(0.95)\frac{kg}{d} \cdot 0.0094\frac{kgNH3}{kgglucose} = 0.12\frac{kg}{d}$$

$$\dot{m}_{biomass,out} = 300 \cdot 0.95\frac{kg}{d} \cdot 0.063\frac{kgbiom}{kgglucose} = 17.9\frac{kg}{d}$$

$$\dot{m}_{citric,out} = 300 \cdot 0.95\frac{kg}{d} \cdot 0.75\frac{kgcitric}{kgglucose} = 212.8\frac{kg}{d}$$

$$\dot{m}_{CO2,outgas} = 300 \cdot 0.95\frac{kg}{d} \cdot 0.32\frac{kgCO2}{kgglucose} = 91\frac{kg}{d}$$

$$\dot{m}_{water,out} = 19{,}697.2\frac{kg}{d} + 300 \cdot 0.95\frac{kg}{d} \cdot 0.30\frac{kgwater}{kgglucose} = 19{,}782.7\frac{kg}{d}$$

The enthalpies of the components per unit mass are:

$$H_{glucose}(T_{IN}) = -7.074 \cdot 10^6\frac{J}{kg} + 1.22 \cdot 10^3\frac{J}{kg \cdot °C}(18-25)°C$$
$$+6.11 \cdot 10^4 = -7.021 \cdot 10^6\frac{J}{kg}$$

$$H_{O2gas}(T_{IN}) = 9.10 \cdot 10^2\frac{J}{kg \cdot °C}(18-25)°C = -6.37 \cdot 10^3\frac{J}{kg}$$

$$H_{N2gas}(T_{IN}) = 1.04 \cdot 10^3\frac{J}{kg \cdot °C}(18-25)°C = -7.28 \cdot 10^3\frac{J}{kg}$$

$$H_{NH3}(T_{IN}) = -2.717\cdot10^6\frac{J}{kg} + 1.95\cdot10^2\frac{J}{kg\cdot°C}(18-25)°C$$

$$+1.794\cdot10^6 = -9.243\cdot10^5\frac{J}{kg}$$

$$H_{water}(T_{IN}) = -1.600\cdot10^7\frac{J}{kg} + 4.19\cdot10^3\frac{J}{kg\cdot°C}(18-25)°C = -1.602\cdot10^7\frac{J}{kg}$$

$$H_{glucose}(T_R) = -7.074\cdot10^6\frac{J}{kg} + 1.22\cdot10^3\frac{J}{kg\cdot°C}(T_R-25)°C + 6.11\cdot10^4$$

$$= -7.043\cdot10^6\frac{J}{kg} + 1.22\cdot10^3 T_R H_{O2gas}(T_R)$$

$$= 9.10\cdot10^2\frac{J}{kg\cdot°C}(T_R-25)°C$$

$$= -2.28\cdot10^4\frac{J}{kg} + 9.10\cdot10^2 T_R$$

$$H_{N2gas}(T_R) = 1.04\cdot10^3\frac{J}{kg\cdot°C}(T_R-25)°C = -2.60\cdot10^4\frac{J}{kg} + 1.04\cdot10^3 T_R$$

$$H_{NH3}(T_R) = -2.717\cdot10^6\frac{J}{kg} + 1.95\cdot10^2\frac{J}{kg\cdot°C}(T_R-25)°C + 1.794\cdot10^6$$

$$= -9.279\cdot10^5\frac{J}{kg} + 1.95\cdot10^2 T_R H_{water}(T_R)$$

$$= -1.600\cdot10^7\frac{J}{kg} + 4.19\cdot10^3\frac{J}{kg\cdot°C}(T_R-25)°C$$

$$= -1.610\cdot10^7\frac{J}{kg} + 4.19\cdot10^3 T_R$$

$$H_{biomass}(T_R) = -8.000\cdot10^6\frac{J}{kg} + 3.00\cdot10^3\frac{J}{kg\cdot°C}(T_R-25)°C$$

$$= -8.075\cdot10^6\frac{J}{kg} + 3.00\cdot10^3 T_R$$

$$H_{citric}(T_R) = -1.021\cdot10^7\frac{J}{kg} + 1.18\cdot10^3\frac{J}{kg\cdot°C}(T_R-25)°C + 9.484\cdot10^4$$

$$= -1.014\cdot10^7\frac{J}{kg} + 1.18\cdot10^3 T_R H_{CO2,outgas}(T_R) = -9.000\cdot10^6\frac{J}{kg}$$

$$+8.71\cdot10^2\frac{J}{kg\cdot°C}(T_R-25)°C = -9.022\cdot10^6\frac{J}{kg} + 8.71\cdot10^2 T_R$$

The enthalpy of the inlet stream is:

$$\dot{m}_{glucose,in}H_{glucose}(T_{IN})+\dot{m}_{O_2,ingas}H_{O_2}(T_{IN})+\dot{m}_{N_2,ingas}H_{N_2}(T_{IN})$$

$$+\dot{m}_{water,in}H_{water}(T_{IN})+m_{NH_3,in}H_{NH_3}(T_{IN})=300\frac{kg}{d}\cdot\left(-7.021\cdot 10^6\frac{J}{kg}\right)+2{,}108\frac{kg}{d}$$

$$\cdot\left(-6.37\cdot 10^3\frac{J}{kg}\right)+6{,}940\frac{kg}{d}\cdot\left(-7.28\cdot 10^3\frac{J}{kg}\right)+2.8\frac{kg}{d}\cdot\left(-9.243\cdot 10^5\frac{J}{kg}\right)$$

$$+19{,}697.2\frac{kg}{d}\cdot\left(-1.602\cdot 10^7\frac{J}{kg}\right)=-3.177\cdot 10^{11}\frac{J}{d}$$

The enthalpy of the outlet stream is:

$$\dot{m}_{glucose,out}H_{glucose}(T_R)+\dot{m}_{O2,outgas}H_{O2}(T_R)+\dot{m}_{N2,outgas}H_{N2}(T_R)+\dot{m}_{NH3,out}H_{NH3}(T_R)$$

$$+\dot{m}_{biom,out}H_{Biom}(T_R)+\dot{m}_{citric,out}H_{citric}(T_R)+\dot{m}_{CO2,outgas}H_{CO2}(T_R)+\dot{m}_{water,out}H_{water}(T_R)$$

$$=15\frac{kg}{day}\cdot(-7.043\cdot 10^6+1.22\cdot 10^3T_R)+1{,}988\frac{kg}{day}\cdot(-2.28\cdot 10^4+9.10\cdot 10^2T_R)$$

$$+6{,}940\frac{kg}{day}\cdot(-2.60\cdot 10^4+1.04\cdot 10^3T_R)+0.12\frac{kg}{day}\cdot(-9.279\cdot 10^5+1.95\cdot 10^2T_R)$$

$$+19.9\frac{kg}{day}\cdot(-8.075\cdot 10^6+3.00\cdot 10^3T_R)+212.8\frac{kg}{day}\cdot(-1.014\cdot 10^7+1.18\cdot 10^3T_R)$$

$$+91\frac{kg}{day}\cdot(-9.022\cdot 10^6+8.71\cdot 10^2T_R)+19{,}782.7\frac{kg}{day}\cdot(-1.610\cdot 10^7+4.19\cdot 10^3T_R)$$

$$=-3.220\cdot 10^{11}+9.232\cdot 10^7T_R$$

T_R can be calculated from:

$$-3.177\cdot 10^{11}=-3.220\cdot 10^{11}+9.232\cdot 10^7T_R$$

which gives $T_R = 46.6°C$.
Using Equation (5.41) with the physical properties of water:

$$Q\rho_{H2O}C_{PH2O}(T_{IN}-T_R)=20\frac{m^3}{d}.1{,}000\frac{kg}{m^3}\cdot 4.19\cdot 10^3\frac{J}{kg\cdot°C}(18-T_R)$$

$$=8.38\cdot 10^7\frac{J}{kg\cdot°C}(18-T_R)$$

$$r_HV\left(\frac{J}{d}\right)=\left[(-r_S)V,\frac{kg}{d}\right]\left(-\Delta H_R,\frac{J}{kg}\right)$$

The rate of substrate removal is:

$$(-r_S)V = (300-15)\frac{kg}{d} = 285\frac{kg}{d}$$

The enthalpy of this reaction under standard conditions was calculated in Example 5.1:

$$\Delta H_R = -1.563\cdot 10^9 \frac{J}{kmol} = -1.563\cdot 10^9 \frac{J}{180kg} = -8.683\cdot 10^6 \frac{J}{kg}$$

So:

$$r_H V\left(\frac{J}{d}\right) = 285\frac{kg}{d}\cdot 8.683\cdot 10^6 \frac{J}{kg} = 2.475\cdot 10^9 \frac{J}{d}$$

So the reactor temperature T_R is calculated by:

$$8.38\cdot 10^7 \frac{J}{kg\cdot °C}(18-T_R) = -2.475\cdot 10^9 \frac{J}{d}$$

Which gives T_R = 47.5°C, in good agreement with the temperature calculated with Equation (5.41).

b. In this case, the enthalpy balance is:

$$\dot{m}_{glucose,in}H_{glucose}(T_{IN}) + \dot{m}_{O2,ingas}H_{O2}(T_{IN}) + \dot{m}_{N2,ingas}H_{N2}(T_{IN})$$
$$+\dot{m}_{water,in}H_{water}(T_{IN}) + \dot{m}_{NH3,in}H_{NH3}(T_{IN}) = \dot{m}_{glucose,out}H_{glucose}(T_R)$$
$$+\dot{m}_{O2,outgas}H_{O2}(T_R) + \dot{m}_{N2,outgas}H_{N2}(T_R) + \dot{m}_{NH3,out}H_{NH3}(T_R)$$
$$+\dot{m}_{biom,out}H_{biom}(T_R) + \dot{m}_{citric,out}H_{citric}(T_R) + \dot{m}_{CO2,outgas}H_{CO2}(T_R)$$
$$+\dot{m}_{water,out}H_{water}(T_R) + UA(T_R - T_J)$$

And the enthalpy balance for the fluid in the jacket is:

$$UA(T_R - T_J) = W_J \cdot C_{PJ}(T_J - T_{JIN})$$

The enthalpy of the inlet stream is the same as calculated at the previous point = -3.177×10^{11}J/d. The enthalpy of components in the outlet stream at 38°C is:

$$H_{glucose}(T_R) = -7.074\cdot 10^6 \frac{J}{kg} + 1.22\cdot 10^3 \frac{J}{kg\cdot °C}(38-25)°C + 6.11\cdot 10^4$$
$$= -6.997\cdot 10^6 \frac{J}{kg}$$

$$H_{O2gas}(T_R) = 9.10\cdot 10^2 \frac{J}{kg\cdot °C}(38-25)°C = 1.183\cdot 10^4 \frac{J}{kg}$$

$$H_{N2gas}(T_R) = 1.04 \cdot 10^3 \frac{J}{kg \cdot °C}(38-25)°C = 1.352 \cdot 10^4 \frac{J}{kg}$$

$$H_{NH3}(T_R) = -2.717 \cdot 10^6 \frac{J}{kg} + 1.95 \cdot 10^2 \frac{J}{kg \cdot °C}(38-25)°C$$

$$+1.794 \cdot 10^6 = -9.205 \cdot 10^5 \frac{J}{kg}$$

$$H_{water}(T_R) = -1.600 \cdot 10^7 \frac{J}{kg} + 4.19 \cdot 10^3 \frac{J}{kg \cdot °C}(38-25)°C = -1.594 \cdot 10^7 \frac{J}{kg}$$

$$H_{biomass}(T_R) = -8.000 \cdot 10^6 \frac{J}{kg} + 3.00 \cdot 10^3 \frac{J}{kg \cdot °C}(38-25)°C = -7.961 \cdot 10^6 \frac{J}{kg}$$

$$H_{citric}(T_R) = -1.021 \cdot 10^7 \frac{J}{kg} + 1.18 \cdot 10^3 \frac{J}{kg \cdot °C}(38-25)°C$$

$$+9.484 \cdot 10^4 = -1.010 \cdot 10^7 \frac{J}{kg}$$

$$H_{CO2,outgas}(T_R) = -9.000 \cdot 10^6 \frac{J}{kg} + 8.71 \cdot 10^2 \frac{J}{kg \cdot °C}(38-25)°C = -8.989 \cdot 10^6 \frac{J}{kg}$$

Therefore, the total enthalpy of the components in the outlet streams is:

$$\dot{m}_{glucose,out} H_{glucose}(T_R) + \dot{m}_{O2,outgas} H_{O2}(T_R) + \dot{m}_{N2,outgas} H_{N2}(T_R)$$

$$+\dot{m}_{NH3,out} H_{NH3}(T_R) + \dot{m}_{biom,out} H_{biom}(T_R) + \dot{m}_{citric,out} H_{citric}(T_R)$$

$$+\dot{m}_{CO2,outgas} H_{CO2}(T_R) + \dot{m}_{water,out} H_{water}(T_R) = 15\frac{kg}{d} \cdot (-6.997 \cdot 10^6)\frac{J}{kg}$$

$$+1{,}988\frac{kg}{d} \cdot (1.183 \cdot 10^4)\frac{J}{kg} + 6{,}940\frac{kg}{d} \cdot (1.352 \cdot 10^4)\frac{J}{kg}$$

$$+0.12\frac{kg}{d} \cdot (-9.205 \cdot 10^5)\frac{J}{kg} + 19.9\frac{kg}{d} \cdot (-7.961 \cdot 10^6)\frac{J}{kg}$$

$$+212.8\frac{kg}{d} \cdot (-1.010 \cdot 10^7)\frac{J}{kg} + 91\frac{kg}{d} \cdot (-8.989 \cdot 10^6)\frac{J}{kg}$$

$$+19{,}782.7\frac{kg}{d} \cdot (-1.594 \cdot 10^7)\frac{J}{kg} = -3.184 \cdot 10^{11} \frac{J}{d}$$

Therefore, from the heat balance for the fluid in the reactor we can calculate T_J:

$$-3.177 \cdot 10^{11}\frac{J}{d} = -3.184 \cdot 10^{11}\frac{J}{d} + \frac{800J}{s°C} 86{,}400\frac{s}{d}(38 - T_J)$$

which gives $T_J = 27.9°C$.

The required flow rate of the cooling fluid is obtained from the heat balance on the cooling fluid:

$$W_J = \frac{UA(T_R - T_J)}{C_{PJ}(T_J - T_{JIN})} = \frac{800\frac{J}{°C}\cdot\frac{86{,}400s}{d}(38-27.9)°C}{4.19\cdot 10^3\frac{J}{kg°C}(27.9-10)°C} = 9{,}308\frac{kg}{d}$$

5.3.2 Heat Balances for Batch Fermentation

We will refer to the general scheme of a batch reactor shown in Figure 5.7.

Using the general form of heat balance (Equation 5.35), the heat balance for the fluid in a batch reactor is:

$$M_R C_{PR}\frac{dT_R}{dt} = r_H V - UA(T_R - T_J) \tag{5.68}$$

In Equation (5.68), M_R is the mass of fluid inside the reactor and C_{PR} is its specific heat. Equation (5.68) corresponds to the analogous heat balance for a continuous system, Equation (5.41), and can be derived with the same methodology. In order to solve Equation (5.68) for T_R, we need an equation for T_J. This equation is the heat balance for the fluid in the jacket:

$$M_J C_{PJ}\frac{dT_J}{dt} = W_J C_{PJ}(T_{JIN} - 25) + UA(T_R - T_J) - W_J C_{PJ}(T_J - 25) \tag{5.69}$$

Equation (5.69) becomes:

$$M_J C_{PJ}\frac{dT_J}{dt} = W_J C_{PJ}(T_{JIN} - T_J) + UA(T_R - T_J) \tag{5.70}$$

Equations (5.68) and (5.70) can be solved simultaneously to calculate the profiles of T_R and T_J as a function of time in a batch reactor. Note that, similarly as for the continuous fermentation, the heat balances (5.68) and (5.70) need to be coupled with the mass balances because the biomass concentration, and therefore the rate of heat generation, changes with time.

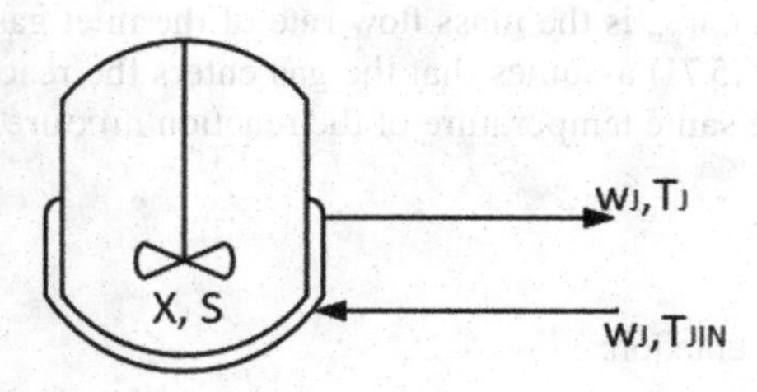

FIGURE 5.7 Scheme of a jacketed batch reactor.

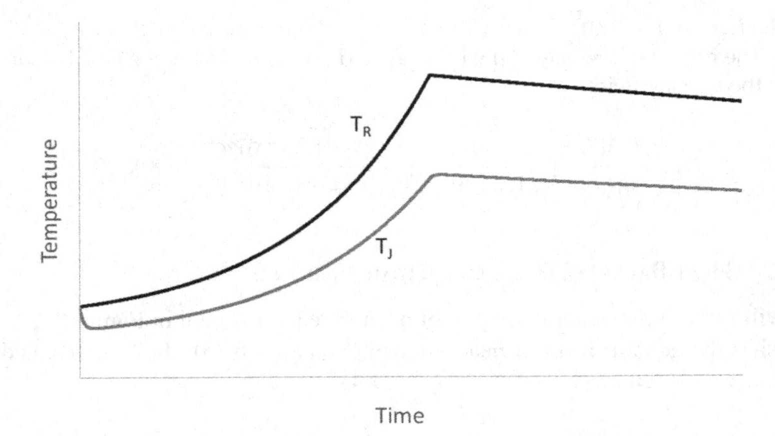

FIGURE 5.8 Typical qualitative time profiles of the reactor and jacket temperatures in batch fermentation.

Typical profiles of T_R and T_J in a batch reactor are shown in Figure 5.8. Typically, for exothermic fermentations, T_R and T_J both increase until the substrate has been completely removed, at which point the heat generation stops and the temperature starts to drop. Note that, when the substrate is present, the rate of temperature rise increases with time, because the biomass concentration increases with time and so the rate of heat generation also increases with time. Figure 5.8 and Equations (5.68) and (5.70) can be used to design the batch fermentation process, to ensure that the process temperature stays within the desired range, without exceeding the maximum allowable temperature for the microorganisms. Similarly to what was discussed for the heat balances in continuous fermenters, for batch reactors the maximum initial substrate concentration can be limited by the heat transfer ability of the vessel.

Note that Equation (5.68) doesn't account for the enthalpy of the inlet gas, if there is any. In case of aerobic fermentations, the enthalpy of the inlet gas is accounted for in Equation (5.71):

$$M_{\text{feed}} C_{P\text{feed}} \frac{dT_R}{dt} + w_{\text{gas}} C_{Pgas} \left(T_R - T_{IN}\right) = r_H \cdot V - UA\left(T_R - T_J\right) \tag{5.71}$$

In Equation (5.71), w_{gas} is the mass flow rate of the inlet gas and $C_{P\text{gas}}$ its specific heat. Equation (5.71) assumes that the gas enters the reactor at temperature T_{IN} and leaves at the same temperature of the reaction mixture, T_R.

Example 5.5

Consider the fermentation:

$$C_6H_{12}O_6 + 0.31NH_3 \rightarrow 0.31C_5H_7O_2N + 1.64C_2H_6O + 1.17CO_2 + 0.464H_2O$$

This fermentation is carried out in a batch reactor. The batch is started with a concentration of microorganisms of 1 kg/m³ and with a glucose concentration of 100 kg/m³. The initial temperature is 25°C. Assume that the density of the fermentation mixture is 1,000 kg/m³, the heat capacity is 4.19 J/kg·°C and the volume is 10 m³. The kinetic parameters for the growth of the microorganism are:

$$\mu_{max} = 5.5\, d^{-1}; K_S = 0.1\,\text{kg/m}^3$$

Calculate the temperature profile of the fermentation fluid vs time,

a. For an adiabatic reaction
b. When a cooling fluid is used in the jacket. The flow rate of the cooling fluid is 1,000 kg/d and its inlet temperature is 10°C. The heat capacity of the cooling fluid is 4.19×10^3 J/kg·°C and UA = 500 W/°C. Assume that the initial temperature in the jacket is 25°C. The volume of the jacket is 50 L.

SOLUTION

a. For the adiabatic case, the heat balance is:

$$\rho C_{P_R} \frac{dT_R}{dt} = r_H$$

The standard enthalpy of the reaction at 25°C is -4.35×10^4 J/mol of glucose. We need to express this per unit of mass of microorganisms produced. According to the reaction stoichiometry, the removal of 1 mol of glucose generates 35.03 g of microorganisms, so the heat of reaction per unit mass of microorganisms formed is:

$$\Delta H_R = -\frac{-4.35 \cdot 10^4 \frac{J}{mol}}{35.03 \frac{g}{mol}} = -1.242 \cdot 10^3 \frac{J}{g} = -1.242 \cdot 10^6 \frac{J}{kg}$$

The heat balance is:

$$1{,}000 \frac{kg}{m^3} 4.19 \cdot 10^3 \frac{J}{kg°C} \frac{dT_R}{dt} = \frac{5.5\; day^{-1} S}{0.1 \frac{kg}{m^3} + S} X \cdot 1.242 \cdot 10^6 \frac{J}{kg}$$

Which simplifies to:

$$\frac{dT_R}{dt}\left(\frac{°C}{d}\right) = 0.296 \frac{5.5S}{0.1 + S} X$$

The heat balance has to be solved together with the mass balances for the substrate and biomass:

$$\frac{dX}{dt} = \frac{5.5S}{0.1 + S} X$$

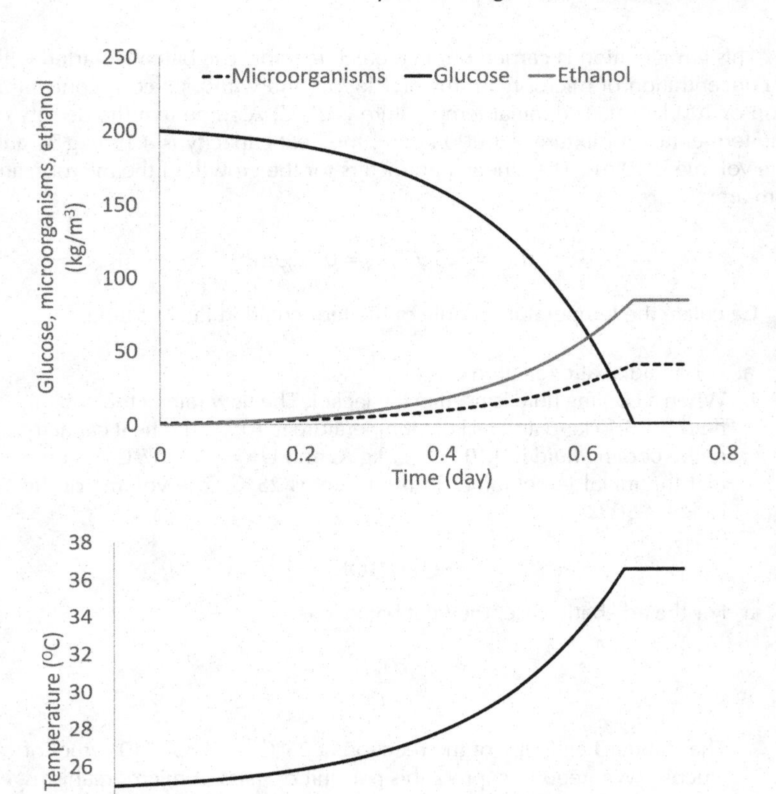

FIGURE 5.9 Profiles for Example 5.5a.

$$\frac{dS}{dt} = -\frac{5.5S}{0.1+S}\frac{X}{0.195}$$

The profiles of S, X, P and T are shown in Figure 5.9.

b. For the case with a jacket, the heat balances for the fluid in the reactor and in the jacket are:

$$\begin{cases} \rho C_{P_R}\dfrac{dT_R}{dt} = r_H - \dfrac{UA}{V}(T_R - T_J) \\ V_J C_{P_J}\dfrac{dT_J}{dt} = W_J \cdot C_{PJ}(T_{JIN} - T_J) + UA(T_R - T_J) \end{cases}$$

In this case, the equations can be written as (using kg, m³, J, °C and d as units):

$$\begin{cases} 1{,}000 \cdot 4.19 \cdot 10^3 \dfrac{dT_R}{dt} = \dfrac{5.5S}{0.1+S} X - \dfrac{500 \cdot 86{,}400}{10}(T_R - T_J) \\ 0.050 \cdot 1{,}000 \cdot 4.19 \cdot 10^3 \dfrac{dT_J}{dt} = W_J \cdot 4.19 \cdot 10^3 \cdot (10 - T_J) + 500 \cdot 86{,}400(T_R - T_J) \end{cases}$$

Which can be simplified to:

$$\begin{cases} \dfrac{dT_R}{dt} = 0.296 \dfrac{5.5S}{0.1+S} X - 1.03(T_R - T_J) \\ \dfrac{dT_J}{dt} = 0.02W_J(10 - T_J) + 206.21(T_R - T_J) \end{cases}$$

The profiles of the reactor and jacket temperature in this case are shown in Figure 5.10.

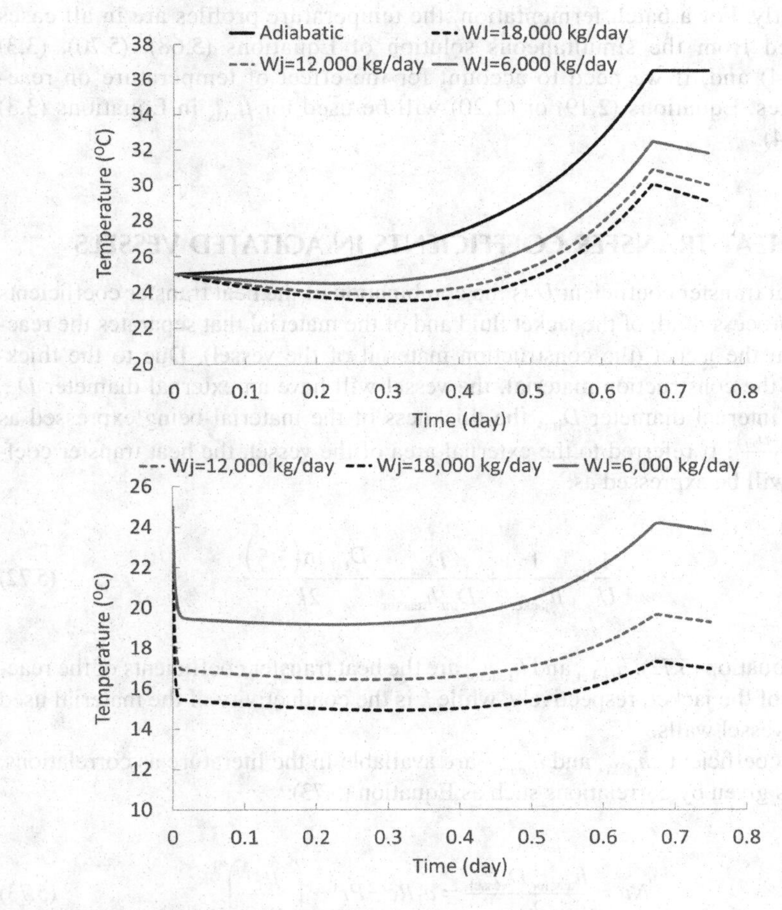

FIGURE 5.10 Profiles for Example 5.5b.

5.4 EFFECT OF VARIABLE TEMPERATURE ON REACTION RATES

So far, we haven't considered explicitly the effect of temperature on reaction rates. This effect can be considered, as seen in Chapter 2, by writing an equation for the dependence of μ_{max} on temperature, e.g. Equations (2.19) and (2.20). Accounting for the effect of temperature on μ_{max} does not change the heat balance equations for continuous and batch reactors, but it changes the mass balance equations which give the substrate, product and biomass profiles. Therefore, the temperatures in continuous and batch processes will be affected by the temperature dependence of kinetic parameters.

For a continuous process, the reactor temperature will still be given by the solution of Equations (5.65), (3.19) and (3.18), but if we take the effect of temperature on μ_{max} into account (e.g. by using Equations 2.19 or 2.20), the three equations will need to be solved simultaneously rather than independently. For a batch fermentation, the temperature profiles are in all cases obtained from the simultaneous solution of Equations (5.68), (5.70), (3.3) and (3.4) and, if we need to account for the effect of temperature on reaction rates, Equations (2.19) or (2.20) will be used for μ_{max} in Equations (3.3) and (3.4).

5.5 HEAT TRANSFER COEFFICIENTS IN AGITATED VESSELS

The heat transfer coefficient U is the combination of the heat transfer coefficients of the process fluid, of the jacket fluid and of the material that separates the reactor from the jacket (the construction material of the vessel). Due to the thickness of the construction material, the vessel will have an external diameter D_{ext} and an internal diameter D_{int}, the thickness of the material being expressed as $x = \frac{(D_{ext} - D_{int})}{2}$. If referred to the external area of the vessel, the heat transfer coefficient will be expressed as:

$$\frac{1}{U} = \frac{1}{h_{jacket}} + \frac{D_{ext}}{D_{int} h_{reactor}} \frac{D_{ext} \ln\left(\frac{D_{ext}}{D_{int}}\right)}{2k} \tag{5.72}$$

In Equation (5.72), h_{jacket} and $h_{reactor}$ are the heat transfer coefficients of the reactor and of the jacket, respectively, while k is the conductivity of the material used for the vessel walls.

The coefficients h_{jacket} and $h_{reactor}$ are available in the literature as correlations. $h_{reactor}$ is given by correlations such as Equation (5.73):

$$Nu = \frac{h_{reactor} D_{vessel}}{k_L} = a_1 Re^{a2} Pr^{0.33} \left(\frac{\mu_L}{\mu_{WL}}\right)^{a_3} \tag{5.73}$$

In Equation (5.73), Nu is the Nusselt number for the fluid in the reactor, and Re and Pr are the Reynolds and Prandtl numbers for the fluid in the reactor, defined as:

$$Re = \frac{ND_{ag}^2 \rho_L}{\mu_L} \tag{5.74}$$

$$Pr = \frac{c_{PL}\mu_L}{k_L} \tag{5.75}$$

In Equations (5.74) and (5.75), N is the agitator stirring speed (revs/s), D_{ag} is the diameter of the agitator (m) and ρ_L, μ_L, c_{PL} and k_L are the density, viscosity, specific heat and conductivity of the fluid in the vessel, respectively. μ_{WL} is the viscosity of the temperature of the vessel wall. Note that the factor $\left(\frac{\mu_L}{\mu_{WL}}\right)^{a_3}$ is only important for very viscous fluids and usually negligible for most fermentation processes. The Reynolds number for the agitator, defined by Equation (5.74), is analogous to the Reynolds number for fluids in pipes, with the fluid velocity taken in this case as the velocity at the tip of the agitator (which is ND_{ag}) and the characteristic dimension as the diameter of the agitator. Equations (5.73) and (5.74) indicate that increasing the agitation speeds increases the heat transfer coefficient for the fluid in the reactor.

The coefficients a_1, a_2, a_3 in Equation (5.73) depend on the type of agitator, on the presence or absence of baffles and other geometrical characteristics of the vessels and, in some cases, on the Reynolds number. For example, for flat-blade disc turbines and baffled vessels, an arrangement often used in fermentation vessels, for Re > 400, $a_1 = 0.74$, $a_2 = 0.67$ and $a_3 = 0.14$.

For the fluid in the jacket, assuming it is a non-viscous liquid, the heat transfer coefficient h_{jacket} is often given by Equation (5.76):

$$Nu_{jacket} = \frac{h_{jacket} D_e}{k_J} = 0.023 Re_J^{0.8} Pr_J^{0.33} \tag{5.76}$$

In Equation (5.76), Nu_{jacket} is the Nusselt number for the fluid in the jacket and k_J is the thermal conductivity of the fluid in the jacket. De is the equivalent diameter defined as:

$$De = \frac{4 \times \text{Cross-sectional area for fluid flow}}{\text{Wetted perimeter}} \tag{5.77}$$

Re_J is the Reynolds number for the fluid in the jacket defined as:

$$Re_J = \frac{\rho_J v D_e}{\mu_J} \tag{5.78}$$

In Equation (5.78), v is the fluid velocity in the jacket and ρ_J and μ_J are the density and viscosity, respectively, of the fluid in the jacket. Pr_J is the Prandtl

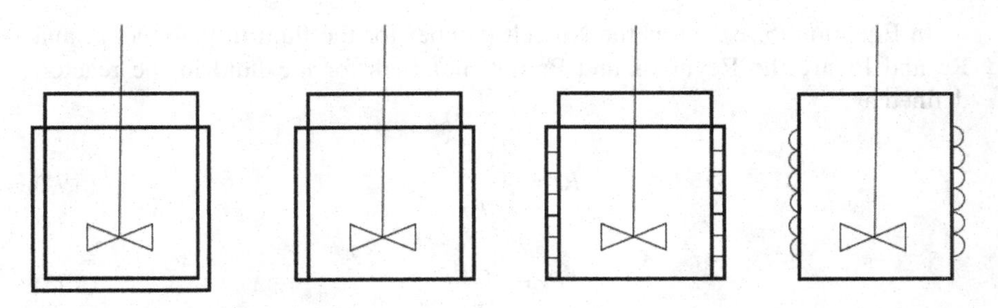

FIGURE 5.11 Scheme of jacketed vessels, from left to right: jacket that covers the whole external area of the vessel; jacket that covers only the cylindrical part; spirally baffled jacket and half-pipe jacket.

number for the fluid in the jacket and is defined in the same way as the Prandtl number for the fluid in the agitator (5.75), using the physical properties for the fluid in the jacket.

Equations (5.76–5.78) show that the heat transfer coefficient for the jacket fluid mainly depends on the fluid velocity in the jacket. In order to increase the fluid velocity, jackets are often equipped with baffles (spirally baffled jackets) or nozzles, or alternatively, half-pipe jackets are also common. Note that the jacket can cover the full area of the vessel (including the base area) or just the cylindrical portion of this area. Figure 5.11 shows a scheme of jacketed vessels with various types of jackets.

Example 5.6

a. Calculate the heat transfer coefficient for a fermentation vessel. The vessel is cylindrical with an internal diameter of 2.1 m and thickness of 10 mm. The vessel is made of stainless steel. The agitator is a flat disc turbine with diameter 0.70 m, rotating at 150 rpm. The jacket covers the cylindrical section only and extends to a height of 1.5 m. The spacing between the jacket and the external walls of the vessel is 50 mm. The jacket fluid has a flow rate of 800 m³/d, enters the jacket from the bottom and leaves at the top. In using Equation (5.73), assume $a_1 = 0.74$, $a_2 = 0.67$, $a_3 = 0.0$ (this corresponds to ignoring the viscosity correction term, as the fluid in the reactor is not very viscous).
b. Repeat the calculations of part (a), this time assuming that the jacket is made of a spiral baffle, with pitch between the spirals of 0.10 m.

Physical properties:

$\rho_L = 1{,}000$ kg/m³ (for both the reactor and the jacket fluid)
$\mu_L = 9 \times 10^{-3}$ Pa·s (for the fluid in the reactor)
$\mu_J = 9 \times 10^{-4}$ Pa·s (for the jacket fluid)
$k = 16$ W/m·°C (for stainless steel)
$k_J = 0.6$ W/m·°C (for both the reactor and the jacket fluid)
$C_P = 4{,}186$ J/kg·°C (for both the reactor and the jacket fluid)

SOLUTION

a. The overall heat transfer coefficient U is given by Equation (5.72), with D_{int} = 2.100 m and D_{ext} = 2.120 m. The heat transfer coefficient for the fluid in the reactor is given by Equation (5.73) with the Reynolds and Prandtl numbers given by Equations (5.74) and (5.75).

$$Re_{ag} = \frac{N D_{ag}^2 \rho_L}{\mu_L} = \frac{\frac{150}{60} \cdot 0.7^2 \cdot 1{,}000}{0.009} = 1.36 \cdot 10^5$$

$$Pr_{ag} = \frac{c_{PL} \mu_L}{k_L} = \frac{4{,}186 \cdot 0.009}{0.6} = 62.8$$

Therefore, $h_{reactor}$ is given by:

$$h_{reactor} = \frac{0.6 \cdot 0.74 \left(1.36 \cdot 10^5\right)^{0.67} 62.8^{0.33}}{2.1} = 2{,}281 \frac{W}{m^2\,^OC}$$

In the jacket, the fluid flows upwards through a rectangular section having sides of $\pi \cdot D_{ext}$ and 0.05 m. Therefore, the cross-sectional area is $\pi \cdot 2.120 \cdot 0.05 = 0.33 m^2$ and the wetted perimeter is $2 \cdot (\pi \cdot 2.120 + 0.05) = 13.41 m$. The equivalent diameter is:

$$De = \frac{4 \times 0.33}{13.41} = 0.098 m$$

The fluid velocity in the jacket is:

$$v = \frac{\frac{800}{86{,}400}}{0.33} = 0.028 \frac{m}{s}$$

The Reynolds number for the jacket is, with Equation (5.78):

$$Re_J = \frac{1{,}000 \cdot 0.028 \cdot 0.098}{0.0009} = 3{,}049$$

The Prandtl umber in the jacket is:

$$Pr_J = \frac{c_{PJ} \mu_J}{k_J} = \frac{4{,}186 \cdot 0.0009}{0.6} = 6.28$$

h_{jacket} is given by re-arranging Equation (5.76):

$$h_{jacket} = \frac{0.6 \cdot 0.023 \cdot 3{,}049^{0.8} 6.28^{0.33}}{0.098} = 158.6 \frac{W}{m^2\,°C}$$

According to Equation (5.72):

$$\frac{1}{U} = \frac{1}{158.6} + \frac{2.12}{2.10 \cdot 2{,}281} + \frac{2.12 \ln\left(\frac{2.12}{2.10}\right)}{2 \cdot 16} = 0.0063$$
$$+ 0.00044 + 0.000627 = 0.0073$$

$$U = 135.7 \frac{W}{m^2 \cdot °C}$$

b. In this case $h_{reactor}$ will be the same of part (a), while h_{jacket} will change. The number of spirals will be equal to the height of the jacket divided by the pitch = 1.5/0.10 = 15 spirals. Each spiral has a rectangular section with sides of 0.10 and 0.05 m, therefore, the cross-sectional area is $0.10 \cdot 0.05 = 0.005 m^2$ and the wetted perimeter is $2 \cdot (0.10 + 0.05) = 0.30 m$. The equivalent diameter is:

$$De = \frac{4 \times 0.005}{0.30} = 0.067 m$$

The fluid velocity in the jacket is:

$$v = \frac{\frac{800}{86,400}}{0.005} = 1.85 \frac{m}{s}$$

The Reynolds number for the jacket is, with Equation (5.78):

$$Re_j = \frac{1,000 \cdot 1.85 \cdot 0.067}{0.0009} = 1.38 \cdot 10^5$$

h_{jacket} is:

$$h_{jacket} = \frac{0.6 \cdot 0.023 \cdot \left(1.37 \cdot 10^5\right)^{0.8} 6.28^{0.33}}{0.067} = 4,859 \frac{W}{m^2 °C}$$

And the overall heat transfer coefficient U:

$$\frac{1}{U} = \frac{1}{4,859} + \frac{2.12}{2.10 \cdot 2,281} + \frac{2.12 \ln\left(\frac{2.12}{2.10}\right)}{2 \cdot 16} = 0.000205$$

$$+ 0.00044 + 0.000627 = 0.00127 \frac{m^2 °C}{W}$$

$$U = 786 \frac{W}{m^2 \cdot °C}$$

5.6 CONSIDERATIONS FOR SCALE-UP AND SCALE-DOWN

As we have seen in Chapter 4, scale-up and scale-down of fermentation processes are often carried out in the industry. In dealing with a scale-up/scale-down problem it is important to ensure that the scaled-up/scaled-down process reproduces the performance of the original reactor. We have seen in Chapter 4 how the scale-up/scale-down affects oxygen transfer. In this section, we will see the effects of scale-up/scale-down on the fermentation temperature and how to deal with these effects.

We will use the same assumptions that we used in Chapter 4, i.e. that the scale-up/scale-down is made maintaining the same residence time, feed concentration and physical properties of the medium. We assume that it is aimed to maintain the temperature of the fermentation constant on scale-up/scale-down. The first equation to consider is the enthalpy balance (5.41). From this equation we see that, if both the original and the scaled-up/scaled-down vessels are adiabatic ($U = 0$) the fermentation temperature T_R will be constant on scale-up/scale-down, as long as the residence time and feed composition are kept constant. Indeed, for constant residence time and feed composition, the terms r_H and $\frac{w_{feed}}{V}$ are constant on scale-up/scale-down. So, the scale-up/scale-down of adiabatic reactors with the aim of maintaining the same temperature is, in principle, straightforward.

The situation is more complicated if the reactor is not adiabatic, but temperature is controlled with an external jacket. In this case, the enthalpy balances are Equations (5.41) and (5.63) (enthalpy balance for the fluid in the jacket). Equation (5.41) indicates, that, in order for the temperature T_R to be constant on a scale-up/scale-down done with the same feed composition and residence time (same r_H and $\frac{w_{feed}}{V}$), the term $\frac{UA(T_R - T_J)}{V}$ needs to be kept constant between the scaled-up/scaled-down and the original reactors. The term $\frac{A}{V}$ is a purely geometrical term and is given, for a cylindrical vessel, where all the external area (including the base) is used for heat transfer, by:

$$\frac{A}{V} = \left(\frac{1}{\alpha} + 4\right)\sqrt[3]{\frac{\alpha\pi}{4V}} \tag{5.79}$$

In Equation (5.79), the parameter α indicates the ratio between the liquid height and the vessel diameter. Equation (5.79) shows that the ratio $\frac{A}{V}$ decreases on scale-up and increases on scale-down. This is an inevitable geometrical effect which needs to be compensated by the term $U(T_R - T_J)$. The enthalpy balance for the jacket fluid (5.63) indicates that, in order to keep the fermentation temperature T_R and, therefore, the term $\frac{UA(T_R - T_J)}{V}$ constant on scale-up/scale-down, the term $\frac{W_J C_{PJ}(T_J - T_{JIN})}{V}$ also needs to be kept constant. If the term $\frac{W_J}{V}$ is kept constant on scale-up/scale-down (i.e. the scale-up/scale-down is made maintaining the flow rate of the jacket fluid proportional to the reactor volume), then the condition of constant $\frac{UA(T_R - T_J)}{V}$ ensures that T_J will also be constant on scale-up/scale-down. Ultimately, therefore, the condition of constant temperature on scale-up/scale-down implies that the term $\frac{UA}{V}$ needs to be kept constant and, therefore, the heat transfer coefficient U needs to compensate for the change of $\frac{A}{V}$. In other words, the condition of constant $\frac{UA}{V}$ on scale-up/scale-down translates, using Equation (5.79), into:

$$\frac{U_{\text{large}}}{U_{\text{small}}} = \sqrt[3]{\frac{V_{\text{large}}}{V_{\text{small}}}} \tag{5.80}$$

The heat transfer coefficient, U, depends, as seen in Section 5.5, on the heat transfer coefficients on the reactor side and on the jacket side. Generally, both these heat transfer coefficients are important, however, a simplified analysis can be made assuming that the controlling heat transfer resistance is the one on the reactor side (this means that the heat transfer coefficient is much lower for the reactor than for the jacket). This generally corresponds to very viscous fermentations (high microorganisms' concentration). In this case, $U \equiv h_{reactor}$ and, using Equation (5.73), with constant physical properties on scale-up/scale-down:

$$\frac{U_{large}}{U_{small}} = \left(\frac{N_{large} D^2_{ag,large}}{N_{small} D^2_{ag,small}}\right)^{a_2} \frac{D_{vessel,small}}{D_{vessel,large}} \tag{5.81}$$

Equation (5.81) can be re-written taking geometrical similarity and the relationship between vessel diameter and volume, obtaining:

$$\frac{U_{large}}{U_{small}} = \left(\frac{N_{large}}{N_{small}}\right)^{a_2} \left(\frac{V_{large}}{V_{small}}\right)^{\frac{2a_2-1}{3}} \tag{5.82}$$

The comparison of Equations (5.82) and (5.80) shows how the agitator speed needs to change on scale-up/scale-down to maintain the same fermentation temperature:

$$\left(\frac{N_{large}}{N_{small}}\right) = \left(\frac{V_{large}}{V_{small}}\right)^{\frac{2}{3}\left(\frac{1}{a_2}-1\right)} \tag{5.83}$$

Since the coefficient a_2 is usually <1, Equation (5.83) indicates that, in order to maintain the fermentation temperature constant, the agitation rate needs to increase in scale-up and decrease in scale-down. This can be a serious problem for scale-up. Indeed, the required increase in agitation speed would cause a significant increase in the required power draw per unit volume (note that the agitator power draw depends on the agitation speed to the third power and to the agitator diameter to the fifth power and the agitator diameter increases on scale-up), which would probably make the scale-up at constant temperature impossible, especially for large scale-up ratios. Figure 5.12 shows a plot of Equation (5.83) for $a_2 = 0.67$.

The change in the agitator power required to maintain the same temperature on scale-up and scale-down is given by:

$$\left(\frac{P_{large}}{P_{small}}\right) = \left(\frac{V_{large}}{V_{small}}\right)^{\left(\frac{2}{a_2}-\frac{1}{3}\right)} \tag{5.84}$$

Equation (5.84) can be derived by combining Equation (5.83) with the definition of power number (4.23), assuming that the power number remains constant on scale-up/scale-down (this corresponds to being in the region of the power curve where the power number is independent of the Reynolds number), and by

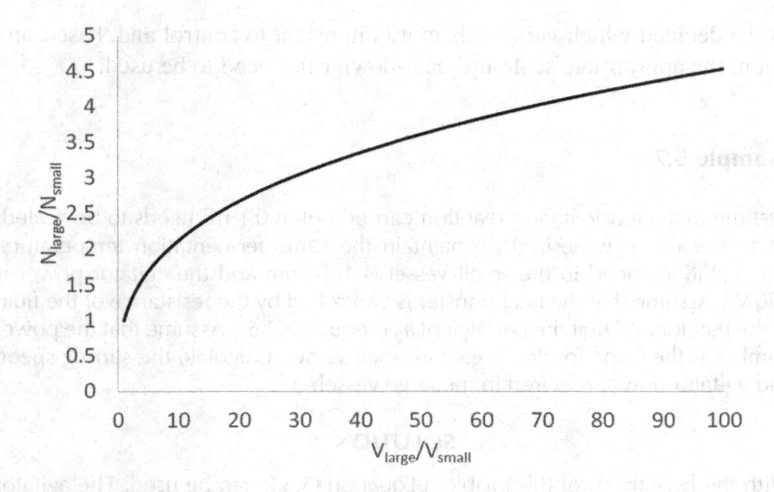

FIGURE 5.12 Plot of Equation (5.83) for $a_2 = 0.67$.

assuming geometric similarity (Equation 4.48). Equation (5.84) can also be re-written to show the change in the required power draw per unit volume:

$$\frac{\left(\frac{P_{large}}{V_{large}}\right)}{\left(\frac{P_{small}}{V_{small}}\right)} = \left(\frac{V_{large}}{V_{small}}\right)^{\left(\frac{2}{a_2}-\frac{4}{3}\right)} \tag{5.85}$$

Equation (5.85) shows that, since a_2 is generally <1, the required power draw per unit volume increases on scale-up.

To try and limit this problem, other strategies are possible: increasing the jacket fluid flow rate more than linearly with the reactor volume, so that T_J decreases and $(T_R–T_J)$ increases; using a jacket fluid with a lower inlet temperature; increasing the heat transfer area in the large vessel, for example, using coils. If the controlling heat transfer resistance is not the one on the reaction side but both the reaction and jacket side resistances are important, the effect of the change in the velocity of the jacket fluid on scale-up/scale-down also becomes important (Equation 5.76). The jacket fluid velocity tends to increase on scale-up, making the scale-up somewhat easier than predicted by Equation (5.83). In general, however, temperature control is easier in smaller vessels because of the higher ratio between the heat transfer area and the reactor volume.

Comparing the findings in this section with the section on scale-up/scale-down of oxygen transfer in Chapter 4, it is evident that the design of the scaled-up/scale-down reactor is different. For example, scaling up to maintain the same oxygen concentration requires a decrease in both the agitation speed and the power draw per unit volume, while scaling-up to maintain the same temperature requires an increase in both variables. In other words, scaling-up/down based on one variable causes changes in other variables. This is typical of all scale-up/scale-down problems. Not all the variables can be kept constant on scale-up/scale-down, it has to be decided which variable is more important to control and, based on this decision, the appropriate scale-up/scale-down rules need to be used.

Example 5.7

Assume that a fermentation reaction carried out at 0.1 m^3 needs to be scaled-up to 0.5 m^3. It is desired to maintain the same fermentation temperature. The agitation speed in the small vessel is 100 rpm and the agitator power is 100 W. Assume that the heat transfer is controlled by the resistance of the fluid in the reactor and that the coefficient a_2 is equal to 0.67. Assume that the power number is the same for the large and small vessel. Calculate the stirring speed and agitator power required in the large vessel.

SOLUTION

With the hypothesis of this problem Equation (5.83) can be used. The agitator speed in the large reactor is:

$$N_{large} = 100(5)^{\frac{2}{3}\left(\frac{1}{0.67}-1\right)} = 170rpm$$

To calculate the required power draw in the large vessel, we use Equation (5.84):

$$P_{large} = 100W \cdot 5^{\left(\frac{2}{0.67}-\frac{1}{3}\right)} = 7{,}116W = 7.1kW$$

This is a very large power. The power per unit volume increases from 1.0 kW/m^3 in the small vessel to 7.1 kW/m^3 in the large vessel. This is very probably impossible to achieve. In this case, the scale-up at constant temperature would require lowering the jacket temperature T_J, using coils to increase the heat transfer area or using multiple vessels to decrease the scale-up ratio.

Questions and Problems

5.1 Describe the main issues in the scale-up of jacketed agitated vessels for fermentation reactions if the reaction is limited by heat transfer.

5.2 With reference to the stoichiometry of glucose fermentation reported below, calculate the heat generated by the reaction (in kJ) when 1 kg of ethanol is produced, if the reaction is carried out at 40°C. Assume that at 40°C all the species are in their reference state and ignore any heat of mixing. Assume that the specific heat is constant in the considered temperature interval. The thermodynamic properties are reported in Table 5.3.

$$C_6H_{12}O_6 + 0.12NH_3 \rightarrow 0.12C_5H_7O_2N + 1.8C_2H_5OH + 1.8CO_2 + 0.36H_2O$$

TABLE 5.3
Thermodynamic Properties for Problem 5.2

Substance	Reference State	Heat of Formation at 25°C (kJ/mol)	Specific Heat (kJ/mol°C)
Glucose	Dissolved in water	−1,270	0.021
Ammonia	Dissolved in water	−80.8	0.078
Biomass	Solid	−770	0.333
Ethanol	Liquid	−278	0.113
Carbon dioxide	Gas	−393	0.037
Water	Liquid	−286	0.075

5.3 A microorganism grows anaerobically on a substrate in a continuous fermenter at steady state. Under the initial operating conditions, the substrate concentration in the feed is S_{0a}, the substrate and biomass concentration in the reactor are S_a and X_a, respectively. The temperature in the reactor is T_{Ra}. The feed flow rate is Q and the reactor volume used for the fermentation is V. Heat is removed via an external jacket where water flows at a flow rate w_J. The influent temperature of the cooling water is T_{JIN} and the effluent temperature is T_{Ja}.

It is proposed to increase the substrate concentration in the feed to S_{0b}, in order to increase the microorganisms' productivity. Determine whether it is possible to operate the reactor with the influent substrate concentration S_{0b} without exceeding the maximum allowable temperature for the microorganisms, $T_{R\max}$.

Assumptions: Assume that the enthalpy of reaction and the heat transfer coefficient U remain unchanged under the new conditions and that the cooling fluid flow rate and inlet temperature w_J and T_{Jin} cannot be changed. In the heat balances, neglect the contribution of the enthalpy of the reactor feed and outlet streams. Assume that the jacket is completely mixed. Assume that the microorganism growth rate follows Monod kinetics and that the values of the kinetic parameters are independent of temperature.

Data:

$$V = 10\,\text{m}^3; Q = 0.001\,\text{m}^3/\text{s}; S_{0a} = 10\,\text{kg/m}^3; S_a = 2\,\text{kg/m}^3;$$
$$X_a = 4\,\text{kg/m}^3; T_{Ra} = 35°\text{C}; S_{0b} = 15\,\text{kg/m}^3;$$
$$w_J = 1.5\,\text{kg/s}; C_{PJ} = 4186\,\text{J/kg.°C}; T_{Jin} = 5°\text{C};$$
$$T_{Ja} = 15°\text{C}; A = 30\,\text{m}^2; T_{R\max} = 50°\text{C}.$$

5.4 A microorganism grows in a continuous fermenter, with a substrate concentration in the feed equal to S_0. The dilution rate is D. The fermenter is cooled via a jacket, where cooling water flows at flow rate w_J. The inlet temperature of the cooling water is T_{JIN} and the heat transfer coefficient is U.

Calculate the temperature in the reactor.

Assumptions: The microorganism follows Monod kinetics. The jacket can be assumed perfectly mixed with a uniform temperature equal to $T_{\text{cool,out}}$. Neglect the enthalpies of the inlet and outlet streams.

Data: $\mu_{max} = 6$ d^{-1}; $K_S = 0.1$ g/L; $S_0 = 20$ g/L; $D = 5$ d^{-1}; $Y_{X/S} = 0.5$ g/g; $Y_H = 3 \times 10^{-8}$ kg/J; $w_J = 2.0$ kg/s; $C_{PJ} = 4186$ J/kg·K; $V = 10$ m^3; $A = 30$ m^2; $T_{Jin} = 5$°C; $U = 300$ W/m^2/K

5.5 The reaction of citric acid ($C_6H_8O_7$) production from glucose by the microorganism *Aspergillus niger* is:

$$C_6H_{12}O_6 + 2.35O_2 + 0.1NH_3 \rightarrow 0.1C_5H_7O_2N + 0.7C_6H_8O_7 + 1.3CO_2 + 3H_2O$$

It is desired to produce 15 t of citric acid per day with a continuous process. Calculate the maximum concentration of glucose (in kg/m^3) that is possible to have in the feed of the reactor without exceeding the maximum operating temperature of 35°C.

The feed of the reactor has a temperature of 25°C. Assume the reactor is adiabatic.

The enthalpy of the reaction is -1.56×10^6 J/mol glucose. Assume that this enthalpy refers to the reactants in their physical state under the reaction conditions. Assume that the enthalpy of reaction is constant in the temperature range considered here. Assume that the density of all the streams is equal to 1,000 kg/m^3 and their heat capacity is 4.19 kJ·kg^{-1}°C^{-1}.

Ignore the enthalpy absorbed by the inlet gas stream.

5.6 The reaction of ethanol production from glucose using the bacterium *Zymomonas mobilis* under anaerobic conditions can be schematised as follows:

$$C_6H_{12}O_6 + 0.30NH_3 \rightarrow 0.30C_5H_7O_2N + 1.5C_2H_5OH + 1.5CO_2 + 0.90H_2O$$

The reaction is carried out in a continuous reactor where the ethanol production rate is 150 kg/h. The volume of the reactor is 56 m^3, the dilution rate is 0.8 d^{-1} (assume that the inlet and outlet flow rates are the same) and the temperature of the feed is 20°C. The reactor is cooled with a jacket, where a fluid having an inlet temperature of 10°C and a flow rate of 10,000 kg/h circulates. The heat transfer area is 40 m^2 and the heat transfer coefficient is 200 W/m^2·°C.

Calculate the temperature in the reactor.

Density of the feed and of the reaction medium: 1,000 kg/m^3. Specific heat of the feed, reaction medium and jacket fluid: 4.19 kJ/kg·°C. The enthalpy of the reaction, assumed constant in the temperature interval considered here, is $\Delta H_R = -118\ kJ/mol\ glucose$.

5.7 A fermentation process is carried out at pilot-scale (50 L) in a cylindrical continuous fermenter. Cooling is achieved with a jacket that covers all the external area (including the bottom base) of the vessel. The fermenter is mixed with an agitator which rotates at 90 rpm.

It is now desired to scale-up the process to a geometrically similar 0.5 m³ vessel, maintaining the same residence time and substrate concentration than in the pilot scale process. It is also desired to maintain all the inlet and outlet temperatures for the reactor and the jacket at the same values as in the pilot scale process. Assuming that in all conditions the heat transfer resistance is controlled by the reaction fluid (i.e. heat transfer coefficients for the vessel walls and for the jacket are much larger than for the fluid in the reactor), calculate the required agitation rate and flow rate of the jacket fluid in the large vessel. Assume that the heat transfer coefficient for the fluid in the reactor is given by the following correlation:

$$\frac{h_{reactor} D_{vessel}}{k_L} = 0.74 Re_{ag}^{0.67} Pr_{ag}^{0.33}$$

5.8 A fermentation reaction is carried out in a continuous process, with the inlet feed at a temperature of 15°C and the reactor temperature of 28°C. The cooling fluid has a flow rate of 100,000 kg/d, enters the jacket at a temperature of 10°C and leaves at 17°C. The fermenter has a working volume of 10 m³ and the feed flow rate is 20 m³/d (density 1,000 kg/m³). The substrate concentration in the feed is 80 kg/m³ and it can be assumed that the substrate is entirely removed by the microorganisms. Calculate the enthalpy of the fermentation (J/kg substrate).

5.9 It is known from experimental data that the fermentation of 1 kg of substrate with the addition of 0.02 kg of ammonia will yield the following: 0.12 kg of microorganisms, 0.40 kg of the desired product, 0.35 kg of carbon dioxide, the balance being produced water. The enthalpies of the species are summarised in Table 5.4 (assume that all the species in

TABLE 5.4
Enthalpy Properties for Problem 5.9: The Specific Heats Are Given for the Substances in Their Reference State and Can Be Assumed Constant within the Considered Temperature Range

Substance	Reference State	ΔH_f^0 (J/kg)	Specific Heat (J/kg·K)
Substrate	Solid	-7.500×10^6	1.50×10^3
Ammonia	Dissolved in water	-4.511×10^6	4.80×10^3
Biomass	Solid	-8.000×10^6	3.00×10^3
Product	Dissolved in water	-1.080×10^7	1.18×10^3
Carbon dioxide	Gas	-9.000×10^6	8.71×10^2
Water	Liquid	-1.600×10^7	4.19×10^3

this reaction are in their reference state). Calculate the maximum substrate concentration (assuming the substrate is completely removed) that is possible to have in an adiabatic continuous fermenter carrying out this reaction without exceeding the maximum allowable temperature of 35°C. Assume that the feed enters the reactor at 22°C. Assume that the feed has a density of 1,000 kg/m^3 and a specific heat of 4.19 J/kg·K.

5.10 A vessel which is being considered for a fermentation process has the following dimensions: volume 25 m^3, diameter 3.2 m, liquid height 3.1 m and agitator diameter 1.0 m. The agitator has a power number (to be assumed constant in all the considered conditions) of 3.5. It is desired not to exceed an agitator power of 15 kW. The jacket is a spiral jacket which covers the cylindrical section only, to the full liquid height. The spacing between the jacket and the external wall is 40 mm and the spiral pitch is 50 mm. The jacket fluid is water entering at 5°C with a maximum flow rate of 50 m^3/d.

It is desired to use this vessel for a fermentation process having a feed of 75 m^3/d at 20°C and a substrate concentration of 70 kg/m^3. The enthalpy of reaction at the fermentation conditions is -9.50×10^5 J/kg substrate. Determine whether this vessel is suitable for this fermentation, considering that the maximum reaction temperature must not exceed 30°C. Assume that the substrate is completely removed in the fermentation and ignore the resistance to heat transfer of the vessel walls. In the calculation of the overall heat transfer coefficient U, assume that the ratio between the external and internal diameter of the vessel is equal to 1.

Assume for all the streams, the following physical properties: conductivity 0.6 W/m·°C, density 1,000 kg/m^3 and specific heat 4.19 kJ/kg·°C. Assume a viscosity of 1.5×10^{-1} Pa·s for the fluid in the reactor and viscosity 8.9×10^{-4} Pa·s for the fluid in the jacket.

5.11 A fermentation process is carried out in batch. The kinetic parameters for the process are the following: mmax = 3.8 d^{-1}, K_S = 0.06 kg/m^3 and $Y_{X/S}$ = 0.14 kg/kg. The enthalpy of the reaction is -9.67×10^5 J/kg substrate. The process is started at 30°C with biomass and substrate concentration of 0.5 and 50 kg/m^3, respectively. It is desired to maintain the fermentation temperature at 30°C throughout the length of the fermentation by manipulating the cooling water flow rate. The cooling water enters the jacket at 15°C. The overall heat transfer coefficient is U = 100 W/m^2·°C.

Calculate and plot the required time profile for the cooling water in this process.

Assume that all the kinetic parameters and physical properties are constant throughout the length of the fermentation. The reactor has a diameter of 1.5 m, a height of 2 m and the jacket width is 50 mm. Assume that the jacket covers only the sides of the vessel and not the bottom. Assume that the density of the reactor and jacket fluid is 1,000 kg/m^3. The specific heat for all streams is 4.19 kJ/kg·°C.

5.12 Calculate the maximum initial substrate concentration and the maximum product productivity (t of product per year) for an adiabatic batch fermentation process characterised by the following parameters: μ_{max} = 4.5 d^{-1}, K_S = 0.07 kg/m^3, $Y_{X/S}$ = 0.12 kg/kg and $Y_{P/S}$ = 0.35 kg/kg. The enthalpy of reaction is –5.50 × 10^5 J/kg. Each batch is started at a temperature of 20°C with 200 mg/L of microorganisms and is stopped when the substrate concentration is below 1 kg/m^3 and the temperature reaches 35°C. Assume that the process runs for 8,000 h per year and the downtime between consecutive batches is 2 h. The volume of the reactor is 10 m^3, the fluid density is 1,000 kg/m^3 and the specific heat is 4.19 kJ/kg·°C.

5.13 A fermentation is carried out in a 10 m^3 continuous reactor. Under the standard fermentation conditions, the feed flow rate is 25 m^3/d and has a substrate concentration of 60 kg/m^3. The effluent substrate and product concentrations are 2 and 25 kg/m^3 respectively, the jacket flow rate is 200 m^3/d and the jacket inlet and outlet temperatures are 10 and 20°C, respectively. The feed of the reactor has a temperature of 22°C and the reactor temperature is 30°C. The reactor is mixed with an agitator rotating at a speed of 80 rpm.

It is now desired to increase the substrate concentration in the feed to 70 kg/m^3, to increase the product productivity, maintaining the same temperature in the reactor. Assuming that the heat transfer rate is controlled by the resistance of the fluid in the reactor, calculate the new required value of the agitation rate. Also calculate the new product productivity and the increase in agitator power draw due to the increased agitation rate. Assume that the power number of the agitator is the same in all cases.

Assume that the heat transfer coefficient for the fluid in the reactor is given by the following correlation:

$$\frac{h_{\text{reactor}} D_{\text{vessel}}}{k_L} = 0.74 Re_{ag}^{0.67} Pr_{ag}^{0.33}$$

6 Design Summary and Examples of Industrial Fermentation Processes

This chapter provides a short summary of the main design considerations covered in previous chapters and shows some examples of industrial fermentation processes. The global market size of chemicals obtained with fermentation processes has been estimated in the order of US$60 billion in 2018. While the market size of fermentation chemicals is large, large are also the challenges in the design and optimisation of these processes. The tools covered in previous chapters (kinetics and stoichiometry, mass balances, mass and heat transfer considerations) should be used to maximise process efficiency and minimise process costs.

6.1 SUMMARY OF DESIGN CONSIDERATIONS

We'll summarise here the main steps for the design of a fermentation process for commercial scale. Here, we assume that the aim of the fermentation is to produce a certain product to be sold on the market.

The main design steps can be summarised as follows:

- *Step 1.* Identify the microorganism that is able to produce the desired product.
- *Step 2.* Gain information from the literature on the metabolism of the microorganism: requirements for substrate and mineral elements, temperature, pH, type of metabolism (product produced as primary or secondary metabolite), suitability for growth under batch, continuous and fed-batch culture.
- *Step 3.* Decide the fermentation mode: batch, continuous and fed-batch. This can be decided based on literature evidence and/or on lab-scale experiments. Considerations on the productivity can underpin the choice of the fermentation mode, as shown in Section 3.7.
- *Step 4.* Decide the operating parameters for the chosen fermentation mode. In all cases, the temperature, pH, concentration of substrate and of mineral elements should be chosen. When choosing the substrate concentration, considerations should be given to the desired productivity and microorganism concentration. Once the substrate concentration has been chosen, the requirements for mineral elements can be estimated from the fermentation stoichiometry (Chapter 2). For batch processes,

DOI: 10.1201/9781003217275-6

the initial microorganism concentration and the length of the batch need to be decided. For continuous and fed-batch processes, the initial microorganism concentration at start-up will not affect the steady state of the fermenter or the process performance. As long as the concentration chosen is high enough to avoid wash-out during start-up. For continuous and fed-batch process, the residence time, or the dilution rate, is the critical parameter to choose as it will greatly affect process performance and productivity. All these considerations, when choosing the operating parameters, should be based on experimental tests, guided by considerations on the reaction stoichiometry and kinetics and by the appropriate mass balances (Chapters 2 and 3).

- *Step 5.* Calculate the required reactor volume. The reactor volume should be calculated based on the desired product productivity and on the chosen residence time. If the required volume is too large, the process may be carried out in multiple fermenters in series.
- *Step 6.* If the fermentation is aerobic, the aeration system should be designed to maintain the desired oxygen concentration. Critical considerations in the design or the aeration system are choosing the agitator type and speed of the aeration rate. Design of the aeration system should be based on the tools presented in Chapter 4 and underpinned by experimental measurements at lab- and/or pilot-scale.
- *Step 7.* Design of the heat transfer system. The heat transfer system should be designed to maintain the desired temperature in the fermentation medium. This designing requires choosing the temperature inlet of the cooling/heating fluid, the type of jacket and the determination of the required heat transfer area. Design considerations for the heat transfer system are reported in Chapter 5.

If the vessel(s) to be used for the fermentation is (are) already available, the design steps above need to be adapted, considering that the reactor volume, agitation type and heat transfer system are available. If a new vessel needs to be purchased, it should be chosen, considering the design steps above, among the available designs from main manufacturers, some of which are reported in the Bibliography.

In the design steps above, we have not mentioned, for conciseness' sake, other important design choices to be made, e.g. pumps and compressors, tubing, filters. Clearly, the economics of the process, considering capital and operating costs, need to be considered to evaluate the viability of the proposed process.

6.2 BIOETHANOL

Bioethanol (Cardona et al., 2010) is the common name for the ethanol molecule (C_2H_5OH) when produced from biomass, rather than from fossil fuels as in conventional processes. Bioethanol is commercially produced using the anaerobic fermentation of the glucose derived from cereal crops or from sugarcane.

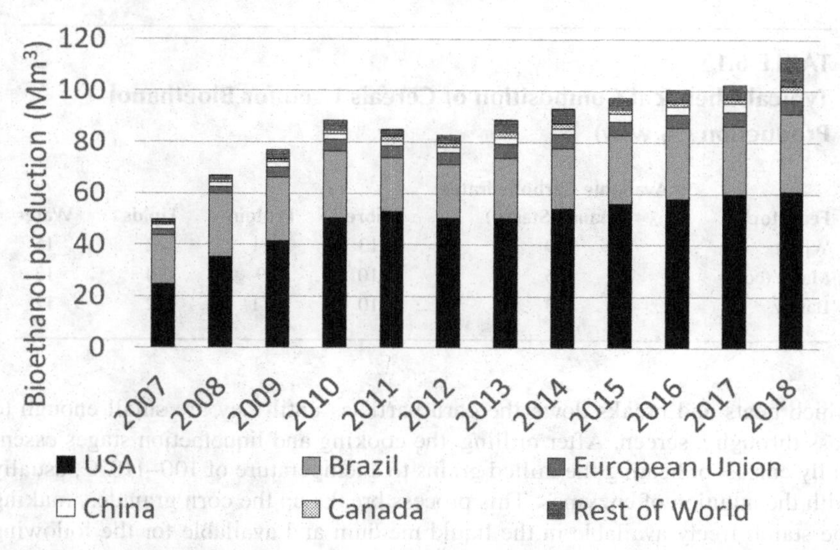

FIGURE 6.1 Production of bioethanol in years 2007–2018. (Adapted from https://www.afdc.energy.gov/data/.)

Bioethanol is an important energy vector in the transition towards renewable fuels and it is mainly used in vehicles since it has similar combustion properties as conventional petrol. Since bioethanol ultimately comes from biomass, which can be harvested at the same rate it is produced, it can be considered a renewable energy source. Furthermore, the carbon dioxide generated by the combustion of bioethanol is absorbed by the crops when they grow, in an ideally zero-carbon process. However, the energy consumption in the bioethanol production process from corn and in the growth of corn can contribute to carbon dioxide emissions, depending on the way this energy is produced.

Figure 6.1 shows the production volumes of bioethanol in different countries in years 2007–2018. The market size of bioethanol has been estimated as US$33 billion in 2020 and is expected to grow to more than US$60 billion by 2025.

The bioethanol production process is partly different depending on whether the feedstock is cereals or sugarcane.

Cereal feedstocks, e.g. corn, wheat and barley, are mainly composed of starch (Table 6.1).

Starch is a polysaccharide made of glucose units. The main units in starch molecules are called amylose (linear polymer in which glucose molecules are joined by α-1,4 bonds) and amylopectin (branched polymer in which linear chains of glucose molecules are joined by α-1,6 bonds). The initial stages in the process (Figure 6.2) aim to generate the free glucose which will be then fermented into ethanol. The first stage is milling, where the particle size of the corn grains is reduced to 3–5 mm. Milling is a mechanical process, usually carried out with hammer mills. In these mills, particle size reduction is obtained by using a hammer,

TABLE 6.1
Typical Chemical Composition of Cereals Used for Bioethanol Production (% w/w)

Feedstock	Available Carbohydrates (Mainly Starch)	Fibre	Protein	Lipids	Water
Wheat	59	13	11	2	15
Maize (corn)	65	10	9	4	12
Barley	63	10	11	2	12

which beats and breaks down the corn particles until they are small enough to pass through a screen. After milling, the cooking and liquefaction stages essentially consist of heating the milled grains to a temperature of 100–110°C, usually with the addition of enzymes. This process breaks up the corn granules, making the starch freely available in the liquid medium and available for the following hydrolysis stage. In the starch hydrolysis stage, enzymes are added which hydrolyse starch to glucose. Table 6.2 reports some common enzymes used for starch hydrolysis, and their optimum conditions of temperature and pH. The reaction of starch hydrolysis can be written as in Equation (6.1):

$$C_6H_{10}O_5 + H_2O \rightarrow C_6H_{12}O_6 \tag{6.1}$$

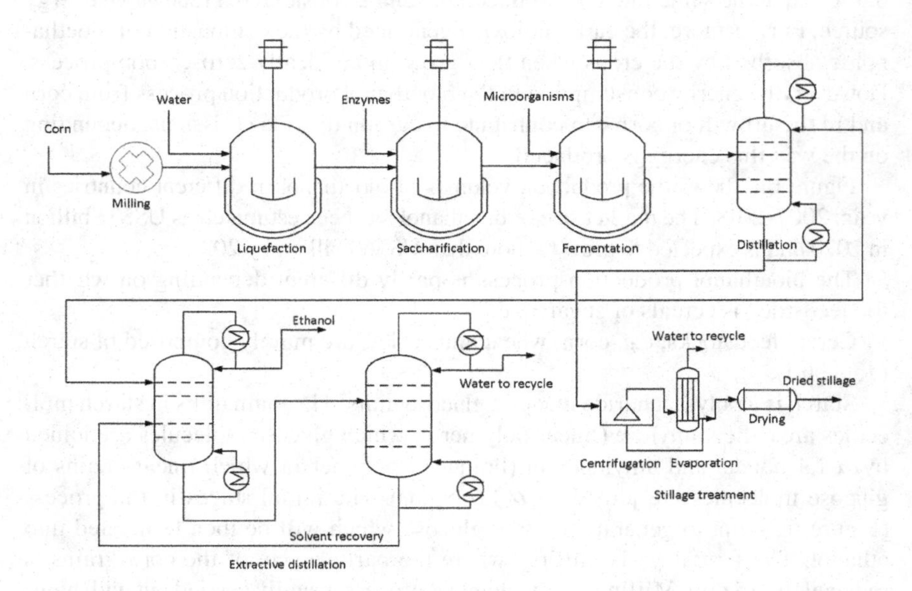

FIGURE 6.2 Simplified scheme of a process for the production of bioethanol from corn.

TABLE 6.2
Examples of Enzymes Used for Starch Hydrolysis

Enzyme	Source	MW	Optimum Temperature (°C)	Optimum pH
α-amylases	*Bacillus subtilis, Bacillus amyloliquefaciens, Bacillus licheniformis*	49,000–62,000	70–90	5.5–6.5
Glucoamylases	*Aspergillus awamori, Aspergillus niger* and many others	38,000–112,000	40–60	4.5–5.0

Once free glucose has been obtained, it is converted into ethanol in the fermentation stage. The fermentation is anaerobic, carried out by one of several species of microorganisms, e.g. the yeast *Saccharomyces cerevisiae*, or the bacterium *Zymomonas mobilis* (Table 6.3). In the fermentation, the microorganisms use ammonia as nitrogen source and generate, in addition to ethanol, new microorganisms, carbon dioxide and water. A typical stoichiometry for the fermentation process is reported in Equation (6.2):

$$C_6H_{12}O_6 + 0.12NH_3 \rightarrow 0.12C_5H_7O_2N + 1.8C_2H_5OH + 1.8CO_2 + 0.36H_2O \quad (6.2)$$

The maximum concentration of ethanol at the end of the fermentation is limited by the fact that ethanol is toxic for microorganisms at high concentrations. The maximum ethanol concentration that can be tolerated by microorganisms is different for different microorganisms, but it is typically in the range 100–160 g/L.

After the fermentation process, the bioethanol needs to be purified to at least 99% w/w to be used as a fuel. Separation of the ethanol from water is in general obtained using distillation. In distillation columns, ethanol, which is more volatile than water, is collected as the distillate at a concentration of up to approximately

TABLE 6.3
Examples of Microorganisms Used for Glucose Fermentation into Ethanol

Microorganisms	Optimum Temperature (°C)	Ethanol Tolerance (g/L)
Saccharomyces cerevisiae	30–37	160
Zymomonas mobilis	30	100
Clostridium thermocellum	55–65	10–30

TABLE 6.4
Chemical Composition (% w/w) of Sugarcane and Molasses

Feedstock	Sucrose	Fibre	Protein	Lipids	Water
Sugarcane	14	13	<1	<1	70
Molasses	37	—	3.5	<1	19

90% w/w. At concentrations higher than this, ethanol and water form an azeotrope and further separation by distillation is no longer possible. In order to increase the ethanol purity to the desired 99% w/w, one of the following technologies is used: (a) vacuum distillation. The azeotrope tends to disappear at reduced pressures; (b) extractive or azeotropic distillation. By adding a third component (often ethylene glycol for extractive distillation), the relative volatility between ethanol and water changes, allowing to reach a higher ethanol purity and (c) adsorption on molecular sieves. Molecular sieves are aluminosilicates that adsorb water but not ethanol, because water molecules are considerably smaller than ethanol molecules and can be trapped in the cage of the sieves, while ethanol is not retained. After the purification processes, ethanol of the desired purity is obtained. Any solid residues from the process are called stillage. Stillage contains all the inorganic and organic matter that is not starch which is present in the feedstock (fibre, lipids, etc.) and the produced microorganisms. Typically, stillage is concentrated by centrifugation, evaporation and drying and is used for animal feed.

The other common feedstock for bioethanol production is sugarcane. In sugarcane, most of the carbohydrates are present as sucrose (Table 6.4). Sucrose is a disaccharide composed of one molecule of glucose and one of fructose. Differently from starch, sucrose can be immediately metabolised by microorganisms, without previous hydrolysis. The pre-treatments of sugarcane for bioethanol production are, therefore, easier than the pre-treatments from cereal feedstocks. In the conventional process (Figure 6.3), bioethanol is produced from

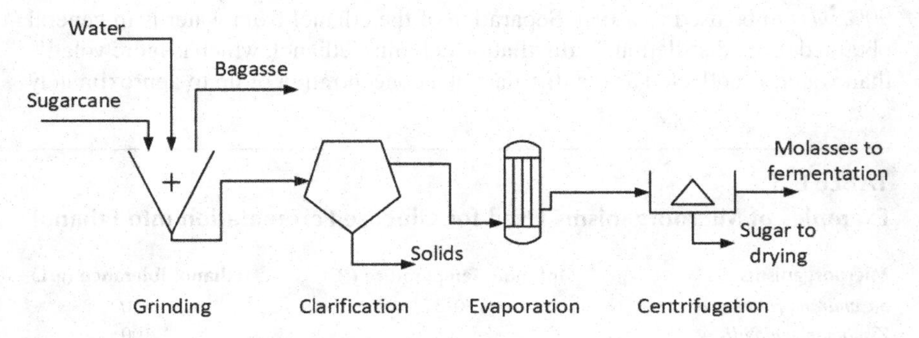

FIGURE 6.3 Simplified scheme of molasses production from sugarcane. Molasses is sent to fermentation for bioethanol production.

sugarcane as a by-product of the sugar industry. Sugarcane is collected manually or mechanically. Then, it is milled with addition of hot water and sugarcane juice, which contains sucrose and all the soluble species, is extracted. Bagasse is the by-product of sugarcane extraction and contains the lignocellulosic (fibre) fraction. Bagasse is typically burned on site to generate electricity. The sugarcane juice is concentrated, by means of evaporation and centrifugation, and in the end sugar crystals are obtained. The residual liquid after centrifugation is called molasses (Table 6.4) and can be used as a substrate for fermentation (without any further pre-treatments) for ethanol production. The fermentation and separation/purification stages are the same as already described for the bioethanol production from cereals.

6.3 PENICILLIN

Penicillin is a group of antibiotics and historically is one of the first antibiotics produced at large scale. Today, penicillin is still used in the treatment of many bacterial infections, e.g. those caused by *Staphylococci* and *Streptococci*. Penicillin is produced by certain species of the *Penicillium* microorganism under aerobic conditions. Current market size of penicillin is in the order of US$2–3 billion.

Fermentation is carried out using sugars (typically glucose or lactose) as substrates, and the fungus *Penicillium* as the microorganism. Also, nutrients containing nitrogen, phosphorus and other salts need to be added.

Penicillin (process flowsheet in Figure 6.4) is not produced during growth of the microorganism but is produced during the stationary phase after growth has stopped (secondary metabolite). When fermentation is complete, biomass is removed by filtration. Typical equipment used for biomass filtration is the rotary

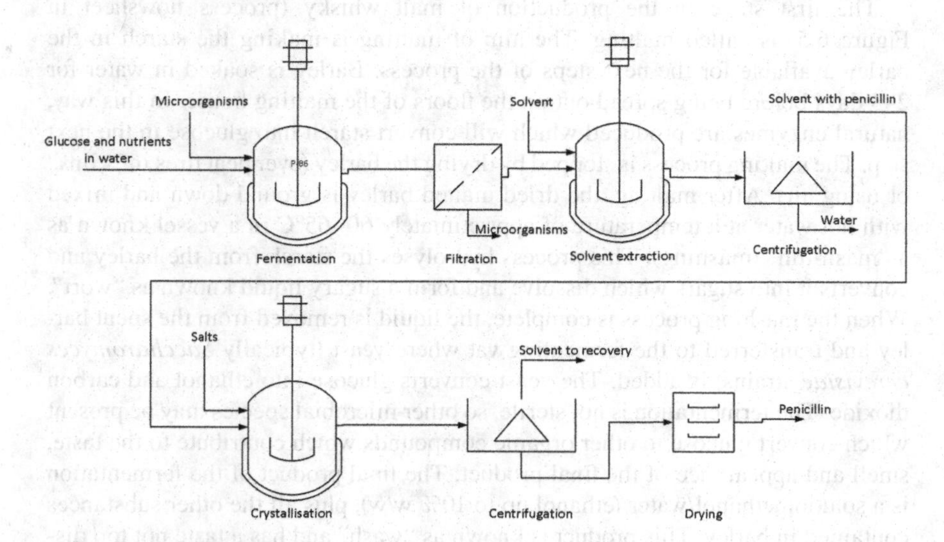

FIGURE 6.4 Simplified scheme of penicillin production.

vacuum filter. After the biomass has been removed, the water phase is mixed with an organic solvent, typically amyl acetate or butyl acetate. The solvent dissolves penicillin but not the many other salts present in the fermentation medium; therefore, purity of penicillin is increased. Solvent extraction is typically carried out in a centrifuge, which also allows for the separation of the water phase. The organic (solvent) phase is sent to the crystallisation stage, where several inorganic reagents are added (e.g. sodium carbonate). The reagents form a salt with penicillin. This allows the precipitation of penicillin (salts are not soluble in organic solvents) as a solid, which will be then further purified. The solid phase (penicillin salt) is separated by the solvent using centrifugation. The penicillin salt is collected from the centrifuge and sent to the dryer, where the residual solvent is removed using hot air.

6.4 WHISKY

Scotch whisky is one of Scotland's biggest exports, second only to oil and gas, as billions of litres are sold each year worldwide. Scotch whisky totals one quarter of all of food and drink exports from the UK making it a very important contributor to both the Scottish and the UK economies. The market size of Scotch whisky was estimated at £4.9 billion in 2019. In 2009, the Scotch Whisky Regulations defined Scotch whisky as a product "that has been distilled at a distillery in Scotland from water and malted barley (to which only whole grains of other cereals may be added)". Malt whisky is the whisky made using barley as only cereal, while grain whisky also uses other cereals (corn, wheat or rye) in addition to barley. The production processes of malt and grain whisky are significantly different.

The first stage in the production of malt whisky (process flowsheet in Figure 6.5) is called malting. The aim of malting is making the starch in the barley available for the next steps of the process. Barley is soaked in water for 2–3 days before being spread out on the floors of the malting house. In this way, natural enzymes are produced which will convert starch into glucose in the next step. The malting process is stopped by drying the barley (over peat fires in "kilns" or using air). After malting, the dried malted barley is ground down and mixed with hot water at a temperature of approximately 60–65°C in a vessel known as a "mash-tun" (mashing). This process hydrolyses the starch from the barley and converts it into sugars which dissolve and form a sugary liquid known as "wort". When the mashing process is complete, the liquid is removed from the spent barley and transferred to the fermenting vat where yeast (typically *Saccharomyces cerevisiae* strains) is added. The yeast converts glucose into ethanol and carbon dioxide. The fermentation is not sterile, so other microbial species may be present which convert glucose to other organic compounds which contribute to the taste, smell and appearance of the final product. The final product of the fermentation is a solution ethanol/water (ethanol up to 10% w/w), plus all the other substances contained in barley. This product is known as "wash" and has a taste not too dissimilar from the taste of unhopped beer. The size of fermentation vessels is in

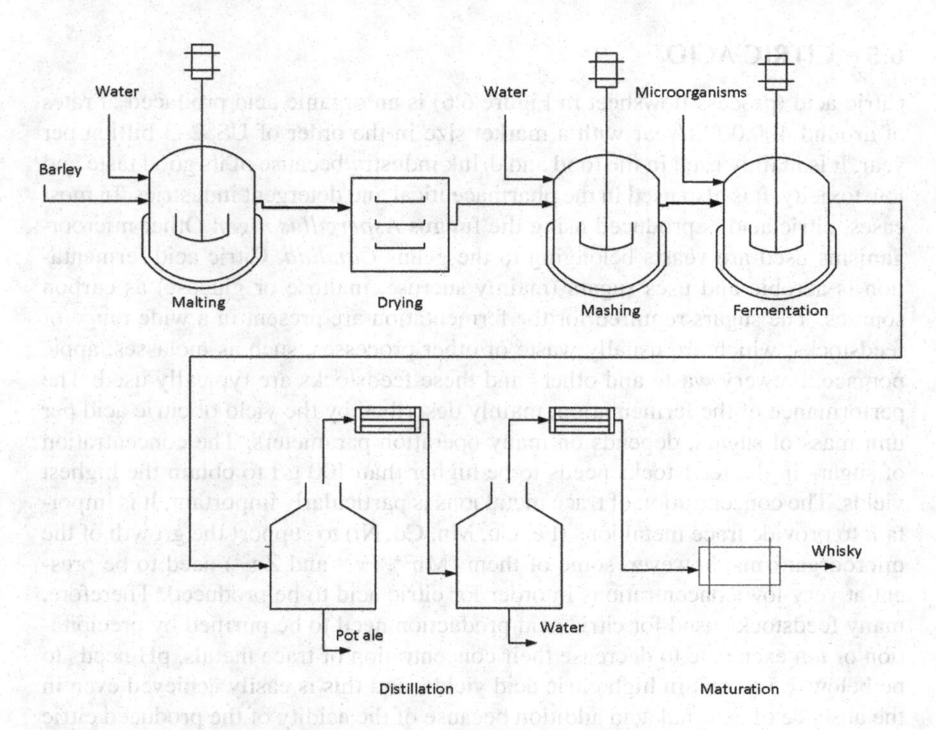

FIGURE 6.5 Simplified scheme of the malt whisky production process.

the range 5–150 m^3. The fermentation is carried out in batch and typically lasts for 2–3 days. The stoichiometry of the fermentation is similar to the one of the bioethanol process.

The wash is then distilled in the characteristic "swan-necked" copper stills. Scottish whisky is traditionally distilled twice, with the second copper still being slightly smaller than the first. After the second distillation the ethanol content is 65–70% (spirit). The final stage in the process is called maturation. The spirit is put into oak casks and stored. The spirit must mature in casks for a minimum of 3 years before it is legally allowed to be called whisky in Scotland. During maturation, the flavours of the spirit combine with natural compounds in the wood cask and this gives the whisky its own characteristic flavour and aroma. During each year of maturation, about 2% of the spirit is lost through natural evaporation.

The grain whisky process uses un-malted cereals, which are finely ground and pressure cooked to release the starch into the suspension. In some cases, malted barley is added to the cooked cereals. The mashing and fermentation processes are similar to the malt whisky process. The distillation of grain whisky is done in continuous tray towers, instead than in the batch copper stills used for malt whisky. Grain whisky has generally lighter aroma than malt whisky because of the different distillation process.

6.5 CITRIC ACID

Citric acid (process flowsheet in Figure 6.6) is an organic acid produced at rates of around 400,000 t/year with a market size in the order of US$2–3 billion per year. It is mainly used in the food and drink industry because of its good taste and low toxicity. It is also used in the pharmaceutical and detergent industries. In most cases, citric acid is produced using the fungus *Aspergillus niger.* Other microorganisms used are yeasts belonging to the genus *Candida*. Citric acid fermentation is aerobic and uses sugars (mainly sucrose, maltose or glucose) as carbon sources. The sugars required for the fermentation are present in a wide range of feedstocks, which are usually waste of other processes, such as molasses, apple pomace, brewery waste and others and these feedstocks are typically used. The performance of the fermentation, mainly described by the yield of citric acid per unit mass of sugars, depends on many operation parameters. The concentration of sugars in the feedstocks needs to be higher than 100 g/l to obtain the highest yields. The concentration of trace metal ions is particularly important. It is important to provide trace metal ions (Fe, Cu, Mn, Co, Ni) to support the growth of the microorganisms, however, some of them (Mn^{2+}, Fe^{3+} and Zn^{2+}) need to be present at very low concentrations in order for citric acid to be produced. Therefore, many feedstocks used for citric acid production need to be purified by precipitation or ion exchange to decrease their concentration of trace metals. pH needs to be below 2.5 to obtain high citric acid yields, and this is easily achieved even in the absence of external acid addition because of the acidity of the produced citric acid. Aeration is also important and oxygen concentration needs to be maintained

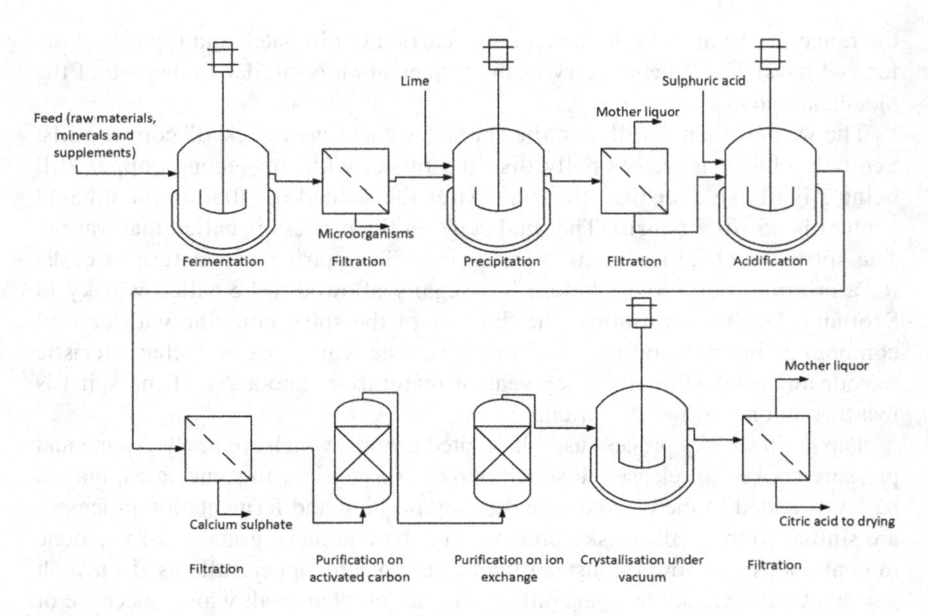

FIGURE 6.6 Simplified scheme of the citric acid production process.

above 25% of saturation to obtain high yields. After the fermentation, the microorganisms are removed by filtration and citric acid is precipitated by the addition of lime, forming tri-calcium citrate tetrahydrate, which has a low solubility in water. The precipitated calcium citrate is separated from water using filtration and the cake is washed several times with water to increase the purity of the salt. After filtration, citric acid is formed again by adding sulphuric acid (this process also forms calcium sulphate which is filtered off). The resulting solution of citric acid is purified using activated carbon and ion exchange resins and citric acid is then crystallised under vacuum. Depending on the temperature of the crystallisation, anhydrous citric acid or citric acid monohydrate is obtained. Citric acid is then dried in the final stage of the process.

6.6 L-GLUTAMIC ACID

L-glutamic acid (or its salt L-glutamate) is the amino acid produced in the largest quantities, with a global production of over 1 Mt/year (process flowsheet in Figure 6.7) and a market size of over US$10 billion per year. It is used by a variety of industries: the food industry uses L-glutamic acid as a flavour enhancer, the pharmaceutical industry uses it as a food supplement and the chemical industry as a building block. L-glutamic acid is also used as additive for animal feed. The production process is based on the fermentation with the bacterium *Corynebacterium glutamicum*, which, under appropriate conditions and using sugars as carbon sources, excretes large concentrations of L-glutamate in the medium. Various operating conditions can trigger in *C. glutamicum* the excretion of L-glutamate, the most common are: (a) growth under biotin limitation; (b) addition of surfactants and (c) addition of penicillin. Furthermore, the concentrations of dissolved oxygen and ammonium and the pH (optimum range 7.0–7.7) need to be controlled in order to maximise the yields. The fed-batch configuration is often the preferred option in industrial processes as it allows us to vary the concentration of ammonium and triggering factors (e.g. surfactants) over the course of the fermentation. At the end of the fermentation, yields of L-glutamate of 60–70% (based on the glucose used) can be obtained and glutamic acid is present mainly as ammonium L-glutamate. After fermentation, the microorganisms are separated by filtration and the liquid phase is passed through an anion

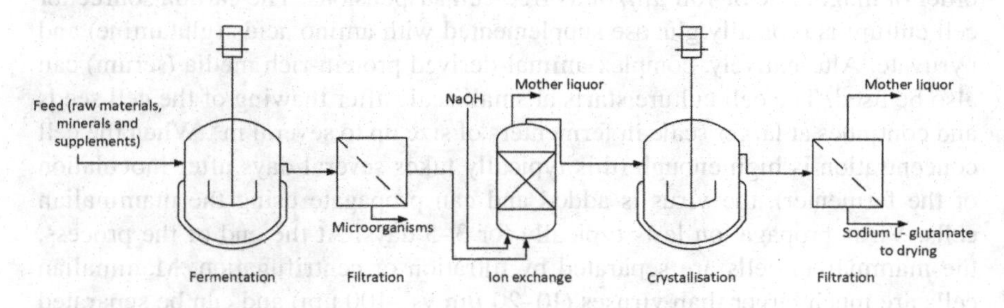

FIGURE 6.7 Simplified scheme of the production process for L-glutamic acid.

exchange resin, where L-glutamate is retained on the resin. The resin is then eluted with sodium hydroxide to form sodium glutamate which is released into solution and to regenerate the resin. Sodium glutamate is then crystallised from the solution and purified.

6.7 CELL-BASED VIRUS VACCINES

Vaccines are used to protect against diseases. In this section, we'll see fermentation processes used to produce vaccines against diseases caused by viruses. Viruses are infective agents made of a nucleic acid molecule (DNA or RNA) coated by proteins and in some cases surrounded by a lipid envelope. Viruses are considered "borderline" between living and non-living, as they can reproduce but only using living cells of other organisms. Viruses can infect any species, microorganisms, plants and animals. When viruses infect the human body, or any other organisms, they bind to and enter the cells of the host body, then they use the cell metabolic machinery to duplicate their genetic material and their protein coating, creating new virus particles and finally leave the cells. Viruses cause diseases in many ways, e.g. by breaking and killing the cells of the host body. Organisms protect themselves against viruses with their immune response, with the use of certain cells and molecules which can kill the infecting agents. The immune response is boosted by vaccination. Vaccines introduce inactivated or weakened viruses in the organism, so that, if the organism is infected with the active virus, it can quickly develop the appropriate immune response. These vaccine viruses are produced using either chicken egg cultures or animal (often mammalian) cells. In this section, we'll see an example of a process flowsheet to produce influenza vaccine viruses from mammalian cells (Figure 6.8). Influenza is a very common disease that can impact hundreds of millions of people every year and is caused by various viruses belonging to the family of *Orthomyxoviridae*.

Influenza vaccines are inactivated or attenuated live viruses, sometimes containing adjuvants (typically aluminium salts) which help the body to develop a stronger immune response. The main mammalian cells used in the production of cell-based influenza vaccines are MDCK (Madine Darby canine kidney cells), Vero (Kidney epithelial monkey cells) or PER.C6 (human retina-derived cells). The cells are grown either attached to microcarriers (e.g. dextran beads, size in the order of magnitude of 100 μm) or as free-cell suspensions. The carbon source for cell culture is typically glucose supplemented with amino acids (glutamine) and pyruvate. Alternatively, complex animal-derived protein-rich media (serum) can also be used. The cell culture starts at small scale after thawing of the cell seeds and continues at larger scale in fermenters of size up to several m^3. When the cell concentration is high enough (this typically takes several days after inoculation of the fermenter), the virus is added and can propagate using the mammalian cells. Virus propagation lasts typically for 3–5 days. At the end of the process, the mammalian cells are separated by filtration or centrifugation. Mammalian cells are much larger than viruses (10–20 μm vs ~100 nm) and can be separated easily by mechanical means. The virus is then inactivated using chemicals. The

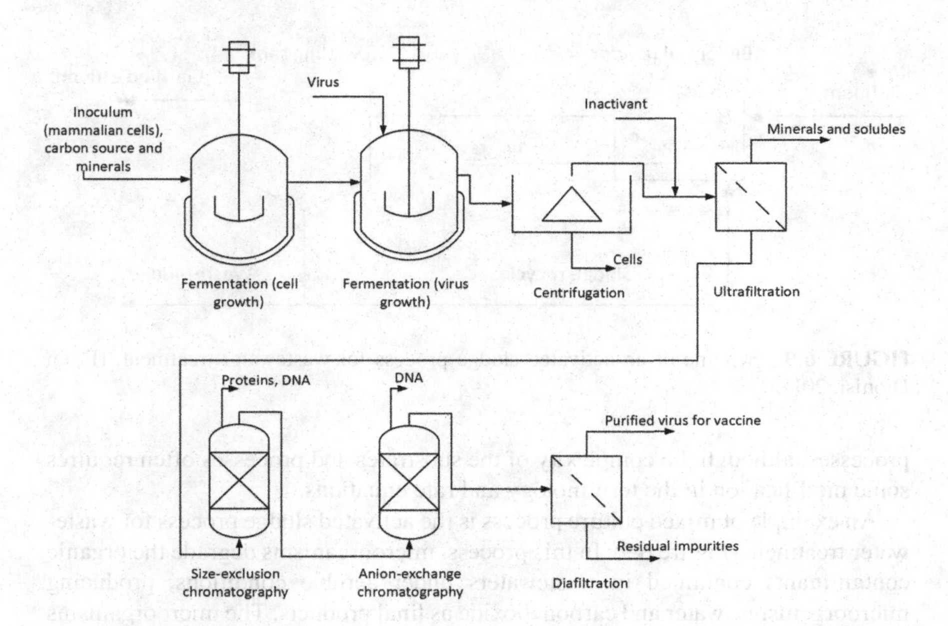

FIGURE 6.8 Example of a process flowsheet to produce influenza vaccine using mammalian cells. (Adapted from Kalbfuss et al., 2007a,b.)

suspension containing the inactivated virus is then flown through ultrafiltration membrane modules to remove small soluble molecules. The concentrated virus is purified using chromatography. Size-exclusion chromatography removes proteins (which have a much smaller size than viruses) and part of the DNA. The rest of the DNA is removed by anion exchange chromatography. Finally, the virus is purified by diafiltration to remove any residual impurities and salts.

6.8 PROCESSES WITH UNDEFINED MIXED CULTURES

Most processes considered so far use pure cultures of selected microbial species, chosen based on the desired product. In the sector of wastewater and organic waste treatment, however, processes which use undefined mixed cultures are more common. Undefined mixed cultures are consortia of many different species of microorganisms, which coexist in the same reactor. Mixed culture processes are usually started up with an inoculum composed of undefined microbial species from another similar plant or using the microorganisms which are naturally present in wastewater or waste. Mixed culture processes are usually better suitable for the treatment of wastewaters and organic waste because these waste streams contain many organic contaminants of different chemical nature, which need a consortium of different microorganisms for their degradation. On the other hand, pure cultures can only grow on a limited range of substrates. Mixed culture processes can be described with the same concepts and models used for pure culture

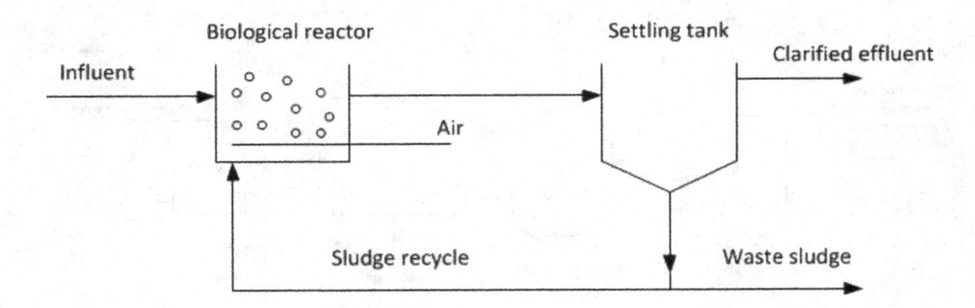

FIGURE 6.9 Scheme of an activated sludge process for wastewater treatment. (From Dionisi, 2017.)

processes, although the complexity of the substrates and processes often requires some modification in the terminology and rate equations.

An example of mixed culture process is the activated sludge process for wastewater treatment (Figure 6.9). In this process, microorganisms degrade the organic contaminants contained in wastewaters under aerobic conditions, producing microorganisms, water and carbon dioxide as final products. The microorganisms are concentrated in a settling tank and recycled to the reactor, while a (usually small) fraction is removed from the system to control their residence time (waste sludge). The settling tank also allows the clarification of the treated wastewater and its discharge to the final treatments.

Another example of mixed culture process is anaerobic digestion of organic waste. In this process (Figure 6.10), a consortium of microorganisms converts the biodegradable components of organic waste into methane and carbon dioxide (biogas), using a complex series of chemical reactions (Figure 6.11), also producing, like in most fermentation reactions, new microorganisms and water. The components of the waste (carbohydrates, proteins and fats) are first hydrolysed by extracellular enzymes into their monomeric constituents (sugars, amino acids and fatty acids), the monomers are then converted into organic acids and

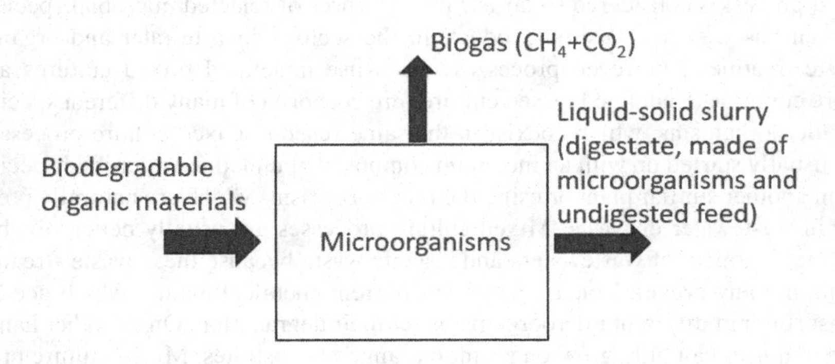

FIGURE 6.10 Scheme of an anaerobic digester for the treatment of organic waste.

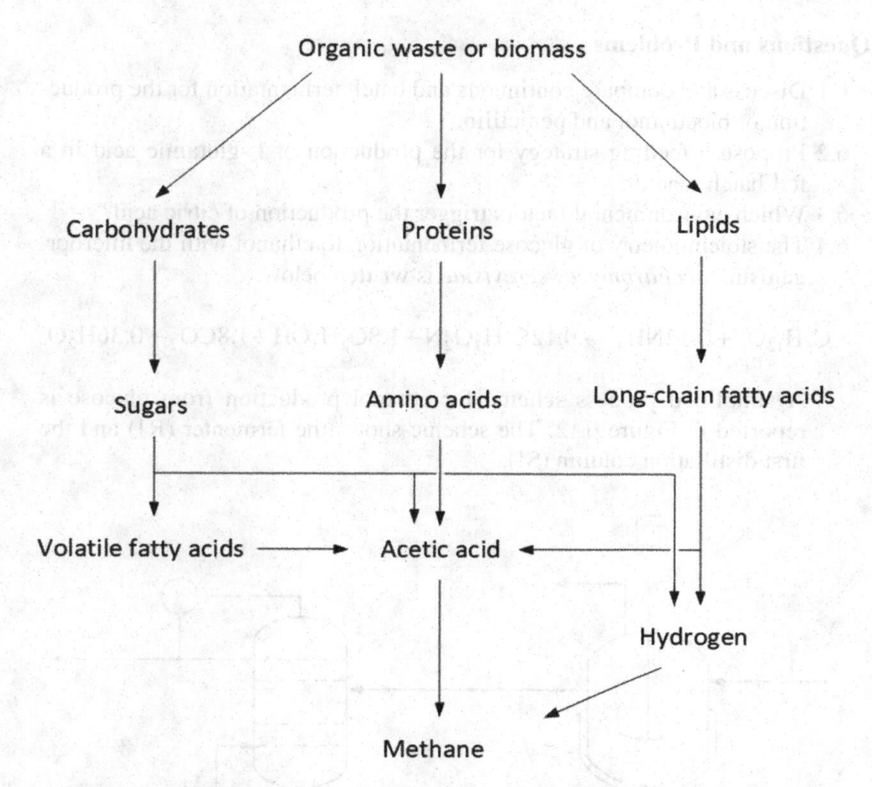

FIGURE 6.11 Simplified scheme of the complex series or reactions occurring in anaerobic digesters.

alcohols, in some cases with the production of hydrogen. Other reactions convert the organic acids and alcohols into acetic acid which is then converted into methane. Hydrogen is also converted to methane by microorganisms which use carbon dioxide as carbon source. Each of the steps in the anaerobic conversion of organic matter into methane is carried out by different microorganisms, so anaerobic digestion requires the coexistence of many different microbial populations which use the products of each other's metabolism as their substrates. The produced methane is usually combusted with air to generate electricity and/or heat; therefore, anaerobic digestion can be seen as a process to produce renewable energy from waste. The solid-liquid slurry (digestate) that leaves the digester is composed of any undigested organics, the microorganisms and the nutrients (particularly nitrogen and phosphorus) generated by the hydrolysis of the organic matter. The digestate can be used in agriculture, partially replacing the need for synthetic fertilisers. The organic acids produced as intermediates in the anaerobic digestion process have also been proposed as carbon source to produce biodegradable plastics (polyhydroxyalkanoates, PHAs) (Villano et al., 2014).

Questions and Problems

6.1 Discuss and compare continuous and batch fermentation for the production of bioethanol and penicillin.

6.2 Propose a feeding strategy for the production of L-glutamic acid in a fed-batch reactor.

6.3 Which environmental factors trigger the production of citric acid?

6.4 The stoichiometry of glucose fermentation to ethanol with the microorganism *Saccharomyces cerevisiae* is written below:

$$C_6H_{12}O_6 + 0.12NH_3 \rightarrow 0.12C_5H_7O_2N + 1.8C_2H_5OH + 1.8CO_2 + 0.36H_2O$$

A simplified process scheme for ethanol production from glucose is reported in Figure 6.12. The scheme shows the fermenter (R1) and the first distillation column (S1).

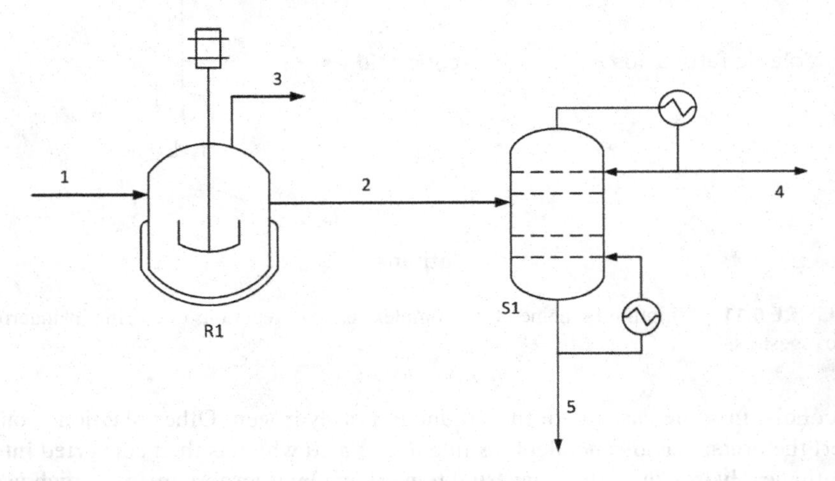

FIGURE 6.12 Simplified scheme of ethanol production for Problem 6.4.

The information available on the mass flow rates of some of the components in the various streams is reported in Table 6.5.

TABLE 6.5
Mass Flow Rates of Components for the Scheme in Figure 6.12

	Streams				
Component	**1**	**2**	**3**	**4**	**5**
Glucose (kg/h)		50	0	0	
Ethanol (kg/h)	0		25		10
Carbon dioxide (kg/h)	0	0	200		

- Calculate the mass flow rate of glucose in the feed (stream 1).
- Calculate the mass flow rate of ethanol in stream 4.

6.5 The stoichiometry of penicillin ($C_9H_{11}O_4N_2S$) production by the microorganism *Penicillium chrysogenum* ($C_5H_7O_2N$) is written below:

$$C_6H_{12}O_6 + 3.24O_2 + 0.55NH_3 + 0.037H_2SO_4 \rightarrow$$
$$0.48C_5H_7O_2N + 0.037C_9H_{11}O_4N_2S + 3.27CO_2 + 4.98H_2O$$

A simplified process flowsheet showing the reactor (R1), the filter (S1) and the solvent extractor-separator (S2) is shown in Figure 6.13.

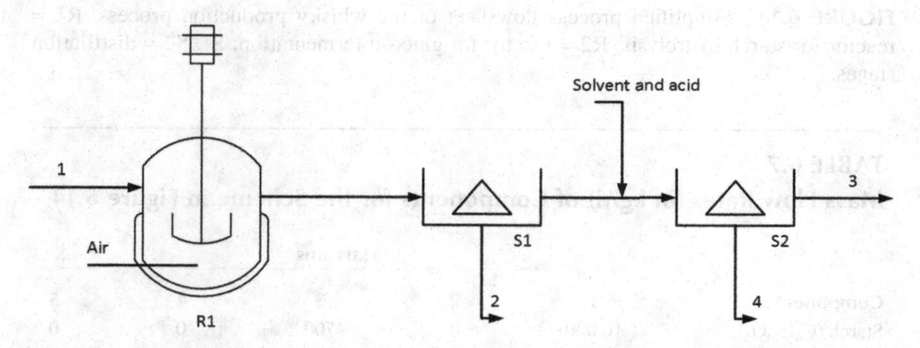

FIGURE 6.13 Simplified process flowsheet of the penicillin production process.

The information available on the mass flow rates of some of the components in the various streams is reported in Table 6.6.

TABLE 6.6
Mass Flow Rates of Components for the Scheme in Figure 6.13

	Streams			
Component	**1**	**2**	**3**	**4**
Penicillin (kg/h)	0	0.1	3.0	0.3
Microorganisms (kg/h)	0		0	1.5

Calculate:

- The oxygen consumption rate by the microorganisms in the reactor.
- The mass flow rate of microorganisms in stream 2.

6.6 Consider the process flowsheet in Figure 6.14, which shows a simplified scheme of whisky production from barley. The simplified stream table is in Table 6.7.

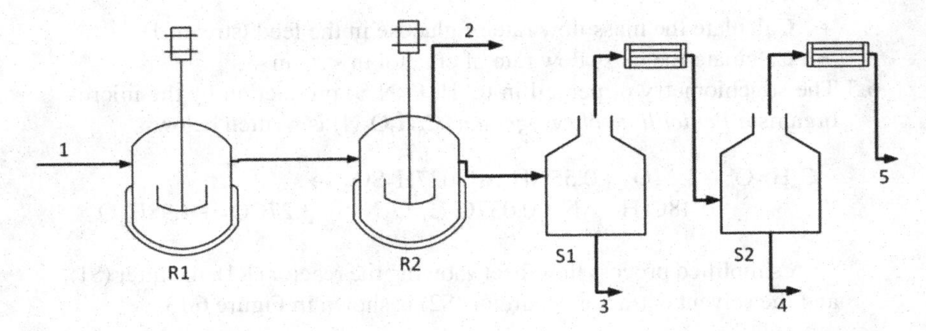

FIGURE 6.14 Simplified process flowsheet of the whisky production process. R1 = reactor for starch hydrolysis; R2 = reactor for glucose fermentation; S1, S2 = distillation stages.

TABLE 6.7
Mass Flow Rates (in kg/d) of Components for the Scheme in Figure 6.14

	Streams				
Component	**1**	**2**	**3**	**4**	**5**
Starch ($C_6H_{10}O_5$)	10,000	0	700	0	0
Glucose ($C_6H_{12}O_6$)	0	0	1,000	0	0
Ethanol (C_2H_6O)	0	0	113	200	
Yeast ($C_5H_7O_2N$)	0	0	1,890	0	0

The reaction of starch hydrolysis is: $C_6H_{10}O_5 + H_2O \rightarrow C_6H_{12}O_6$

The fermentation of glucose uses ammonia and produces microorganisms, ethanol, carbon dioxide and water. Assume that all the produced carbon dioxide leaves in stream 2.

Calculate:

- The mass flow rate of ethanol in stream 5 (as kg/d).
- The mass flow rate (kg/d) of carbon dioxide in stream 2.

Appendix A: Numerical Solutions of Differential Equations in Excel

The mass balances during batch fermentation processes, as for any batch processes, are described by differential equations. The differential equations that describe the mass balances for batch fermentation were reported in Chapter 3. This section describes a procedure to solve differential equations using Microsoft Excel.

Consider the generic differential equation:

$$\frac{dx}{dt} = f(x) \tag{A.1}$$

with the initial condition $x(t=0)=x_0$.

Solving this differential equation means finding the function $x(t)$ that satisfies Equation (A.1). A numerical solution of the equation can be found by noting that, if the time increment dt is very small, we have:

$$x\big|_{t+dt} = x\big|_t + \frac{dx}{dt}\bigg|_t dt = x\big|_t + f(x)\big|_t \, dt \tag{A.2}$$

In this way, the values of the function $x(t)$ can be obtained as a series of data points starting from the initial condition $x(0)=x_0$, e.g.,

$$\begin{aligned} x\big|_{dt} &= x_0 + f(x)\big|_0 \, dt \\ x\big|_{2dt} &= x_{dt} + f(x)\big|_{dt} \, dt \\ &\dots \\ x\big|_{n \cdot dt} &= x_{(n-1)dt} + f(x)\big|_{(n-1)dt} \, dt \end{aligned} \tag{A.3}$$

This numerical integration of the differential Equation (A.1) gives correct results only if dt is sufficiently small. Equations (A.3) can be easily solved using Excel.

Example A.1

Solve numerically the differential equation

$$\frac{dx}{dt} = 1.5 \cdot x^2 \tag{A.4}$$

over the time interval $t = \begin{bmatrix} 0 & 0.5 \end{bmatrix}$ with the initial condition $x(0) = 1$. The units of time and x in this example are arbitrary.

SOLUTION

In order to solve this equation numerically using Excel, we need to generate a column of time values using small increments dt, e.g. $dt = 0.01$. Then we need to create columns for x and for $\frac{dx}{dt} = 1.5 \cdot x^2$. The first value in the x column will be $x(0) = 1$ and the following values will be given by $x(0.01) = x(0) + \left.\frac{dx}{dt}\right|_0 \cdot 0.01$, $x(0.02) = x(0.01) + \left.\frac{dx}{dt}\right|_{0.01} \cdot 0.01$ and so on. This is shown in the screenshot below (Figure A.1).

The values of $x(t)$ obtained with this procedure are shown in Figure A.2.

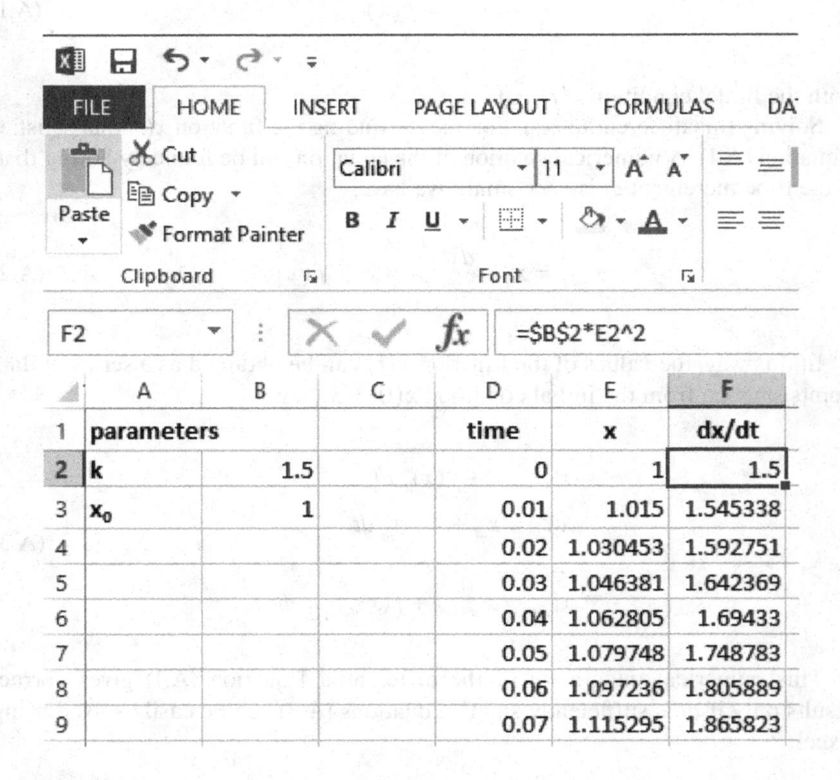

	A	B	C	D	E	F
1	**parameters**			**time**	**x**	**dx/dt**
2	**k**	1.5		0	1	1.5
3	**x_0**	1		0.01	1.015	1.545338
4				0.02	1.030453	1.592751
5				0.03	1.046381	1.642369
6				0.04	1.062805	1.69433
7				0.05	1.079748	1.748783
8				0.06	1.097236	1.805889
9				0.07	1.115295	1.865823

FIGURE A.1 Excel screenshot exemplifying the numerical solution of the differential equation in Example A.1.

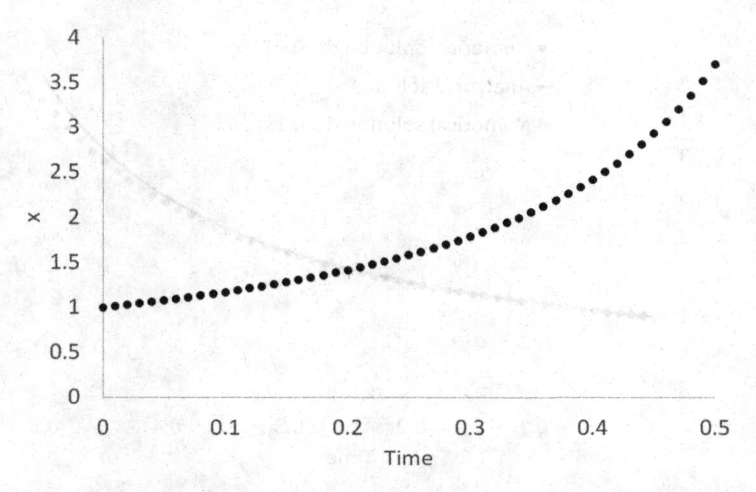

FIGURE A.2 Plot of $x(t)$ generated by numerical integration of the differential equation in Example A.1. Values of x obtained with $dt = 0.01$.

The numerical solution obtained in this way can be compared to the analytical solution which, in this particular case, can be obtained very easily:

$$x = \frac{x_0}{1 - 1.5 \cdot t \cdot x_0}$$

The comparison of the analytical and numerical solution is shown in Figure A.3. It can be seen that the agreement between the numerical and the analytical solution is very good, this means that the numerical error is very small.

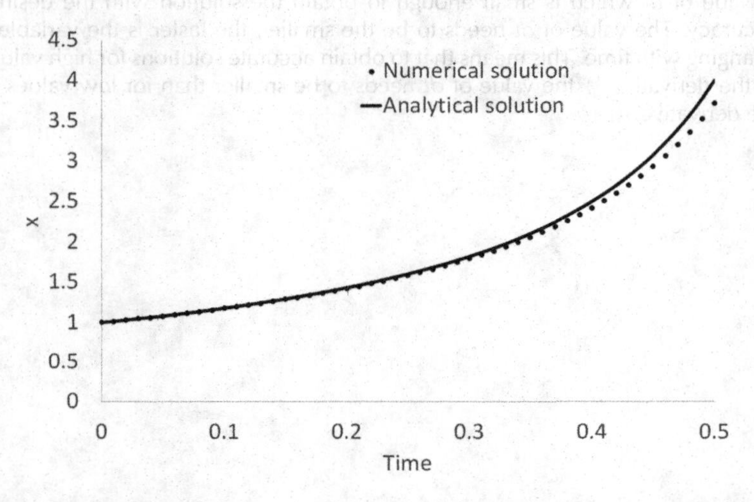

FIGURE A.3 Comparison of the numerical solution (obtained with $dt = 0.01$) with the analytical solution.

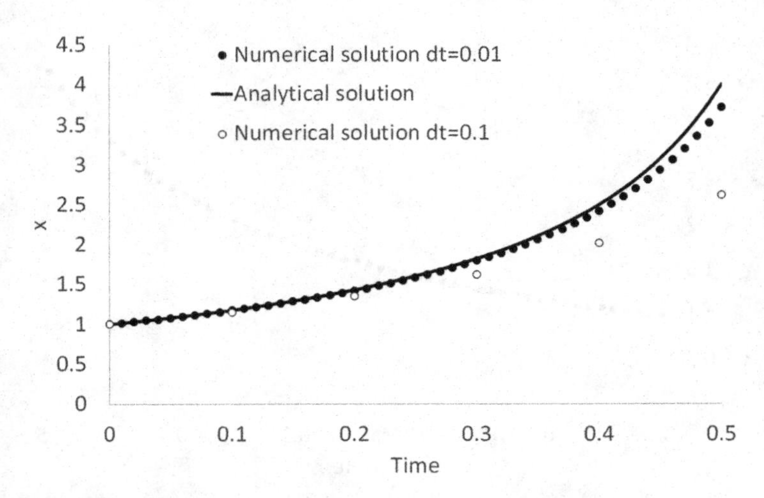

FIGURE A.4 Comparison of the numerical solution obtained with $dt = 0.1$ with the numerical solution obtained with $dt = 0.01$ and the analytical solution.

It is important to observe that the critical parameter that determines the accuracy of the numerical solution is the time interval used for discretisation, dt, which needs to be as small as possible, compatibly with the computational power available. For example, let us see what happens if we choose a time interval $dt = 0.1$, i.e. ten times larger than the dt value used so far. Figure A.4 compares the values of the function x with $dt = 0.1$, $dt = 0.01$ and the analytical solution.

From Figure A.4, it is evident that the numerical solution with $dt = 0.1$ is largely incorrect (large numerical error) and this shows the need to choose a value of dt which is small enough to obtain the solution with the desired accuracy. The value of dt needs to be the smaller, the faster is the variable x changing with time. This means that to obtain accurate solutions for high values of the derivative $\frac{dx}{dt}$ the value of dt needs to be smaller than for low values of the derivative.

Appendix B: Numerical Solutions of Systems of Equations in Excel

Systems of equations describe the steady state of continuous-flow processes like the chemostat. In many cases, these equations can be solved manually by substitution; however, this section presents a method to solve them using Excel.

Consider a system of equations:

$$\begin{cases} f_1\left(x_1, x_2, \dots, x_n\right) = 0 \\ f_2\left(x_1, x_2, \dots, x_n\right) = 0 \\ \dots \dots \\ f_n\left(x_1, x_2, \dots, x_n\right) = 0 \end{cases} \tag{B.1}$$

This is a system of n equations in the n unknowns x_1, x_2, …, x_n. Solving this system of equations means finding the values of x_1, x_2, … x_n that satisfy all the equations $f_1\left(x_1, x_2, \dots, x_n\right) = 0$, $f_2\left(x_1, x_2, \dots, x_n\right) = 0$, …, $f_n\left(x_1, x_2, \dots, x_n\right) = 0$.

The procedure to solve the system of equations (B.1) in Excel is quite simple. First, initial guesses have to be given to the unknowns. Then the equations have to be written in the same form as for Equations (B.1), i.e. with the left-hand side equal to 0. The left-hand sides of the equations have to be written into Excel as a function of the given initial guesses for the unknowns. Excel will calculate a numerical value for each equation (or rather for the left-hand side of each equation) which will be in general different than 0, because the initial guesses given are not the solution of the system of equations. We will call these calculated values "residuals". Then each residual is squared and all the squared residuals are added up together. At this point, Solver is used to find the minimum of the sum of the squared residuals, by manipulating the values of the unknowns. If Solver is successful and the sum of squared residuals is very close to 0, the values found for the unknown are the solutions of the system of equations.

Note that the procedure is quite sensitive (or very sensitive in some cases) to the given values of the initial guesses. Therefore, it is important to have an idea at least of the order of magnitude of the values of the unknowns. The procedure is shown in the example below.

Example B.1

Solve the following system of equations:

$$\begin{cases} x+2y+z=3 \\ 2x+y-z=5 \\ x-y+2z=2 \end{cases}$$

SOLUTION

This is a system of three equations in the three unknowns x, y and z. This is a simple system that can be easily solved manually by substitution; however, we will use it as an example of the Excel method. First of all, initial guesses have to be given to x, y and z; we will give the value of 1 to all of them. Then we need to write the equations in Excel in the form:

$$\begin{cases} x+2y+z-3 \\ 2x+y-z-5 \\ x-y+2z-2 \end{cases}$$

Excel will calculate a value for each equation (residual). We need to square each of the residuals and then add all the squared residuals together. Then use Solver to find the values of x, y and z than make the sum of the squared residuals minimum.

This procedure is shown in the screenshot (see Figure B.1).

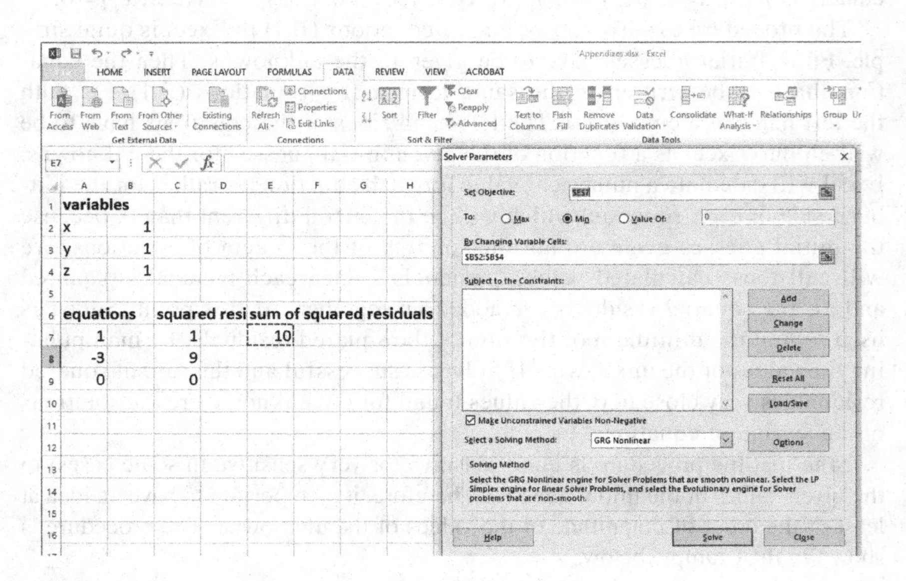

FIGURE B.1 Screenshot showing the procedure to solve Example B.1.

In this case, solver gives the correct values of the unknowns:

$$\begin{cases} x = 2.33 \\ y = 0.33 \\ z = 0 \end{cases}$$

With these values, the sum of the squared residuals is of the order of 10^{-12}, i.e. a very low value which ensures we have found the solution of the system of equations.

Appendix C: Answers and Solutions

CHAPTER 1

1.1 Main cell components and their functions:

- Membranes are the external cell boundaries and allow for the retention of the intracellular materials and for the in-out flow of the substrate, nutrient and products.
- Enzymes are proteins which catalyse all the metabolic reactions in the cell.
- DNA contains the genetic information which is used to synthesise proteins, including the enzymes which are used to synthesise all the main cell components. RNA is used for the transcription of the genetic information in the DNA and for the protein synthesis.
- Cytoplasm includes water and all the intracellular components of the cell.
- ATP and ADP are used as energy vectors for chemical reactions in the cell. ATP transfers energy to the reagents to allow their conversion into products and in this process is converted to ADP. ADP is converted back to ATP by catabolic reactions.
- NADH and NAD^+ are used in the oxidation and reduction reactions in the cell. In aerobic metabolism, NADH is converted back to NAD^+ using oxygen in a process that generates ATP from ADP. In anaerobic metabolism, NADH is converted back to NAD^+ using the substrate or metabolic intermediates, generating metabolic products which have often commercial value but with no intracellular energy generation.

1.2 Primary metabolites are produced during cell growth and duplication. Examples of primary metabolites are the products of anaerobic fermentations, for example, methane, ethanol and organic acids. Secondary metabolites are produced after cell growth has stopped or when it has slowed down considerably. The production of secondary metabolites is often triggered by the lack of essential nutrients or metabolites in the medium or by a non-balanced composition of the medium. Examples of secondary metabolites are organic products of aerobic metabolism, e.g. many amino acids and certain organic acids.

1.3 In aerobic metabolism usually much more energy is produced per unit of substrate removed from the medium than in anaerobic metabolism.

The reason for the higher energy production is the oxidation of NADH molecules by oxygen, which generates up to 3 moles of ATP per mol of NADH converted to NAD^+ (this process is called oxidative phosphorylation). Under anaerobic conditions, there is no oxygen and oxidative phosphorylation cannot happen. Therefore, the energy production in anaerobic metabolism is only due to substrate-level phosphorylation, and is much lower, per unit of metabolised substrate, than under aerobic conditions. The energy required for the synthesis of cell components is similar for aerobic and anaerobic metabolism. Therefore, under anaerobic conditions a higher fraction of the substrate needs to be used in the catabolic pathways than under aerobic conditions, giving a lower production of microorganisms per unit of substrate removed.

1.4 Batch processes are easy to operate and allow for an easy traceability of the product back to the production batch, which is good for safety and regulatory purposes, especially for the pharmaceutical industry. Continuous processes don't have downtime and have usually higher microorganism concentration than batch processes, giving, therefore, higher reaction rates. However, chemostat processes require a careful control of the fermentation conditions with risks of process instability and contamination. Fed-batch processes combine the easy traceability of batch processes with the high microorganism concentration of continuous processes. Furthermore, fed-batch processes are highly flexible because they allow for a change in the feed composition during the feeding; therefore, they are particularly suitable for the production of secondary metabolites, whose production is triggered by a change in the medium composition. However, fed-batch reactors have downtime due to the removal of the reaction volume and their volumetric productivity is lower than for continuous processes because part of the fermentation volume remains in the reactor and does not contribute to the productivity.

1.5 Stirred tank reactors have the advantages of maintaining the same composition in any points of the reactors. In addition, mass and heat transfer rates are enhanced by mixing. Packed- and fluidised-bed fermenters allow for microorganism retention in the reactor, which facilitates product purification and can give higher microorganism concentration and so higher reaction rates. However, mass and heat transfer rates may be lower than for stirred tank fermenters due to the absence of mechanical agitation. Airlift reactors give good mixing similar as for stirred tank fermenters and are better for shear-sensitive cultures, however, the mass and heat transfer rates may be lower than in agitated vessels. Bubble columns are cheap to manufacture and are suitable for shear-sensitive cultures, however, the mass and heat transfer rates are typically low because of the absence of mechanical agitation.

CHAPTER 2

2.1 The generic stoichiometry for this fermentation can be written as:

$$C_2H_5OH + aO_2 + bNH_3 \rightarrow cC_5H_7O_2N + dCO_2 + eH_2$$

We know that 0.20 kg of microorganisms are produced per kg of ethanol consumed, so:

$$\frac{c \cdot 113}{46} = 0.20 \quad \Rightarrow \quad c = 0.08$$

The other coefficients can be calculated from the elemental balances for C, H, N and O:

$$\begin{cases} C: 2 = 5 \cdot 0.08 + d \\ H: 6 + 3b = 7 \cdot 0.08 + 2e \\ O: 1 + 2a = 2 \cdot 0.08 + 2d + e \\ N: b = 0.08 \end{cases}$$

The solution of the system of equations gives $b = 0.08$; $d = 1.60$; $e = 2.84$ and $a = 2.60$. So the stoichiometry for this process is:

$$C_2H_5OH + 2.60O_2 + 0.08NH_3 \rightarrow 0.08C_5H_7O_2N + 1.60CO_2 + 2.84H_2O$$

2.2 Glucose converted is 30,000 × 0.95 = 28,500 t/year.

- From the reaction stoichiometry, 0.60 kg of glutamic acid are produced per kg of glucose, so the glutamic acid produced is 28,500 × 0.60 = 17,100 t/year.
- 0.45 kg of carbon dioxide are produced per kg of glucose, so 12,888 t/year of carbon dioxide are produced.
- 0.39 kg of oxygen are consumed per kg of glucose, so 11,115 t/year of oxygen are consumed.

2.3 From the stoichiometry, we need 0.1 × 14 = 1.4 g of N per 0.7 × 192 = 134.4 g of citric acid, so we need 0.0104 tonnes of N per tonne of citric acid. For 10,000 tonnes of citric acid per year, we need 104.17 tonnes of N, i.e. 398 tonnes of ammonium chloride.

- From the stoichiometry, we need 0.56 tonnes of oxygen per tonne of citric acid produced, so to produce 10,000 tonnes of citric acid we need 5,600 tonnes of oxygen.
- 0.43 tonnes of carbon dioxide are produced per tonne of citric acid produced, so 4,300 tonnes of carbon dioxide will be produced for the production of 10,000 tonnes of citric acid per year.

2.4 The generic stoichiometry for this fermentation can be written as:

$$C_6H_{12}O_6 + aO_2 + bNH_3 \rightarrow cC_5H_7O_2N + dC_3H_5O_2N + eCO_2 + fH_2O$$

We know that 0.20 kg of microorganisms are produced per kg of glucose consumed, so:

$$\frac{c \cdot 113}{180} = 0.20 \quad \Rightarrow \quad c = 0.32$$

We also know that 0.60 kg of protease are produced per kg of glucose consumed, so:

$$\frac{d \cdot 87}{180} = 0.60 \quad \Rightarrow \quad d = 1.24$$

The remaining coefficients are calculated from the elemental balances:

$$\begin{cases} C: 6 = 5 \cdot 0.32 + 3 \cdot 1.24 + e \\ H: 12 + 3b = 7 \cdot 0.32 + 5 \cdot 1.24 + 2f \\ O: 6 + 2a = 2 \cdot 0.32 + 2 \cdot 1.24 + 2e + f \\ N: b = 0.32 + 1.24 \end{cases}$$

The solution of the system of equations gives b = 1.56; e = 0.68; f = 4.12 and a = 1.30. So, the stoichiometry for this process is:

$$C_6H_{12}O_6 + 1.30O_2 + 1.56NH_3 \rightarrow 0.32C_5H_7O_2N$$

$$+ 1.24C_3H_5O_2N + 0.68CO_2 + 4.12H_2O$$

2.5 The coefficient x can be determined using the C, H or O balance. The C balance is:

$$6 = 5 \cdot 0.15 + 2x + 2.1$$

from which x = 1.575.

2.6 The rate of lactic acid (the product) production is proportional to the rate of lactose (the substrate) removal according to:

$$r_P = Y_{P/S} \cdot (-r_S)$$

With the data available, to determine $Y_{P/S}$ we need to calculate the stoichiometry of the fermentation. The general stoichiometry for this fermentation is:

$$C_{12}H_{22}O_{11} + aNH_3 \rightarrow bC_5H_7O_2N + cC_3H_6O_3 + dH_2O$$

It is known that the initial lactose concentration is $0.80 \times 50 = 40$ g/L. The removal of 70% of the initial lactose corresponds to a removal of

28 g/L of lactose. This coincides with the formation of 2 g/L of microorganisms, so:

$$\frac{b\cdot 113}{342}=\frac{2}{28}\Rightarrow b=0.22$$

The remaining coefficients can be calculated from the elemental balances:

$$\begin{cases} C: 12=5\cdot 0.22+3c \\ H: 22+3a=7\cdot 0.22+6c+2d \\ O: 11=2\cdot 0.22+3c+d \\ N: a=0.22 \end{cases}$$

The solution of the system of equations gives $a = 0.22$; $c = 3.63$ and $d = -0.33$ (this means that water is a reagent and not a product). So, the stoichiometry for this process is:

$$C_{12}H_{22}O_{11}+0.22NH_3+0.33H_2O\rightarrow 0.22C_5H_7O_2N+3.63C_3H_6O_3$$

From this stoichiometry, $Y_{P/S} = 0.96$ kg/kg. Therefore, the rate of lactic acid production when the rate of lactose removal is 0.4 g/L.h is:

$$r_P=0.96\cdot 0.4=0.38\frac{g}{L\cdot h}$$

Note that in this case the general fermentation equation has only four unknown coefficients because there is no carbon dioxide production. However, it is not possible to determine the four coefficients using the elemental balances alone because the balances of H and O are the combination of each other, so they are not independent equations. The information on the microorganism yield is, therefore, still needed. Related to this, also note that, in the solution of the four elemental balances the coefficient d can be obtained either from the H or from the O balance obtaining the same result.

2.7 According to Equation (2.19):

$$\mu_{\max}=Ae^{-\frac{Ea}{RT}}$$

Taking the natural logarithm of both sides:

$$\ln\mu_{\max}=\ln A-\frac{E_a}{RT}$$

Therefore, if we generate a series of values of $\mu_{\max}$ at different temperatures, the plot of ln $\mu_{\max}$ vs $1/T$ should give a straight line with intercept equal to ln A and slope equal to –Ea/R, allowing for the calculation of A and Ea. For the determination of each μ_{max} value at each temperature, experiments should be carried out by measuring the biomass and substrate concentration as a function of time, as shown in Section 2.4.

2.8 We use the Monod model with maintenance metabolism. In this case:

$$r_S = -\frac{\mu \cdot X}{Y_{X/S}} - m_S X$$

$$r_X = \mu \cdot X$$

Therefore:

$$r_S = -\frac{r_X}{Y_{X/S}} - m_S X = -\frac{1.7\frac{kg}{m^3 \cdot h} \cdot 24\frac{h}{d}}{0.2\frac{kg}{kg}} - 0.4d^{-1} \cdot 2.5\frac{kg}{m^3} = -205\frac{kg}{m^3 \cdot d}$$

2.9 Equation (2.17), which describe the inhibition of microbial growth by a substance I, can be re-written as:

$$\mu = \frac{\mu_{max} S}{K_S + S} \frac{K_I}{K_I + I}$$

The rate of substrate removal is, therefore:

$$r_S = -\frac{\mu \cdot X}{Y_{X/S}} = -\frac{\mu_{max} S}{K_S + S} \frac{K_I}{K_I + I} \frac{X}{Y_{X/S}}$$

If $S \gg K_S$, this equation becomes:

$$r_S = -\mu_{max} \frac{K_I}{K_I + I} \frac{X}{Y_{X/S}}$$

which can be re-arranged as:

$$\frac{X}{(-r_S)} = \frac{Y_{X/S}}{\mu_{max}} + \frac{Y_{X/S}}{\mu_{max} K_I} I$$

Therefore, a series of batch experiments can be carried out, under the condition that at the start of each experiment $S \gg K_S$, measuring the initial substrate removal rate at different concentrations of the inhibitor I. Then $X/(-r_S)$ should be plotted vs I obtaining a linear plot with slope equal to $Y_{X/S}/\mu_{max} \cdot K_I$, from which K_I can be calculated since $Y_{X/S}$ and μ_{max} are known.

2.10 According to Equation (2.19):

$$\mu_{max} = Ae^{-\frac{Ea}{RT}}$$

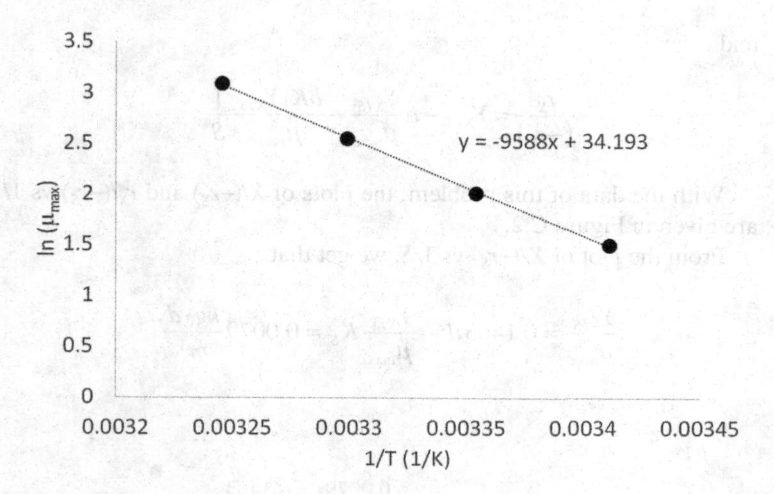

FIGURE C.1 Plot for Problem 2.10.

This equation can be linearised as:

$$\ln \mu_{\max} = \ln A - \frac{E_a}{RT}$$

Therefore, the values of A and E_a can be obtained from a linear plot of $\ln \mu_{\max}$ vs $1/T$. With the given experimental data, the linear plot is in Figure C.1.

From the regression line, we find that:

$$A = e^{34.193} = 7.077 \cdot 10^{14}$$

and

$$\frac{E_a}{R} = 9{,}588K \quad \Rightarrow \quad E_a = 79.68 \frac{kJ}{mol}$$

At 28°C, the value of $\mu_{\max}$ is:

$$\mu_{\max} = 7.077 \cdot 10^{14} e^{-\frac{79{,}670}{8.31 \cdot 301.15}} = 10.6 d^{-1}$$

2.11 If the metabolism is described by the Monod model with endogenous metabolism, the linearisation of the rate equations for substrate removal and biomass growth gives:

$$\frac{X}{(-r_S)} = \frac{Y_{X/S}}{\mu_{\max}} + \frac{Y_{X/S}}{\mu_{\max}} K_S \frac{1}{S}$$

and

$$\frac{r_X}{(-r_S)} = Y_{X/S} - b\frac{Y_{X/S}}{\mu_{max}} - \frac{bK_S Y_{X/S}}{\mu_{max}}\frac{1}{S}$$

With the data of this problem, the plots of $X/(-r_S)$ and $r_X/(-r_S)$ vs $1/S$ are given in Figure C.2.

From the plot of $X/(-r_S)$ vs $1/S$, we get that:

$$\frac{Y_{X/S}}{\mu_{max}} = 0.1463\,d; \quad \frac{Y_{X/S}}{\mu_{max}}K_S = 0.0079\frac{kg \cdot d}{m^3}$$

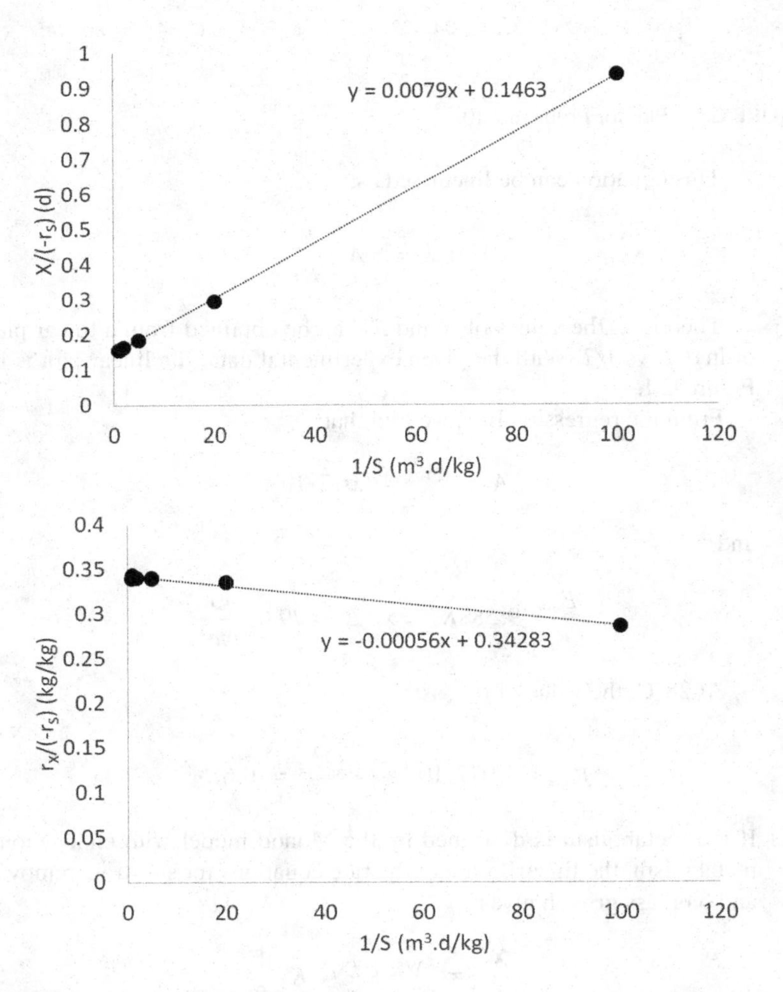

FIGURE C.2 Plot for Problem 2.11.

Therefore, $K_S = 0.054$ kg/m³.
From the plot of $r_X/(-r_S)$ vs $1/S$, we get that:

$$Y_{X/S} - b\frac{Y_{X/S}}{\mu_{max}} = 0.3428\frac{kg}{kg}; \quad \frac{bK_S Y_{X/S}}{\mu_{max}} = 0.00056\frac{kg}{m^3}$$

Therefore, $b = 0.07$ d^{-1}, $Y_{X/S} = 0.35$ kg/kg and $\mu_{max} = 2.4$ d^{-1}.

2.12 The total rate of oxygen consumption by the microorganisms is given by:

$$\left(-r_{O2biom}\right) = \frac{\mu X}{Y_{X/O2}} + \frac{bX}{Y_{X/O2,end}}$$

We know that in this case the net growth rate of the microorganisms is:

$$r_X = \mu X - bX = 0.3\frac{kg}{m^3 \cdot h} \text{ and } bX = 0.045\frac{kg}{m^3 d}$$

Therefore, $\mu X = 0.255\frac{kg}{m^3 d}$ and $\left(-r_{O2biom}\right) = 0.38\frac{kg}{m^3 d}$

CHAPTER 3

3.1 From the reaction stoichiometry, we calculate the parameters $Y_{X/S}$ and $Y_{P/X}$. These are:

$$Y_{X/S} = \frac{0.1 \cdot 113}{180} = 0.063\frac{kg}{kg}$$

$$Y_{P/X} = \frac{0.66 \cdot 192}{0.1 \cdot 113} = 11.21\frac{kg}{kg}$$

$$\mu_{max} = \frac{\ln 2}{t_d} = 0.315 h^{-1}$$

The maximum growth rate can be calculated from the doubling time:
The differential equations that give the profiles for the substrate, microorganisms and product are, in this case (with concentrations in kg/m³ and time in *h*):

$$\frac{dX}{dt} = \frac{0.198S}{0.07 + S}X$$

$$\frac{dS}{dt} = -\frac{0.198S}{0.07 + S}\frac{X}{0.063}$$

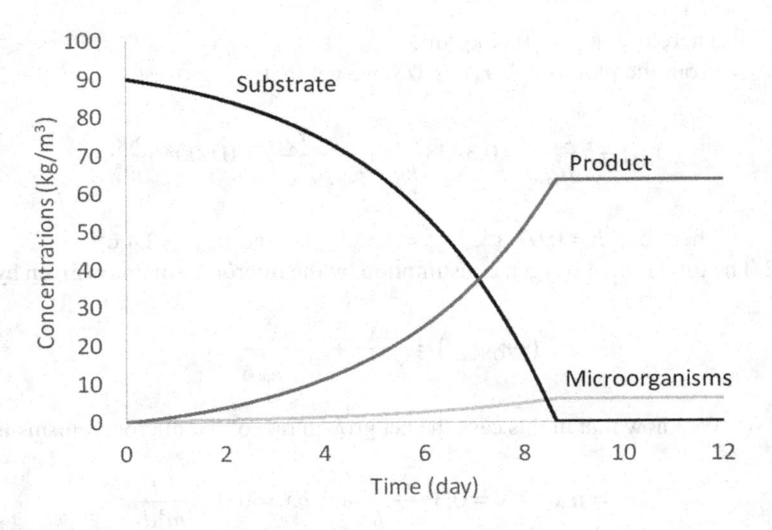

FIGURE C.3 Plot for Problem 3.1.

$$\frac{dP}{dt} = \frac{0.198S}{0.07 + S} X \cdot 11.21$$

These equations can be solved with $S_0 = 90$ kg/m³, $X_0 = 0.2$ kg/m³ and $P_0 = 0$. The obtained profiles are reported in Figure C.3.

3.2 The effluent substrate concentration in a chemostat is given by:

$$S = \frac{DK_S}{\mu_{max} - D}$$

This equation can be applied to the two operating conditions given in the problem (where the time is in *d* and the concentration in kg/m³):

$$0.05 = \frac{1.0K_S}{\mu_{max} - 1.0}$$

$$0.15 = \frac{1.3K_S}{\mu_{max} - 1.3}$$

These are two equations in the two unknowns: μ_{max} and K_S. The solution gives $\mu_{max} = 1.53$ d^{-1} and $K_S = 0.026$ kg/m³.

3.3 For a fed-batch reactor at the "quasi-steady state" the substrate concentration during feeding is:

$$S = \frac{DK_S}{\mu_{max} - D}$$

Imposing that $S = 0.1$ kg/m^3, we obtain: $D = 5.33$ d^{-1}. This is the maximum dilution rate that we can have at any time during feeding so that the substrate concentration is always lower than 0.1 kg/m^3. With an initial volume of 1 m^3, the maximum flow rate during feeding is:

$$Q = V \cdot D = 5.33\frac{m^3}{d}$$

3.4 From the reaction stoichiometry:

$$Y_{P/X} = 9.50\frac{kg}{kg}$$

$$Y_{X/S} = 0.063\frac{kg}{kg}$$

$$Y_{X/O2} = 0.16\frac{kg}{kg}$$

The concentrations of biomass, substrate and product and the oxygen consumption rate as a function of the dilution rate at steady state are given by:

$$S = \frac{DK_S}{\mu_{\max} - D}$$

$$X = Y_{X/S}\left(S_0 - S\right)$$

$$P = X \cdot Y_{P/X}$$

$$r_{O2biom} = -D\frac{X}{Y_{X/O_2}}$$

The results are plotted in Figure C.4.

3.5 The yields $Y_{X/S}$ and $Y_{P/X}$ are, in this case, from the reaction stoichiometry:

$$Y_{P/X} = 6.11\frac{kg}{kg}$$

$$Y_{X/S} = 0.075\frac{kg}{kg}$$

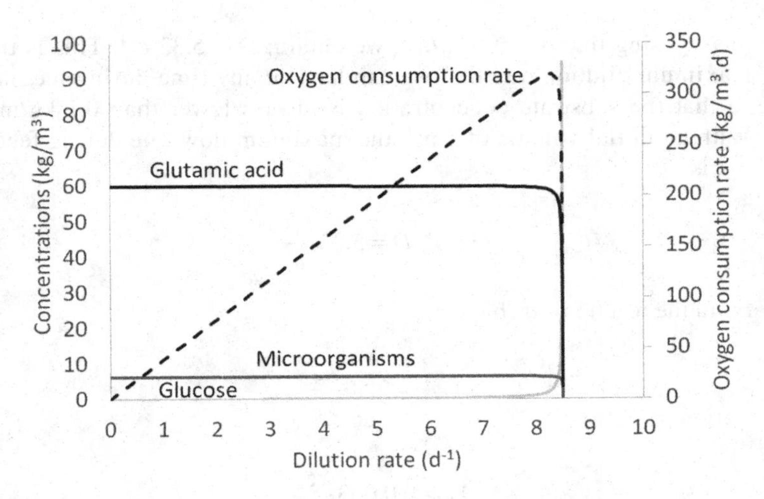

FIGURE C.4 Plot for Problem 3.4.

For the batch process, the time profiles of microorganisms, glucose and ethanol are given by the following equations, where time is in d:

$$\frac{dX}{dt} = \frac{12.1S}{0.1+S}X$$

$$\frac{dS}{dt} = -\frac{12.1S}{0.1+S}\frac{X}{0.075}$$

$$\frac{dP}{dt} = \frac{12.1S}{0.1+S}X \cdot 6.11$$

The initial conditions are $S_0 = 120$ kg/m^3, $X_0 = 0.5$ kg/m^3 and $P_0 = 0$. The time profiles are shown in Figure C.5.

From the graph, we see that the time required to obtain 99% glucose removal is 0.242 d. The downtime is 2 h = 0.083 d, so the total time for one batch is approximately 0.33 d. At the end of each batch, the concentration of ethanol is 53 kg/m^3. So, the ethanol productivity is 53/0.33 = 161 kg/m^3·d = 58.6 t/m^3·year. Therefore, the reactor volume required to produce 20,000 t of ethanol per year is 20,000/58.6 = 341.2 m^3.

For the chemostat, the dilution rate required for a concentration of 1.2 kg glucose/m^3 (which corresponds to 99% conversion of the glucose in the feed) is given by:

$$D = \frac{\mu_{max} S}{K_S + S} = 11.2 d^{-1}$$

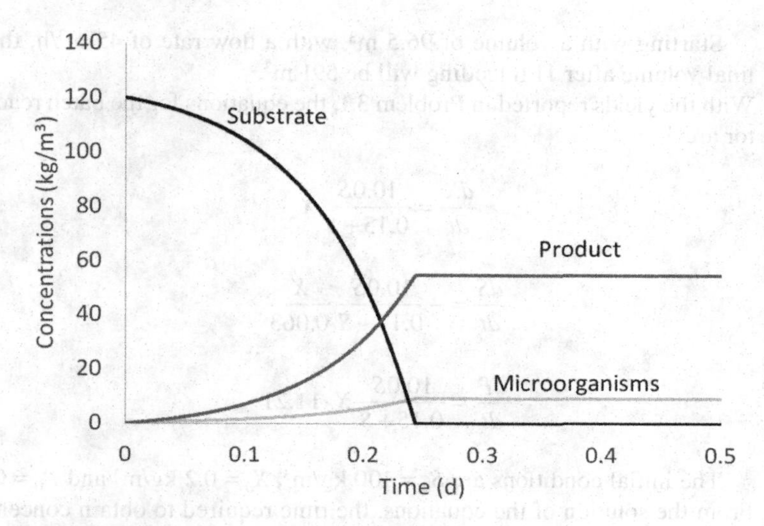

FIGURE C.5 Plot for Problem 3.5.

From the reaction stoichiometry, we consume 2.17 t of glucose per 1 t of ethanol, so to produce 20,000 t of ethanol per year we need to feed 43,400 t of glucose per year. This allows us to calculate the required flow rate to the reactor:

$$Q = \frac{43,400 \frac{t}{year}}{0.12 \frac{t}{m^3}} = 361,700 \frac{m^3}{year}$$

The required volume of the chemostat is, therefore:

$$V = \frac{Q}{D} = \frac{361,700 \frac{m^3}{year}}{\frac{11.2 \frac{1}{d}}{365 \frac{d}{year}}} = 88.5 m^3$$

For the fed-batch reactor, the annual feed flow rate is the same as calculated for the chemostat, i.e. 361,700 m^3/year. However, the reactor is fed over 11 h per cycle, with 1 h downtime between batches. Therefore, the actual feed flow rate over the 11 h feeding time will be 361,700 × 12/11 = 394,582 m^3/year = 1,081 m^3/d = 45.0 m^3/h. The maximum dilution rate which is required to maintain the glucose concentration always below 0.1 kg/m^3 is the same as calculated for the chemostat, i.e. 11.2 d^{-1}. The initial volume of the fed-batch reactor is, therefore:

$$V_0 = \frac{Q}{D} = 96.5 m^3$$

Starting with a volume of 96.5 m³, with a flow rate of 45 m³/h, the final volume after 11 h feeding will be 591 m³.

3.6 With the yields reported in Problem 3.1, the equations for the batch reactor are:

$$\frac{dX}{dt} = \frac{10.0S}{0.15+S}X$$

$$\frac{dS}{dt} = -\frac{10.0S}{0.15+S}\frac{X}{0.063}$$

$$\frac{dP}{dt} = \frac{10.0S}{0.15+S}X \cdot 11.21$$

The initial conditions are $S_0 = 100$ kg/m³, $X_0 = 0.2$ kg/m³ and $P_0 = 0$. From the solution of the equations, the time required to obtain concentration of glucose of 0.5 kg/m3 is 0.35 d. By adding the downtime (2 h = 0.083 d), the total batch time will be 0.43 d. When the glucose concentration is 0.5 kg/m³, the citric acid concentration is 70 kg/m³ (this can be obtained from the graph or directly from the product yield or reaction stoichiometry). Therefore, the citric acid productivity in the batch process is 70/0.43 = 162.8 kg/m³.d.

For the chemostat, the dilution rate required to achieve a glucose concentration of 0.5 kg/m³ is given by:

$$D = \frac{10.0 \cdot 0.5}{0.15+0.5} = 7.7d^{-1}$$

The effluent citric acid concentration is 70 kg/m³ (same as for the batch reactor), therefore, the productivity per unit volume of reactor is: 70 × 7.7 = 539 kg/m³.d.

For the fed-batch reactor, the maximum dilution rate at the start of the cycle is still 7.7 d⁻¹.

The product productivity is given by:

$$\frac{dP_{TOT}}{dt}\left(\frac{kgproduct}{d}\right) = \frac{Y_{P/S}S_0D}{1+Dt_{feed}} = 119.0\frac{kg}{m^3d}$$

This is the citric acid productivity during the feed period. However, we need to consider that there is no production during the downtime, so the actual productivity will be 119 × 11/12 = 109.1 kg/m³.d.

3.7 The parameters $Y_{X/S}$ and m_S can be obtained by plotting $\frac{(S_0-S)}{X}$ vs $\frac{1}{D}$. With the data of this problem, the plot is shown in Figure C.6.

From the slope, we obtain $m_S = 1.47$ d⁻¹, from the intercept we obtain $Y_{X/S} = 0.30$ kg/kg.

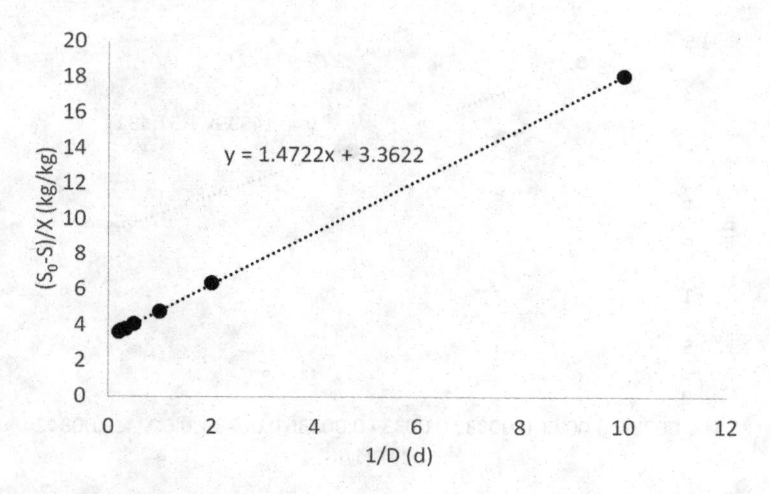

FIGURE C.6 Plot for Problem 3.7.

3.8 At $T = 30°C$, we calculate the μ_{max} from the doubling time.
From Equation (2.19), the ratio between the μ_{max} at two temperatures T_1 (30°C = 303.15 K) and T_2 (35°C = 308.15 K) is:

$$\frac{\mu_{max\,2}}{\mu_{max\,1}} = e^{-\frac{Ea}{R}\left(\frac{1}{T_2}-\frac{1}{T_1}\right)} = e^{-\frac{90{,}000}{8.31}\left(\frac{1}{308.15}-\frac{1}{303.15}\right)}$$

We obtain $\mu_{max2} = 0.62\ d^{-1}$.

3.9 For each data series, we can plot $\frac{1}{S}$ vs $\frac{1}{D}$ (Figure C.7) to calculate μ_{max}.

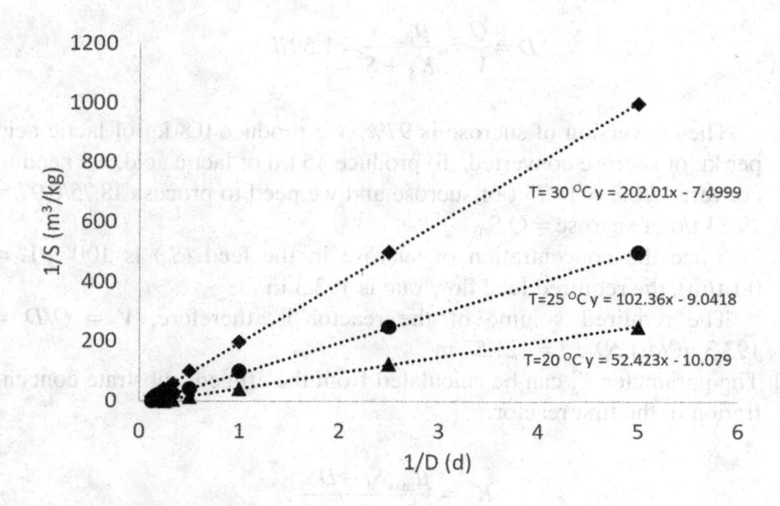

FIGURE C.7 Plot for Problem 3.9.

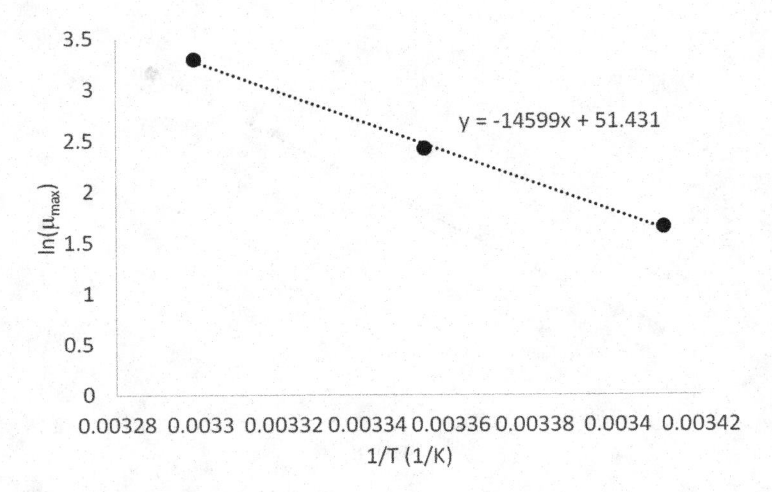

FIGURE C.8 Plot for Problem 3.9.

From the plot, we obtain the μ_{max} at the three temperatures: T = 20°C, μ_{max} = 5.2 d^{-1}; T = 25°C, μ_{max} = 11.3 d^{-1} and T = 30°C, μ_{max} = 26.9 d^{-1}.

To obtain the parameters A and E_a of Equation (2.19), we need to plot $\ln(\mu_{max})$ vs $\frac{1}{T}$. The plot is reported in Figure C.8.

From the slope of the plot, we calculate E_a = 121,318 J/mol and $A = 2.17 \times 10^{22}$.

3.10 From the microorganisms balance, the required dilution rate is:

$$D = \frac{Q}{V} = \frac{\mu_{max} S}{K_S + S} = 1.59\,d^{-1}$$

The conversion of sucrose is 97%. We produce 0.8 kg of lactic acid per kg of sucrose converted. To produce 15 t/d of lactic acid, we need to convert 15/0.8 = 18.75 t of sucrose and we need to process 18.75/0.97 = 19.33 t/d of sucrose = $Q \cdot S_0$.

Since the concentration of sucrose in the feed (S_0) is 100 g/L = 0.1 t/m^3, the required feed flow rate is 193.3 m^3/d.

The required volume of the reactor is, therefore, $V = Q/D$ = 193.3 m^3/d/1.59 d^{-1} = 121.57 m^3.

3.11 The parameter K_S can be calculated from the effluent substrate concentration in the first reactor:

$$K_S = \frac{\mu_{max} S_1 - DS_1}{D_1}$$

In the first reactor $D = 2\ d^{-1}$, so $K_S = 9\ kg/m^3$. The parameters $Y_{X/S}$ and $Y_{P/S}$ can also be calculated from the effluent substrate concentration:

$$X_1 = Y_{X/S}\left(S_0 - S_1\right)$$

$$P_1 = Y_{P/S}\left(S_0 - S_1\right)$$

In this case, we obtain: $Y_{X/S} = 0.052$ kg/kg and $Y_{P/S} = 0.36$ kg/kg.

The feed to the second reactor is the effluent of the first reactor, so the mass balance for the substrate, biomass and product are:

$$\frac{\mu_{max} S_2}{K_S + S_2} X_2 \cdot V_2 + QX_1 = Q \cdot X_2$$

$$Q \cdot S_1 = Q \cdot S_2 + \frac{\mu_{max} S_2}{K_S + S_2} \frac{X_2}{Y_{X/S}} \cdot V_2$$

$$\frac{\mu_{max} S_2}{K_S + S_2} \frac{X_2}{Y_{X/S}} Y_{P/S} \cdot V_2 + QP_1 = Q \cdot P_2$$

We have that $Q/V_2 = D_2 = 1.33\ d^{-1}$. The equations can be therefore written as follows:

$$\frac{\mu_{max} S_2}{K_S + S_2} X_2 + D_2 X_1 = D_2 X_2$$

$$D_2 S_1 = D_2 S_2 + \frac{\mu_{max} S_2}{K_S + S_2} \frac{X_2}{Y_{X/S}}$$

$$\frac{\mu_{max} S_2}{K_S + S_2} \frac{X_2}{Y_{X/S}} Y_{P/S} + D_2 P_1 = D_2 P_2$$

This is a system of three equations in the three unknowns, S_2, X_2 and P_2, which can be solved analytically or numerically. The solution gives $S_2 = 0.045\ kg/m^3$, $X_2 = 5.15\ kg/m^3$ and $P_2 = 36.06\ kg/m^3$. If we wanted to achieve the same final substrate effluent concentration ($S_2 = 0.045\ kg/m^3$) using one single CSTR, with sterile feed and inlet substrate concentration of 100 kg/m³, the required dilution rate would be:

$$D = \frac{\mu_{max} S_2}{K_S + S_2} = 0.040\, d^{-1}$$

And the required volume would be $V = Q/D = 250\ m^3$. Note that the volume of one single reactor would be much higher than the combined volumes of two reactors in series. This is because in this case the parameter K_S is larger than the substrate concentration in the reactors, and the reaction rate is limited by the substrate concentration. In two reactors in series, the reaction rate is larger than in one single reactor because the substrate concentration in the first reactor is larger than the one in one single reactor.

3.12 At the quasi-steady state, the biomass balance in a fed-batch reactor can be written as:

$$D = \frac{Q}{V} = \frac{\mu_{max}}{K_S + S}$$

In a fed-batch reactor at quasi-steady state, the substrate concentration, although it can be assumed very low and much lower than the feed concentration, is the highest at the beginning of the feed and lowest at the end of the feed. Therefore, in the chemostat process which achieves the same effluent substrate concentration as the fed-batch, the dilution rate should be equal to the dilution rate of the fed-batch at the start of the cycle (this is the maximum dilution rate in the fed-batch cycle). This will ensure that the substrate concentration in the chemostat is always much lower than the feed while ensuring that the chemostat volume is kept to a minimum.

For the fed-batch, the dilution rate at the start of the cycle is:

$$D = \frac{Q}{V} = \frac{1.6\frac{m^3}{h}}{2m^3} = 0.8h^{-1}$$

This is the required dilution rate for the chemostat. The residence time in the chemostat is therefore $\tau = \frac{1}{D} = 1.25d$.

To calculate the effluent substrate and biomass concentrations in the chemostat, we need the coefficients $Y_{P/S}$ and $Y_{X/S}$, which can be obtained from the fed-batch data:

$$X = Y_{X/S} S_0$$

$$P = Y_{P/S} S_0$$

We obtain: $Y_{X/S} = 0.10$ kg/kg and $Y_{P/S} = 0.33$ kg/kg. Therefore, the effluent concentrations of a chemostat with 80 kg/m^3 feed will be $X = 8$ kg/m^3 and $P = 26.4$ kg/m^3.

3.13 The biomass concentration throughout the length of the feed phase is:

$$X = Y_{X/S} S_0 = 7.5\frac{kg}{m^3}$$

At the end of the feed phase, after the pH shift, the rate of product generation is:

$$r_P = 8.9X = 66.75\frac{kg}{m^3 \cdot d}$$

The product concentration at the end of the cycle is:

$$P = r_P \cdot t = 5.56\frac{kg}{m^3}$$

In each cycle, we generate 3 m³ of effluent, so we produce 16.7 kg of product per cycle. The cycle consists of 3 h feeding, plus 2 h product production and 1 h downtime, so the total length is 6 h. In 8,000 h/year, we can run 1,333 cycles, so we produce 22,261 kg of product per year or 61.0 kg/d.

3.14 The data on the two lab-scale chemostat experiments can be used to calculate the parameters μ_{max}, K_S, $Y_{X/S}$ and $Y_{P/S}$ for the fermentation.

The parameters μ_{max} and K_S can be obtained from the effluent substrate concentration in the two experiments:

$$D_1 = \frac{\mu_{max} S_1}{K_S + S_1} \quad D_2 = \frac{\mu_{max} S_2}{K_S + S_2}$$

where $D_1 = 3\ d^{-1}$, $D_2 = 5\ d^{-1}$, $S_1 = 0.5\ kg/m^3$ and $S_2 = 1.1\ kg/m^3$. Solving this system of two equations in the two unknowns, μ_{max} and K_S, we obtain: $\mu_{max} = 11.25\ d^{-1}$ and $K_S = 1.375\ kg/m^3$.

From the biomass and product production in the first experiment, we obtain $Y_{X/S}$ and $Y_{P/S}$:

$$X_1 = Y_{X/S}\left(S_0 - S_1\right)$$

$$P_1 = Y_{P/S}\left(S_0 - S_1\right)$$

We obtain: $Y_{X/S} = 0.051$ kg/kg and $Y_{P/S} = 0.28$ kg/kg.

The dilution rate that gives the optimum product productivity is obtained by plotting the volumetric productivity DP vs D. The graph is shown in Figure C.9 (for a feed of 50 kg/m³).

From this graph, we see that the optimum volumetric product productivity is 113.2 kg/m³.d and is obtained for $D_{opt} = 9.4\ d^{-1}$. The biomass concentration that corresponds to this dilution rate is 2.2 kg/m³. If the vessel volume is 2 m³, the product productivity is 226.2 kg/d.

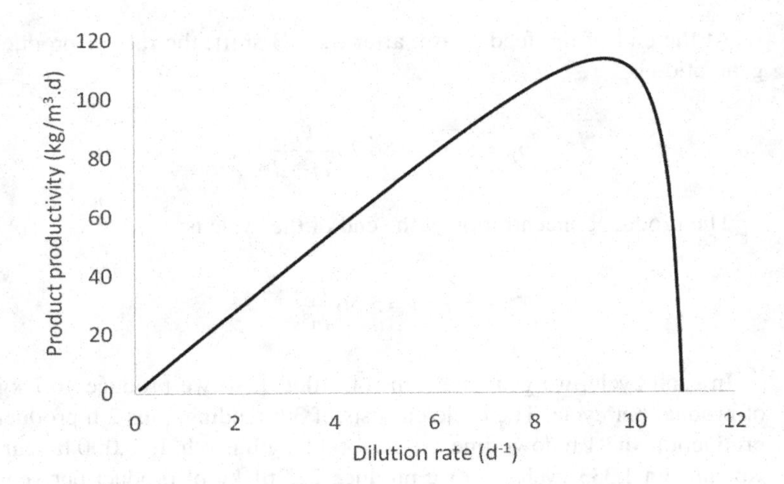

FIGURE C.9 Plot for Problem 3.14.

3.15 For a chemostat where microbial metabolism is described by the Monod equation with maintenance, the biomass and substrate concentrations at steady state are:

$$X = \frac{Y_{X/S}\left(S_0 - S\right)}{1 + \frac{m_S Y_{X/S}}{D}}$$

$$D = \frac{\mu_{max}}{K_S + S}$$

Using the biomass and substrate concentration obtained at the two dilution rates, we have a system of four equations with the four unknowns: μ_{max}, K_S, $Y_{X/S}$ and m_s. After re-arrangements, we obtain: $\mu_{max} = 0.75\ h^{-1}$, $K_S = 0.2\ kg/m^3$, $Y_{X/S} = 0.65$ kg/kg and $m_S = 0.11\ h^{-1}$.

3.16 In a fed-batch reactor at the quasi-steady state, the substrate concentration is linked to the dilution rate by the equation:

$$D = \frac{Q}{V} = \frac{\mu_{max}}{K_S + S}$$

The dilution rate, and so the substrate concentration, is highest at the start of the feed, when the volume is minimum. The substrate concentration that corresponds to 1% of the feed concentration is $S = 0.2\ kg/m^3$. The maximum feed flow rate that can be used is, therefore:

$$Q = \frac{\mu_{max} S}{K_S + S} V = \frac{7.3 \cdot 0.2}{0.1 + 0.2} 0.2 = 0.97\, m^3/d$$

3.17 For maintenance metabolism, the mass balances for biomass and product are the same as for the Monod model without maintenance, and they lead to the same quasi-steady state equations for biomass and product, respectively:

$$S = \frac{DK_S}{\mu_{\max} - D}$$

$$P = S_0 \cdot Y_{P/S} \text{ (assuming } S \ll S_0\text{)}$$

The substrate balance at the quasi-steady state with maintenance metabolism is:

$$Q(S_0 - S) = \frac{\mu_{\max}}{K_S + S} \frac{X \cdot V}{Y_{X/S}}$$

Which gives (assuming $S \ll S_0$)

$$X = \frac{Y_{X/S} S_0}{1 + \frac{m_S Y_{X/S}}{D}}$$

For the endogenous metabolism, the mass balances for substrate and product are the same as for the simple Monod model, giving (assuming $S \ll S_0$):

$$X = Y_{X/S} S_0$$

and

$$P = S_0 \cdot Y_{P/S}$$

The biomass balance at quasi-steady state with endogenous metabolism is:

$$XQ = \frac{\mu_{\max} S}{K_S + S} X \cdot V - bXV$$

which gives:

$$S = \frac{K_S (D + b)}{\mu_{\max} - (D + b)}$$

Note that the quasi-steady state equations for the fed-batch reactor with maintenance and endogenous metabolism are the same as for the corresponding chemostat models.

3.18 The maximum allowable substrate concentration in the fed-batch cycle is 0.3 kg/m³ (1% of 30 kg/m³). This means that the maximum feed flow rate is:

$$Q=\frac{15\cdot 0.3}{0.08+0.3}\cdot 1=11.8\frac{m^3}{d}$$

We need to feed 9 m³ per cycle, so the time required for each feed is 0.76 d = 18.2 h; including the downtime, the total cycle time is 18.7 h.

The product concentration throughout all the cycle can be approximated by (assuming $S \ll S_0$):

$$P=30\cdot 0.3=9\frac{\text{kg}}{\text{m}^3}$$

Each batch produces 9 m³ of effluent at a product concentration of 9 kg/m³, so each batch produces 81 kg of product. In 8,000 h/year, we can have 427 batches, with a total product production of 34,587 kg/year or 94.8 kg/d.

Biomass concentration is given by:

$$X=\frac{0.2\cdot 30}{1+\frac{0.8\cdot 0.2}{11.8}}=5.9\frac{\text{kg}}{\text{m}^3}$$

Each batch produces 53.1 kg of biomass, with a biomass productivity of 62.1 kg/d.

CHAPTER 4

4.1 Factors that affect the oxygen transfer rate in agitated vessels are:

- Mixing conditions (type, speed and dimensions of the agitator, geometry of the vessel, presence of baffles, etc.). The k_La, in particular the specific area a, increases as the turbulence in the vessel increases because gas bubbles are broken into smaller bubbles, with a larger area per unit volume of reactor. Increase in turbulence is usually obtained by increasing the size and stirring speed of the agitator. The geometry of the vessel and the presence of baffles also affect the mixing conditions and, therefore, the mass transfer coefficient.
- Temperature. Higher temperatures increase the film coefficient k_L because the diffusivity of oxygen in water increases. However, the oxygen solubility in water decreases as the temperature increases, and this causes a reduction in the driving force.
- Gas flow rate. Increasing the gas flow rate increases the number of bubbles per unit volume of the vessel; this increases the area available for mass transfer and in turn increases the oxygen transfer rate.

- Composition of the inlet gas. Higher oxygen concentrations in the inlet gas correspond to higher saturation concentration of oxygen in water, increasing the driving force and the oxygen transfer rate.
- Concentration of solids and of microorganisms. Higher concentrations of solids or microorganisms increase the viscosity of the liquid phase, decreasing the coefficient k_L and decreasing the oxygen transfer rate.
- Total pressure. For a given mass flow rate of the inlet gas, increasing the pressure in the vessel, increases the oxygen saturation concentration in the liquid phase, increasing the driving force. However, higher pressures will cause the gas bubbles to be smaller, decreasing the area available for mass transfer per unit volume (parameter a).

4.2 The slope of the plot $\ln(CO_2^* - CO_2)$ *vs* t is equal to $-k_La$. From the graph we see that $k_La = 0.18$ min^{-1}. If $CO_2^* = 9.1\frac{mg}{l}$ and $CO_2 = 2.1\frac{mg}{l}$ the oxygen transfer rate is

$$r_{O2transf} = 0.18\cdot(9.1-2.1) = 1.26\frac{mg}{l\cdot\min} = 1.26\cdot10^{-3}\frac{kg}{m^3\min}$$

4.3 For geometrically similar vessels:

$$\frac{D_{\text{small}}}{H_{\text{small}}} = \frac{D_{\text{large}}}{H_{\text{large}}}$$

and

$$\frac{D_{ag,\text{small}}}{D_{\text{small}}} = \frac{D_{ag,\text{large}}}{D_{\text{large}}}$$

For the small vessel $\frac{D_{\text{small}}}{H_{\text{small}}} = 1$, so this ratio will be the same for the large vessel. For the large vessel, the volume is:

$$V_{\text{large}} = \frac{\pi}{4}D_{\text{large}}^2 H_{\text{large}} = \frac{\pi}{4}D_{\text{large}}^3$$

which gives:

$$D_{\text{large}} = 2.67\ m$$

Therefore, $H_{\text{large}} = 2.67$ m.

For the small vessel, the ratio between the agitator diameter and the vessel diameter is 0.3; this ratio will be the same for the large vessel, which will have, therefore, an agitator of a diameter equal to $D_{ag,\text{large}} = 0.3D_{\text{large}} = 0.80m$.

4.4 We need to find the dimensions of the large vessel. For the small vessel:

$$V_{small} = \frac{\pi}{4} D_{small}^2 H_{small} \quad \Rightarrow \quad H_{small} = \frac{4V_{small}}{\pi D_{small}^2} = 1.57m$$

Therefore, $\frac{D_{small}}{H_{small}} = \frac{D_{large}}{H_{large}} = 0.57$.
For the large vessel:

$$V_{large} = \frac{\pi}{4} D_{large}^2 H_{large} = \frac{\pi}{4} \frac{D_{large}^3}{0.57} \quad \Rightarrow \quad D_{large} = \sqrt[3]{\frac{4 \cdot 0.57 V_{large}}{\pi}} = 1.94m$$

and $H_{large} = \frac{D_{large}}{0.57} = 3.40\ m$.
Furthermore, the ratio between agitator diameter and vessel diameter is:

$$\frac{D_{ag,small}}{D_{small}} = \frac{D_{ag,large}}{D_{large}} = 0.33$$

Therefore, $D_{ag,large} = 0.65$ m.
The ratio between the mass transfer coefficients in both vessels is:

$$\frac{k_L a_{small}}{k_L a_{large}} = \left(\frac{P_{Gsmall}}{P_{Glarge}} \frac{V_{large}}{V_{small}} \right)^{0.62} \left(\frac{Q_{Gsmall}}{Q_{Glarge}} \frac{A_{large}}{A_{small}} \right)^{0.34}$$

$$= \left(\frac{N_{small}^3 D_{ag,small}^5}{N_{large}^3 D_{ag,large}^5} \frac{V_{large}}{V_{small}} \right)^{0.62} \left(\frac{Q_{Gsmall}}{Q_{Glarge}} \frac{D_{large}^2}{D_{small}^2} \right)^{0.34}$$

Replacing the known numerical values this becomes:

$$1 = \left(\frac{2.5^3 0.3^5}{N_{large}^3 0.65^5} \frac{10}{1} \right)^{0.62} \left(\frac{250}{2{,}500} \frac{1.94^2}{0.9^2} \right)^{0.34}$$

From this we obtain $N_{large} = 1.29\ s^{-1} = 77$ rpm.

4.5 The oxygen mass balance for the chemostat is ignoring the oxygen concentration in the feed:

$$k_L a \left(C_{O2}^* - C_{O2} \right) V = Q \cdot C_{O2} + \left(-r_{O2biom} \right) \cdot V$$

which becomes

$$k_L a \left(C_{O2}^* - C_{O2} \right) = D C_{O2} + \left(-r_{O2biom} \right)$$

The substrate consumption rate in this chemostat is:

$$(-r_S)V = Q(S_0 - S) \;\Rightarrow\; (-r_S) = D(S_0 - S) = 0.5(40-4) = 18\frac{kg}{m^3 d}$$

From the fermentation stoichiometry, we calculate the ratio between oxygen and substrate consumption:

$$Y_{O2/S} = \frac{32 \cdot 3}{180} = 0.53\frac{kg}{kg}$$

Therefore, the rate of oxygen consumption is:

$$(-r_{O2biom}) = (-r_S)Y_{O2/S} = 9.5\frac{kg}{m^3 d}$$

The $k_L a$ in this system is:

$$k_L a = \frac{DC_{O2} + (-r_{O2biom})}{(C^*_{O2} - C_{O2})} = 1{,}583 d^{-1}$$

4.6 In the small reactor (reactor 1):

$$(-r_{O2,max1})V_1 = Q_{O2max1} = k_L a_1 (C^*_{O2} - C_{O2min})V_1$$

From this equation, we obtain $k_L a_1 = 7.1\ \text{min}^{-1}$.
In the large reactor (reactor 2):

$$(-r_{O2,max2})V_2 = Q_{O2max2} = k_L a_2 (C^*_{O2} - C_{O2min})V_2$$

$k_L a_2$ can be calculated from the given correlation for oxygen transfer rate:

$$\frac{k_L a_2}{k_L a_1} = \frac{\alpha\left(\frac{P_G}{V}\right)_2^{0.7}\left(\frac{Q_G}{A}\right)_2^{0.2}}{\alpha\left(\frac{P_G}{V}\right)_1^{0.7}\left(\frac{Q_G}{A}\right)_1^{0.2}} = \gamma^{0.7}$$

Therefore, $k_L a_2 = 4.4\ \text{min}^{-1}$ and

$$Q_{O2max2} = k_L a_2 (C^*_{O2} - C_{O2min})V_2 = 308\frac{g_{O2}}{\text{min}}$$

Q_{O2max2} is the maximum oxygen transfer rate achievable in the large reactor.

We need now to calculate how much biomass is produced per unit of oxygen consumed, this can be obtained by calculation of the growth stoichiometry. In this case, the general growth stoichiometry is:

$$C_6H_{12}O_6 + aO_2 + bNH_3 \rightarrow cCO_2 + dH_2O + eC_5H_7O_2N$$

The coefficients *a, b, c, d, e* can be calculated from the elemental balances and from the given growth yield, $Y_{X/S}$. We obtain the following stoichiometry:

$$C_6H_{12}O_6 + 3.6O_2 + 0.48NH_3 \rightarrow 3.6CO_2 + 5.04H_2O + 0.48C_5H_7O_2N$$

From this stoichiometry we have:

$$Y_{O2/X} = \frac{32 \cdot 3.6}{0.48 \cdot 113} = 2.12\frac{kg}{kg}$$

The maximum biomass production rate in the large vessel is:

$$r_{X\max}V_2 = (QX)_{\max} = \frac{Q_{O2\max 2}}{Y_{O2/X}} = 209\frac{kg}{d}$$

4.7 From the batch test, we can calculate the growth stoichiometry and the kinetic parameters. The generic growth stoichiometry in this case is:

$$C_6H_{12}O_6 + aO_2 + bNH_3 \rightarrow cC_5H_7O_2N + dH_2O + eCO_2$$

We know that $Y_{X/S}$ = 0.3 kg/kg. From the elemental balances, we obtain all the coefficients:

$$C_6H_{12}O_6 + 1.2O_2 + 0.16NH_3 \rightarrow 0.16C_5H_7O_2N + 1.68H_2O + 1.2CO_2$$

From the stoichiometry, we obtain $Y_{X/O2} = 0.47$ kg/kg.

The graph gives the kinetic parameters. From the linearisation of the substrate removal rate according to the Monod equation, we obtain:

$$\frac{X}{(-r_S)} = \frac{Y_{X/S}}{\mu_{\max}}K_S\frac{1}{S} + \frac{Y_{X/S}}{\mu_{\max}}$$

From the graph, the intercept is

$$\frac{Y_{X/S}}{\mu_{\max}} = 0.088\frac{kg}{kg}d$$

From which we calculate $\mu_{\max} = 3.41$ d^{-1}.

The slope from the graph is:

$$\frac{Y_{X/S}}{\mu_{\max}}K_S = 0.0111\frac{kg \cdot d}{m^3} \quad \Rightarrow \quad K_S = 0.13\frac{kg}{m^3}$$

The chemostat will work with $D = 0.9 \cdot \mu_{max} = 3.07\ d^{-1}$. Under these conditions, the effluent substrate concentration will be:

$$S = \frac{DK_S}{\mu_{max} - D} = 1.18 \frac{kg}{m^3}$$

And the biomass concentration will be:

$$X = Y_{X/S}(S_0 - S) = 5.65 \frac{kg}{m^3}$$

The oxygen balance in the chemostat is:

$$k_L a(C_{O2}^* - C_{O2}) = D \cdot C_{O2} + \frac{DX}{Y_{X/O2}}$$

The required $k_L a$ is, therefore:

$$k_L a = \frac{D \cdot C_{O2} + \frac{DX}{Y_{X/O2}}}{(C_{O2}^* - C_{O2})} = 5{,}678 d^{-1}$$

4.8 The mass balance for oxygen during the batch fermentation is:

$$\frac{dC_{O2}}{dt} = k_L a(C_{O2}^* - C_{O2}) - \frac{\mu_{max}}{K_S + S} \frac{X}{Y_{X/O2}}$$

From the problem we know that it needs to be $dC_{O2}/dt = 0$, so the required $k_L a$ during the fermentation will be given by:

$$k_L a = \frac{\mu_{max}}{K_S + S} \frac{X}{Y_{X/O2}} \frac{1}{(C_{O2}^* - C_{O2})}$$

With the assumption that $S \gg K_S$

$$k_L a = \mu_{max} \frac{X}{Y_{X/O2}} \frac{1}{(C_{O2}^* - C_{O2})}$$

From the given stoichiometry: $Y_{X/O2} = 0.15$ kg biom/kg O_2, μ_{max} can be calculated from the doubling time:

$$t_d = \ln \frac{2}{\mu_{max}}$$

Which gives $\mu_{max} = 0.76\ h^{-1}$

We need to calculate the biomass concentration after 4.5 h:

$$X = X_0 e^{\mu_{max} t} \text{ which gives } X = 3.06\ g/L.$$

Therefore, the k_La required after 4.5 h will be equal to 3,365 h^{-1} = 0.935 s^{-1}.

From the given correlations, we obtain:

$k_La = 0.020 \cdot 2.40^{0.8} N^{2.4}$ from which N can be calculated:

$$N = 3.67 \text{ revs/s.}$$

4.9 a From the reaction stoichiometry, 0.98 g of acetic acid are produced per g of ethanol converted. The concentration of acetic acid in the reactor is 48 g/L, so 48/0.98 = 49 g/L of ethanol are converted. The ethanol concentration in the feed is equal to 60 g/L so the effluent ethanol concentration will be 60 – 49 = 11 g/L.

Under the new conditions, an acetic acid productivity of 1,500 kg/d is desired. With the same flow rate and reactor volume, this means that the acetic acid concentration in the reactor needs to be 1,500/20 = 75 kg/m^3. Therefore, under the new conditions 75/0.98 = 76.5 kg/m^3 of ethanol need to be converted. The effluent ethanol concentration will remain equal to 11 g/L (the residence time does not change), so the glucose concentration in the feed under the new conditions needs to be equal to 76.5 + 11 = 87.5 kg/m^3.

b The k_La for oxygen under the initial conditions can be calculated using a mass balance for oxygen:

$$r_{O2\,transf} V = Q \cdot C_{O2} + (-r_{O2\,biom}) \cdot V$$

$r_{O2\,biom}$ is the oxygen consumption rate by the microorganisms and can be calculated from the stoichiometry of the reaction and from the given mass flow rates:

$$(-r_{O2biom}) = \frac{Q \cdot C_{CH3COOH}}{V} Y_{O2/CH3COOH}$$

$$= \frac{20\ m^3/\text{d} \cdot 48\ \text{kg}/m^3 \cdot 0.71\ \text{kgO2/kg acetic acid}}{10 m^3}$$

$$= 68.2\ \text{kgO2}/m^3.\text{d}$$

The equation for the rate of oxygen transfer is:

$$r_{O2\,transf}\left(\frac{kg_{O_2}}{m^3 \cdot day}\right) = k_La \cdot (C^*_{O2} - C_{O2})$$

So, the only unknown in the oxygen balance is the k_La, which can be calculated as equal to 17,950 d^{-1}.

Under the new conditions, we aim to produce 1,500 kg of acetic acid per day. This means that the oxygen consumption rate by the microorganisms will be equal to:

$$\left(-r_{O2biom}\right)=\frac{1{,}500\ kg/d \cdot 0.71\ kgO_2/kg\ acetic\ acid}{10m^3}=106.5kgO_2/m^3.d$$

If the k_La stays the same we can use the oxygen balance to calculate the new oxygen concentration in the reactor. This will be equal to 2.1 mg/L, so, yes, the current aeration system can work well under the new conditions and can provide the required oxygen concentration.

4.10 From the oxygen balance in a chemostat:

$$k_La=\frac{Q\cdot\left(C_{O2}-C_{O20}\right)+D\frac{X}{Y_{X/O2}}\cdot V}{\left(C_{O2}^*-C_{O2}\right)V}$$

In this case, we ignore the contributions of the inlet and outlet streams, so:

$$k_La=\frac{D\frac{X}{Y_{X/O2}}\cdot V}{\left(C_{O2}^*-C_{O2}\right)V}$$

In the scale-up, the residence time and the inlet substrate concentration are kept constant, this means that D and X are the same in the small and large reactors, therefore:

$$\frac{k_La_{\text{large}}}{k_La_{\text{small}}}=\frac{\left(C_{O2}^*-C_{O2}\right)_{\text{small}}}{\left(C_{O2}^*-C_{O2}\right)_{\text{large}}}=\frac{8.0-4.0}{8.0-2.0}=0.67$$

For geometrically similar vessels:

$$\frac{A_{\text{small}}}{A_{\text{large}}}=\left(\frac{D_{\text{small}}}{D_{\text{large}}}\right)^2 \quad \frac{V_{\text{small}}}{V_{\text{large}}}=\left(\frac{D_{\text{small}}}{D_{\text{large}}}\right)^3$$

Combining these equations, with the condition that the gas flow rate per unit volume is the same in the scale-up:

$$\frac{\left(\frac{Q_G}{A}\right)_{\text{large}}}{\left(\frac{Q_G}{A}\right)_{\text{small}}}=\left(\frac{V_{\text{large}}}{V_{\text{small}}}\right)^{0.333}$$

Using the correlation for k_La:

$$\frac{k_La_{large}}{k_La_{small}} = \frac{\left(\frac{P_G}{V}\right)_{large}^{0.75}\left(\frac{Q_G}{A}\right)_{large}^{0.45}}{\left(\frac{P_G}{V}\right)_{small}^{0.75}\left(\frac{Q_G}{A}\right)_{small}^{0.45}} = \left(\frac{P_{Glarge}}{P_{Gsmall}}\right)^{0.75}\left(\frac{V_{small}}{V_{large}}\right)^{0.75}\left(\frac{V_{large}}{V_{small}}\right)^{0.33 \cdot 0.45}$$

$$= \left(\frac{P_{Glarge}}{P_{Gsmall}}\right)^{0.75}\left(\frac{V_{small}}{V_{large}}\right)^{0.60}$$

which becomes:

$$0.67 = \left(\frac{P_{Glarge}}{P_{Gsmall}}\right)^{0.75}(0.20)^{0.60}$$

and so $P_{Glarge}/P_{Gsmall} = 2.12$. Since the ratio P_G/P is the same for both vessels, the ratio between P_{large} and P_{small} will also be equal to 2.12.

Comparing the power numbers for the large and small vessels:

$$\frac{\left(\frac{P}{\rho_L N^3 D_{ag}^5}\right)_{large}}{\left(\frac{P}{\rho_L N^3 D_{ag}^5}\right)_{small}} = 1 \Rightarrow \frac{P_{large}}{P_{small}} = \frac{\left(N^3 D_{ag}^5\right)_{large}}{\left(N^3 D_{ag}^5\right)_{small}} \Rightarrow \left(\frac{N_{large}}{N_{small}}\right) = \left[\frac{\frac{P_{large}}{P_{small}}}{\left(\frac{D_{aglarge}}{D_{agsmall}}\right)^5}\right]^{0.33}$$

Since the vessels are geometrically similar $D_{ag,small}/D_{ag,large} = (V_{small}/V_{large})^{0.33}$ and so:

$$\left(\frac{N_{large}}{N_{small}}\right) = \left(\frac{2.12}{5^{0.33 \cdot 5}}\right)^{0.33} = 0.53$$

Therefore, the agitation speed required in the large vessel is $N = 0.53 \times 180 = 95.4$ rpm.

4.11 The oxygen balance in the chemostat is:

$$k_La\left(C_{O2}^* - C_{O2}\right) = D \cdot C_{O2} + \frac{DX}{Y_{X/O2}}$$

Since the feed, residence time and oxygen saturation concentration are the same for the small and the large reactor, the steady state biomass concentration will also be the same, so the requirement to have the same dissolved oxygen concentration implies that the k_La will have to be the same.

From the given correlation for k_La:

$$\frac{k_La_{large}}{k_La_{small}} = \frac{0.03\left(\frac{P_G}{V}\right)_{large}^{0.69}\left(\frac{Q_G}{A}\right)_{large}^{0.39}}{0.03\left(\frac{P_G}{V}\right)_{small}^{0.69}\left(\frac{Q_G}{A}\right)_{small}^{0.39}}$$

Re-arranging and considering that in this case the superficial gas velocity remains the same in the two reactors:

$$\frac{k_L a_{\text{large}}}{k_L a_{\text{small}}} = \frac{\left(\frac{P_G}{V}\right)_{\text{large}}^{0.69}}{\left(\frac{P_G}{V}\right)_{\text{small}}^{0.69}} = \left(\frac{P_{G,\text{large}}}{P_{G,\text{small}}}\frac{V_{\text{small}}}{V_{\text{large}}}\right)^{0.69} = \left(\frac{N_{\text{large}}^3 D_{ag,\text{large}}^5}{N_{\text{small}}^3 D_{ag,\text{small}}^5}\frac{V_{\text{small}}}{V_{\text{large}}}\right)^{0.69}$$

Under conditions of geometric similarity:

$$\frac{D_{ag,\text{large}}}{D_{ag,\text{small}}} = \left(\frac{V_{\text{large}}}{V_{\text{small}}}\right)^{\frac{1}{3}}$$

Therefore, the condition of equal $k_L a$ in the large and small vessels becomes:

$$1 = \left(\frac{N_{\text{large}}^3}{N_{\text{small}}^3}\left(\frac{V_{\text{small}}}{V_{\text{large}}}\right)^{\frac{2}{3}}\right)^{0.69} = \frac{N_{\text{large}}^{2.07}}{N_{\text{small}}^{2.07}}\frac{V_{\text{small}}^{0.46}}{V_{\text{large}}^{0.46}} \Rightarrow N_{\text{large}} = 120 \cdot 6^{0.22} = 179 rpm$$

4.12 The volumetric oxygen mass transfer rate is:

$$r_{\text{O2transf}} = k_L a \cdot \left(C_{\text{O2}}^* - C_{\text{O2}}\right)$$

The correlation for k_l is:

$$Sh_L = \frac{k_l \cdot d_{\text{bubbles}}}{D_{g,L}} = 2.0 + 0.31\left(\frac{d_{\text{bubbles}}^3 \rho_L g}{\mu_L D_{g,L}}\right)^{\frac{1}{3}}$$

From this correlation:

$$k_l = \left[2.0 + 0.31\left(\frac{d_{\text{bubbles}}^3 \rho_L g}{\mu_L D_{g,L}}\right)^{\frac{1}{3}}\right]\frac{D_{g,L}}{d_{\text{bubbles}}}$$

And therefore:

$$\frac{k_{l,35}}{k_{l,25}} = \frac{\left[2.0 + 0.31\left(\frac{d_{\text{bubbles}}^3 \rho_L g}{\mu_L D_{g,L}}\right)_{35}^{\frac{1}{3}}\right]\left(\frac{D_{g,L}}{d_{\text{bubbles}}}\right)_{35}}{\left[2.0 + 0.31\left(\frac{d_{\text{bubbles}}^3 \rho_L g}{\mu_L D_{g,L}}\right)_{25}^{\frac{1}{3}}\right]\left(\frac{D_{g,L}}{d_{\text{bubbles}}}\right)_{25}} = \frac{\left[2.0 + 0.31\left(\frac{0.002^3 1000 \cdot 9.8}{0.00072 \cdot 3.0 \cdot 10^{-9}}\right)_{35}^{\frac{1}{3}}\right]\left(\frac{3.0 \cdot 10^{-9}}{0.002}\right)_{35}}{\left[2.0 + 0.31\left(\frac{0.002^3 1000 \cdot 9.8}{0.00089 \cdot 2.2 \cdot 10^{-9}}\right)_{25}^{\frac{1}{3}}\right]\left(\frac{2.2 \cdot 10^{-9}}{0.002}\right)_{25}} = 1.33$$

For the specific area a:

$$a = 1.44\left[\left(\frac{P_G}{V}\right)^{0.4}\left(\frac{\rho_L}{\sigma^3}\right)^{0.2}\right]\left(\frac{v_g}{v_t}\right)^{0.5}$$

With the assumptions of this problem, all the physical properties and variables in this equation don't change between 25°C and 35°C, so $a_{25} = a_{35}$.

Therefore:

$$\frac{k_L a_{35}}{k_L a_{25}} = 1.33$$

and

$$\frac{r_{O2\text{transf},35}}{r_{O2\text{transf},25}} = \frac{k_L a_{35}\cdot(7.5-2.5)}{k_L a_{25}\cdot(9.3-2.5)} = 0.98 \;\Rightarrow\; r_{O2\text{transf},35} = 0.137\frac{kg}{m^3 s}$$

Note that the rate of oxygen transfer at 35°C is approximately the same as at 25°C because the increase in the k_l is compensated by the decrease in the driving force.

4.13 The mass balance for oxygen in a chemostat is:

$$k_L a\left(C^*_{O2} - C_{O2}\right)V = QC_{O2} + \frac{r_X V}{Y_{X/O2}}$$

Since we know the maximum $k_L a$ that we can have in this vessel, we can calculate the maximum biomass production rate:

$$r_X = \left[k_L a\left(C^*_{O2} - C_{O2}\right) - DC_{O2}\right]Y_{X/O2}$$

$$= \left[313.2(0.009 - 0.002) - 0.125\cdot 0.002\right]0.18$$

$$= 0.39\frac{kg}{m^3 h}$$

The maximum product volumetric production rate is, therefore:

$$r_P = r_X Y_{P/X} = 2.0\frac{kg}{m^3 h}$$

and the product productivity is:

$$r_P V = 4.0\frac{kg}{h} = 48\frac{kg}{d}$$

The maximum possible substrate concentration in the feed is obtained from the substrate balance, which, with the hypothesis that $S \ll S_0$, is:

$$QS_0 = \frac{r_X V}{Y_{X/S}} \quad \Rightarrow \quad S_0 = \frac{r_X}{DY_{X/S}} = 31.2\frac{kg}{m^3}$$

4.14 The mass balance for oxygen in a chemostat is:

$$k_L a\left(C^*_{O2} - C_{O2}\right)V = QC_{O2} + \frac{r_X V}{Y_{X/O2}}$$

which can be re-arranged as:

$$k_L a\left(C^*_{O2} - C_{O2}\right) = DC_{O2} + \frac{DX}{Y_{X/O2}}$$

The value of $k_L a$ for this chemostat is:

$$k_L a = \frac{DC_{O2} + \frac{DX}{Y_{X/O2}}}{\left(C^*_{O2} - C_{O2}\right)} = \frac{0.083\frac{1}{h}\cdot 0.0035\frac{kg}{m^3} + 0.083\frac{1}{h}\frac{4}{0.25}\frac{kg}{m^3}}{(0.0092 - 0.0035)\frac{kg}{m^3}} = 233h^{-1}$$

Under the new proposed conditions of residence time 10 h, which corresponds to $D = 0.10\ h^{-1}$, the biomass concentration should stay the same, because the biomass concentration in a chemostat is given by:

$$X = Y_{X/S}\left(S_0 - S\right)$$

and, with the assumptions of this problem, $S \ll S_0$ in all conditions.

The oxygen concentration with the new proposed residence time can be calculated again from the oxygen balance under the new conditions, which will have the same $k_L a$ and biomass concentration:

$$C_{O2} = \frac{k_L a C^*_{O2} - \frac{DX}{Y_{X/O2}}}{D + k_L a} = 0.0023\frac{kg}{m^3} = 2.3\frac{mg}{L}$$

Therefore, the residence time can be reduced to 10 h without reducing the oxygen concentration below the minimum acceptable level.

4.15 Experiments should be carried out in a range of agitation speed and inlet gas flow rate. For each condition, the $k_L a$ should be measured with the oxygen absorption method (or with any other method).

In a first series of experiments, the gas flow rate will be held constant while the agitation speed is varied. By varying the agitation speed the power draw will vary and we will collect a series of data of $k_L a$ vs P_G (we assume that adequate correlations are available for the calculation of the power draw of the agitator in the vessel, or that this can be measured directly). We will be able to calculate the parameter a of Equation (4.22):

$$k_L a = k\left(\frac{P_G}{V}\right)^a v_g^b$$

Indeed, Equation (4.22) can be linearised as:

$$ln\left(k_L a\right) = a ln\left(\frac{P_G}{V}\right) + ln\left(k v_g^b\right)$$

and the plot of $ln(k_L a)$ vs $ln\ (P_G/V)$ should give a straight line with slope equal to a and intercept equal to $\ln\left(k v_g^b\right)$.

The next series of experiments should be carried out at constant agitation speed (i.e. constant P_G) and variable Q_G (i.e. variable v_g). The values of $k_L a$ collected in this series of experiments will be correlated with the linearised equation:

$$ln\left(k_L a\right) = b ln\left(v_g\right) + ln\left(k\left(\frac{P_G}{V}\right)^a\right)$$

which will allow the calculation of b from the slope of the line $ln(k_L a)$ vs $ln(v_g)$.

Once the values of a and b are known, the value of k can be calculated from either of the linearised form of the correlation [$ln(k_L a)$ vs $ln\ (P_G/V)$ or $ln(k_L a)$ vs $ln(v_g)$]. The best procedure is to calculate k from both correlations and calculate the average of the two values obtained.

4.16 If we assume the model with maintenance, we need to take into account the oxygen consumed by the oxidation of the substrate for maintenance in the oxygen balance. The oxygen consumption due to maintenance is (Equation 2.36):

$$r_{O2biom,m} = \frac{-m_S X}{Y_{S/O2,m}}$$

Therefore, the oxygen balance for a chemostat becomes:

$$Q \cdot C_{O20} + k_L a \cdot \left(C_{O2}^* - C_{O2}\right)V = Q \cdot C_{O2} + \frac{\mu_{max} S}{K_S + S}\frac{XV}{Y_{X/O2}} + \frac{m_S XV}{Y_{S/O2,m}}$$

Therefore, Equation (4.34) modifies into:

$$k_L a = \frac{D \cdot \left[C_{O2} - C_{O20} + \frac{X}{Y_{X/O2}}\right] + \frac{m_S X}{Y_{S/O2,m}}}{\left(C_{O2}^* - C_{O2}\right)}$$

For the model with endogenous metabolism, the oxygen consumption due to endogenous metabolism is (Equation 2.42):

$$r_{O2biom,end} = \frac{-bX}{Y_{X/O2,end}}$$

and the oxygen balance in a chemostat becomes:

$$Q \cdot C_{O20} + k_L a \cdot \left(C^*_{O2} - C_{O2}\right)V = Q \cdot C_{O2} + \frac{\mu_{max} S}{K_S + S} \frac{XV}{Y_{X/O2}} + \frac{bXV}{Y_{X/O2,end}}$$

Equation (4.34) modifies into:

$$k_L a = \frac{D \cdot \left[C_{O2} - C_{O20} + \frac{X}{Y_{X/O2}}\right] + \frac{bX}{Y_{S/O2,m}}}{\left(C^*_{O2} - C_{O2}\right)}$$

4.17 The mass balance for oxygen in a batch reactor is:

$$\frac{dC_{O2}}{dt} = k_L a\left(C^*_{O2} - C_{O2}\right) - \frac{\mu_{max}}{K_S + S} \frac{X}{Y_{X/O2}}$$

Under the requirement of this problem, the oxygen concentration is constant ($\frac{dC_{O2}}{dt} = 0$), $S \gg K_S$, therefore, the required $k_L a$ during the batch can be calculated from:

$$k_L a = \frac{\mu_{max} \frac{X}{Y_{X/O2}}}{\left(C^*_{O2} - C_{O2}\right)}$$

The time profile of the biomass concentration X vs time is obtained from the biomass balance in the batch reactor:

$$\frac{dX}{dt} = \mu_{max} X \quad \Rightarrow \quad \int_{X0}^{X} \frac{dX}{X} = \mu_{max} \int_0^t dt \quad \Rightarrow \quad X = X_0 e^{\mu_{max} t}$$

Therefore, the time profile of the required $k_L a$ is given by:

$$k_L a = \frac{\mu_{max} \frac{X_0 e^{\mu_{max} t}}{Y_{X/O2}}}{\left(C^*_{O2} - C_{O2}\right)}$$

From the correlation for $k_L a$ given in this problem, we calculate the following equation for the required gas flow rate as a function of the $k_L a$:

$$Q_G = A\left(\frac{k_L a}{k\left(\frac{P_G}{V}\right)^{0.60}}\right)^{\frac{1}{0.42}}$$

and introducing the previous equation for the time profile of k_La (note that in the equation for Q_G, k_La needs to be in s^{-1}, therefore, the units of μ_{max} need to be converted accordingly):

$$Q_G = A\left(\frac{\mu_{max} X_0 e^{\mu_{max} t}}{Y_{X/O2}\left(C^*_{O2} - C_{O2}\right)0.045\left(\frac{P_G}{V}\right)^{0.60}}\right)^{\frac{1}{0.42}}$$

$$= 2.8\left(\frac{\frac{4.8}{86{,}400}\cdot 0.1 e^{4.8t}}{0.25(0.009 - 0.003)0.045\left(\frac{1{,}000}{3}\right)^{0.60}}\right)^{\frac{1}{0.42}}$$

This equation gives the required time profile for the gas flow rate vs time in this batch process, ensuring that the oxygen concentration remains at 3.0 mg/L. The fermentation needs to be carried out until the substrate concentration drops to 0.2 kg/m³, from the initial 20 kg/m³. The substrate balance in this case is:

$$\frac{dS}{dt} = -\frac{\mu_{max} X}{Y_{X/S}} \Rightarrow \int_{S_0}^{S} dS = -\frac{\mu_{max} X_0}{Y_{X/S}} \int_0^t e^{\mu_{max} t} dt \Rightarrow S_0 - S = \frac{X_0 e^{\mu_{max} t}}{Y_{X/S}}$$

Therefore, the end time of the fermentation t_{final} is:

$$t_{final} = \frac{\ln\left(\frac{(S_0 - S)Y_{X/S}}{X_0}\right)}{\mu_{max}} = 0.77\,d = 18.4\,h$$

The time profile of the required gas flow rate during the fermentation is shown in Figure C.10. Note that the required gas flow rate is very low

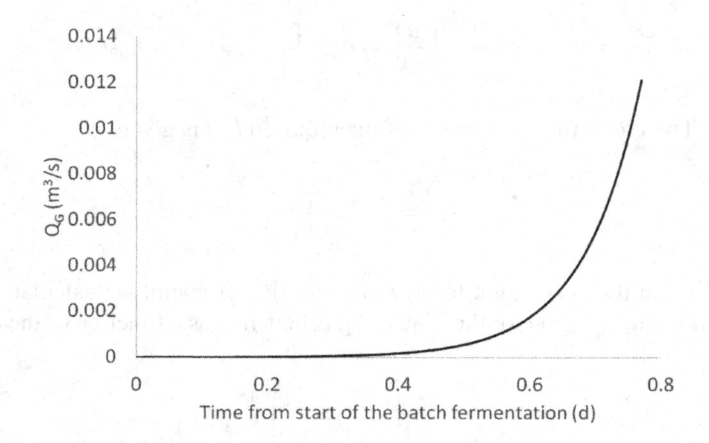

FIGURE C.10 Plot for Problem 4.17.

at the beginning of the fermentation and then increases exponentially because of the exponential increase in biomass concentration.

4.18 With the assumptions of ignoring the oxygen mass flow rates in the inlet and outlet streams, the mass balance for oxygen becomes:

$$k_L a\left(C_{O2}^* - C_{O2}\right) = \frac{DX}{Y_{X/O2}}$$

The required $k_L a$ to maintain the oxygen concentration C_{O2} in the chemostat is:

$$k_L a = \frac{DX}{Y_{X/O2}\left(C_{O2}^* - C_{O2}\right)}$$

In the initial conditions (subscript "old"), the oxygen concentration is 2.0 mg/L, in the new conditions (subscript "new") the oxygen concentration is 4.0 mg/L. The ratio between the required $k_L a$ in the two conditions is:

$$\frac{k_L a_{new}}{k_L a_{old}} = \frac{\frac{DX}{Y_{X/O2}\left(C_{O2}^* - C_{O2new}\right)}}{\frac{DX}{Y_{X/O2}\left(C_{O2}^* - C_{O2old}\right)}} = \frac{\left(C_{O2}^* - C_{O2old}\right)}{\left(C_{O2}^* - C_{O2new}\right)} = \frac{8.5 - 4.0}{8.5 - 2.0} = 0.69$$

From the given correlation for $k_L a$, we obtain:

$$\frac{k_L a_{\text{new}}}{k_L a_{\text{old}}} = \left(\frac{P_{\text{Gnew}}}{P_{\text{Gold}}}\right)^{0.70} = \left(\frac{N_{\text{new}}}{N_{\text{old}}}\right)^{2.1} \Rightarrow N_{\text{new}} = N_{\text{old}} \cdot \left(\frac{k_L a_{\text{new}}}{k_L a_{\text{old}}}\right)^{\frac{1}{2.1}} = 151 rpm$$

The reduction in the required power draw can be calculated using the same equation:

$$\frac{P_{\text{Gnew}}}{P_{\text{Gold}}} = \left(\frac{k_L a_{\text{new}}}{k_L a_{\text{old}}}\right)^{\frac{1}{0.70}} = 0.59$$

From the equation for the agitator power draw as a function of the power number, we can calculate P_{Gold}:

$$P_{Gold} = 0.43P = 0.43 P_0 \rho N_{old}^3 D_{ag}^5 = 0.43 \cdot 1{,}000 \cdot 4.2 \cdot \left(\frac{180}{60}\right)^3 1.3^5 = 181.05 kW$$

Therefore, P_{Gnew} = 106.82 kW and the savings in power are 74.23 kW. Over 1 year (8,000 h), the energy saved will be 2.14×10^{12} J = 593,840 kWh.

4.19 The oxygen mass balance in the chemostat, ignoring the oxygen flow with the inlet and outlet streams becomes:

$$r_{O2transf}V = k_L a\left(C^*_{O2} - C_{O2}\right)V = \frac{DXV}{Y_{X/O2}} = \frac{1.9d^{-1} \cdot 3.5\frac{kg}{m^3} \cdot 30m^3}{0.29\frac{kg}{kg}} = 687.9\frac{kg}{d}$$

This is the rate of oxygen transfer from the gas phase to the liquid phase.

We need to calculate the rate at which oxygen is provided to the reactor with the inlet gas phase. The volumetric flow rate of the inlet gas phase is 3,000 m³/h, therefore, the volumetric flow rate of inlet oxygen is 3,000 × 0.21 = 630 m³/h. The corresponding mass flow rate of inlet oxygen is:

$$\frac{630\frac{m^3}{h} \cdot 1{,}000\frac{l}{m^3}\, 1atm}{0.0821\frac{l \cdot atm}{mol \cdot K} \cdot 298.15K} = 25{,}737\frac{mol}{h} = 19{,}766\frac{kg}{d}$$

The efficiency of oxygen transfer is, therefore, 687.9/19,766 = 3.5%.

4.20 The first step is to calculate the required dilution rate for a substrate concentration of 0.1 kg/m³. This is:

$$D = \frac{\mu_{max} S}{K_S + S} = 3.22\, d^{-1}$$

The influent flow rate is, therefore: $Q = V \cdot D = 32.2$ m³/d.

The product productivity is:

$$QP = Q\left(S_0 - S\right)Y_{P/S} = 578.3\frac{kg}{d}$$

The required $k_L a$ can be calculated from the oxygen balance:

$$k_L a\left(C^*_{O2} - C_{O2}\right)V = \frac{DXV}{Y_{X/O2}} \Rightarrow k_L a = \frac{DY_{X/S}\left(S_0 - S\right)}{Y_{X/O2}\left(C^*_{O2} - C_{O2}\right)} = 14{,}082 d^{-1}$$

CHAPTER 5

5.1 In the scale-up of agitated vessels where fermentation reactions take place, the first criterion is to keep the residence time and the temperature the same in the small and in the large vessels so that the performance of the process remains the same.

The condition that the residence time remains the same means that the Q/V ratio remains the same for the small and large vessels.

The heat balance for a continuous process is described by the heat balances for the fluid in the reactor and in the jacket:

$$w_{\text{feed}} C_{P\text{feed}} (T_{IN} - T_R) + r_H V = UA(T_R - T_J)$$

$$UA(T_R - T_J) = W_J \cdot C_{PJ} (T_J - T_{JIN})$$

The two heat balances can be re-arranged as:

$$\frac{w_{\text{feed}}}{V} C_{P\text{feed}} (T_{IN} - T_R) + \frac{DX}{Y_H} = \frac{UA(T_R - T_J)}{V}$$

$$\frac{UA(T_R - T_J)}{V} = \frac{W_J \cdot C_{PJ} (T_J - T_{JIN})}{V}$$

If the residence time and feed properties (density, composition and temperature) stay constant with scale-up, the heat balance for the fluid in the reactor indicates that, in order to maintain the reactor temperature T_R constant with scale-up, the ratio $\frac{UA(T_R - T_J)}{V}$ will have to be constant with scale-up. The heat balance for the fluid in the jacket indicates that, if the flow rate of the jacket fluid is scaled up to maintain W_J/V constant, the jacket temperature T_J will stay the same on scale-up.

The previous considerations indicate that, in order to maintain the reaction temperature constant with scale-up, the ratio UA/V needs to be kept constant. The ratio A/V is fixed by geometric considerations, i.e.:

$$A = \frac{\pi D^2}{4} + \pi DH$$

$$V = \frac{\pi D^2}{4} H$$

Calling α the ratio H/D, which remains constant on scale-up for geometrically similar vessels:

$$A = \frac{\pi D^2}{4} + \pi \alpha D^2 = D^2 \left(\frac{\pi}{4} + \pi \alpha \right)$$

$$V = \frac{\pi \alpha D^3}{4}$$

$$\frac{A}{V} = \frac{\left(4 + \frac{1}{\alpha}\right)}{D}$$

The ratio A/V decreases as the size of the vessel increases. Therefore, in order to keep the ratio UA/V constant with scale-up, U needs to be larger for the large vessel than for the small vessel. If the overall heat transfer coefficient U is controlled by the reactor-side coefficient, the heat transfer coefficient for the reactor h_{reactor} needs to be larger for the larger vessel. Using the common correlation for h_{reactor}:

$$Nu = \frac{h_{\text{reactor}} D}{k_L} = a_1 Re_{ag}^{a2} Pr_{ag}^{0.33} \left(\frac{\mu_L}{\mu_{WL}} \right)^{a_3}$$

and noting that the physical properties of the fluid will stay the same on scale-up, we see that h_{reactor} is proportional to $\frac{N^{a2} D_{ag}^{2a2}}{D}$, i.e. is proportional to $N^{a_2} D^{(2a_2-1)}$. Therefore, in the scale-up the agitation rate N needs to be adjusted to obtain an increased h_{reactor}, in order to obtain the same UA/V ratio with a lower A/V. In practice, depending on the scale-up ratio, this may or may not be possible, and it may be, therefore, need, on scale-up, to use multiple vessels in parallel, rather than one big vessel, in order to have a more favourable A/V ratio.

5.2 The enthalpy of the reaction is: $\Delta H_R \left(\frac{J}{mol\,glucose} \right) = 0.12 H_{\text{biom}} + 1.8 H_{\text{ethanol}} + 1.8 H_{CO2} + 0.36 H_{H2O} - H_{GLu} - 0.12 H_{NH3}$

For all the species, the enthalpy (J/mol) is expressed as: $H_{fi}(25°C) + C_{pi} \cdot (40 - 25)$.

So we have (all values in kJ/mol):

$$\text{Biomass:} -770 + 0.333 * 15 = -765$$

$$\text{Ammonia:} -79.6$$

$$\text{Ethanol:} -276$$

$$\text{Carbon dioxide:} -392$$

$$\text{Water:} -285$$

$$\text{Glucose:} -1270$$

And the enthalpy of reaction is –118.45 kJ/mol glucose.

To produce 1 kg of ethanol, we need 2.17 kg of glucose to react. This corresponds to 12.05 mol of glucose. Therefore, the heat generated by the reaction when 1 kg of ethanol is produced is 1,427 kJ.

5.3 By heat balance on reactor:

$$\textit{Heat generated} = \textit{heat removed}$$

Heat balance under condition a:

$$Q_{\text{generated},a} = \frac{r_{Xa} V}{Y_H} = Q_{\text{removed},a} = w_J \cdot C_{PJ} \cdot \left(T_{Ja} - T_{Jin} \right)$$

$$\frac{r_{Xa}V}{Y_H} = w_J \cdot C_{PJ} \cdot \left(T_{Ja} - T_{Jin}\right)$$

Now r_{xa} can be calculated by mass balance for substrate in the chemostat:

$$QS_{0a} = \left(-r_{S,a}\right)V + QS_a$$

$$\left(-r_S\right) = \frac{r_X}{Y_{X/S}}$$

By putting value of rate, we get:

$$QS_{0a} = \frac{r_{X,a}}{Y_{X/S}}V + QS_a$$

Now by biomass's mass balance:

$$r_{X,a}V = QX_a$$

$$r_{X,a} = \frac{QX_a}{V}$$

$$r_{X,a} = 0.001 \times \frac{4}{10} = 0.0004 \frac{kg}{m^3 \cdot s}$$

To calculate the value of $Y_{X/S}$, we obtain after re-arranging:

$$Y_{X/S} = \frac{r_{X,a}V}{Q(S_{0a} - S_a)}$$

$$Y_{X/S} = \frac{X_a}{(S_{0a} - S_a)}$$

$$Y_{X/S} = \frac{4}{10-2} = 0.5 \frac{kg\ biomass}{kg\ substrate}$$

Now Y_H can be calculated using the heat balance for condition a:

$$Y_H = \frac{r_{Xa}V}{w_J \cdot C_{PJ} \cdot \left(T_{Ja} - T_{Jin}\right)}$$

$$Y_H = \frac{0.0004 \times 10}{1.5 \times 4186 \times (15-5)}$$

$$Y_H = 6.37 \times 10^{-8}\, kg/J$$

Now to calculate the heat transfer coefficient (U):

$$\frac{r_{Xa}V}{Y_H} = UA\left(T_{Ra} - T_{Ja}\right)$$

$$U = \frac{r_{Xa}V}{Y_H A\left(T_{Ra} - T_{Ja}\right)}$$

$$U = \frac{0.0004 \times 10}{6.37 \times 10^{-8} \times 30 \times (35 - 15)}$$

$$U = 104.65 \frac{W}{m^2 \cdot K}$$

Now for condition b, the heat balance for reactor:

$$\frac{r_{Xb}V}{Y_H} = w_J \cdot C_{PJ} \cdot \left(T_{Jb} - T_{Jin}\right)$$

Now we need r_{xb} and for that we will be doing the substrate mass balance:

$$QS_{0b} = \left(-r_{Sb}\right)V + QS_b$$

$$\left(-r_S\right) = \frac{r_X}{Y_{X/S}}$$

Therefore,

$$QS_{0b} = \frac{r_{X,b}}{Y_{X/S}} V + QS_b$$

$$r_{Xb} = \frac{\left(Q\left(S_{0b} - S_b\right)\right)}{V} Y_{X/S}$$

Since the residence time is the same, it will be:

$$S_a = S_b$$

Now $r_{x,b}$ will be:

$$r_{Xb} = \frac{(0.001(15-2))}{10} \times 0.5 = 0.00065 \frac{kg}{m^3 \cdot s}$$

From heat balance equation of condition b:

$$T_{Jb} = T_{Jin} + \frac{r_{Xb}V}{Y_H \cdot w_J \cdot C_{PJ}}$$

$$T_{cool,out,b} = 5 + \frac{0.00065 \times 10}{6.37 \times 10^{-8} \times 1.5 \times 4186}$$

$$T_{Jb} = 21.25°C$$

Now to calculate the T_{Rb}, we'll be using the other equation of heat balance, i.e.:

$$\frac{r_{Xb}V}{Y_H} = UA\left(T_{Rb} - T_{Jb}\right)$$

$$T_{Rb} = T_{Jb} + \frac{r_{Xb}V}{Y_H UA}$$

$$T_{Rb} = 21.25 + \frac{0.00065 \times 10}{6.37 \times 10^{-8} \times 104.65 \times 30}$$

$$T_{Rb} = 53.75°C$$

Hence,

$$T_{reactor,b} > T_{reactor,\max}$$

Which is not possible, therefore, reactor can't be operated with $S_0 = S_{0b}$.

5.4 The effluent substrate concentration is:

$$S = \frac{DK_S}{\mu_{\max} - D} = 0.50\frac{kg}{m^3}$$

The biomass concentration is:

$$X = Y_{X/S}\left(S_0 - S\right) = 9.75\frac{kg}{m^3}$$

The rate of heat generation is:

$$r_H V = \frac{DXV}{Y_H} = 188kW$$

With the assumptions of this problem, the heat balance for the fluid in the reactor is:

$$r_H V = UA(T_R - T_J)$$

The heat balance for the fluid in the vessel is:

$$w_J c_{PJ}(T_J - T_{JIN}) = UA(T_R - T_J)$$

Combining these equations, we calculate $T_J = 27°C$. T_R is obtained from:

$$T_R = \frac{r_H V}{UA} + T_J = 48\,^oC$$

5.5 The heat balance for the fluid in the reactor is, in this case:

$$w_{\text{feed}} C_{P\text{feed}}(T_{IN} - T_R) + r_H V = 0$$

The enthalpy of the reaction is $-1.56 \cdot 10^6$ J/mol glucose. We need to produce 15 t of citric acid per day. From the reaction stoichiometry, we need to convert 1.34 kg glucose per kg of citric acid produced. So, we need to convert 20.1 t = 111.67 kmol of glucose per day. The rate of heat generation is, therefore, $r_H V = 1.56 \times 10^9$ J/kmol × 111.67 kmol/d = 1.74×10^{11} J/d. From the heat balance, the feed flow rate is:

$$w_{\text{feed}} = \frac{-r_H V}{C_{Pfeed}(T_{IN} - T_R)} = \frac{-1.74 \cdot 10^{11}\, J/d}{4.19 \cdot 10^3\, J/kg.^oC(25-35)\,^oC} = 4.15 \cdot 10^6 \frac{kg}{d}$$

Since the density of the feed is 1,000 kg/m³, the flow rate of the feed is Q = 4,150 m³/d. It has to be that $Q \times S_0$ = 20,100 kg of glucose/d, therefore, $S_0 = 4.84$ kg/m³.

5.6 We need to write the enthalpy balances for the fluid in the reactor and for the fluid in the jacket. Under the hypotheses of the problem, these two equations are the following:

$$Q\rho c_P((T_{\text{feed}} - T) + r_{react}(-\Delta H_R)V = UA(T - T_J)$$

$$w_J c_{PJ}(T_J - T_{JIN}) = UA(T - T_J)$$

These equations have the two unknowns: T and T_J.

The term $r_{\text{react}}V$ represents the mol of glucose which react (kmol/h). The ethanol production rate is 150 kg/h = 3.26 kmol/h. From the reaction stoichiometry, this corresponds to a glucose reaction rate of 2.17 kmol/h. Therefore, the heat generation rate by this reaction is equal to $r_{\text{react}}(-\Delta H_R)V$ = 256 MJ/h = 71.1 kW.

The flow rate through the reactor is $Q = D \times V = 44.8$ m³/d.

All the other terms in the heat balances are given and the two equations can be solved by substitution. The solution is $T_J = 17.6$°C and $T = 37.3$°C.

5.7 The heat balances for the fluid in the reactor and in the jacket can be re-arranged as:

$$\frac{W_{\text{feed}}}{V} C_{P\text{feed}} \left(T_{IN} - T_R\right) + \frac{DX}{Y_H} = \frac{UA\left(T_R - T_J\right)}{V}$$

$$\frac{UA\left(T_R - T_J\right)}{V} = \frac{W_J \cdot C_{PJ}\left(T_J - T_{JIN}\right)}{V}$$

Since the residence time, feed properties and reactor temperature will be kept constant on scale-up, the heat balance for the fluid in the reactor indicates that the ratio $\frac{UA(T_R - T_J)}{V}$ will be constant on scale-up. Furthermore, T_J will need to stay the same on scale-up, so the term UA/V will have to stay constant on scale-up.

The ratio A/V will change on scale-up:

$$\frac{A}{V} = \frac{\left(4 + \frac{1}{\alpha}\right)}{D}$$

In this case $\alpha = 1$, so $\frac{A}{V} = \frac{5}{D}$.

The diameter of a cylindrical vessel is related to the volume by the equation:

$$D = \sqrt{\frac{4V}{\pi\alpha}}$$

Therefore, we have:

$$\frac{D_{\text{small}}}{D_{\text{large}}} = \left(\frac{V_{\text{small}}}{V_{\text{large}}}\right)^{0.33}$$

The change in the A/V ratio on scale-up will be:

$$\frac{\left(\frac{A}{V}\right)_{\text{large}}}{\left(\frac{A}{V}\right)_{\text{small}}} = \frac{\frac{5}{D_{\text{large}}}}{\frac{5}{D_{\text{small}}}} = \frac{D_{\text{small}}}{D_{\text{large}}} = \left(\frac{V_{\text{small}}}{V_{\text{large}}}\right)^{0.33} = 0.46$$

So, in order for the UA/V ratio to be constant on scale-up, it will have to be:

$$\frac{U_{\text{large}}\left(\frac{A}{V}\right)_{\text{large}}}{U_{\text{small}}\left(\frac{A}{V}\right)_{\text{small}}} = 1 \Rightarrow \frac{U_{\text{large}}}{U_{\text{small}}} = \frac{\left(\frac{A}{V}\right)_{\text{small}}}{\left(\frac{A}{V}\right)_{\text{large}}} = 2.15$$

Since we are assuming that $U \approx h_{\text{reactor}}$, this ratio is also equal to $h_{\text{reactor,large}}/h_{\text{reactor,small}}$.

With the given correlation for the heat transfer coefficients, we get:

$$\frac{h_{\text{reactor,large}}}{h_{\text{reactor,small}}} = \frac{N_{\text{large}}^{0.67} D_{\text{large}}^{0.34}}{N_{\text{small}}^{0.67} D_{\text{small}}^{0.34}} = \frac{N_{\text{large}}^{0.67}}{N_{\text{small}}^{0.67}} \cdot 1.30$$

Therefore, it will be:

$$N_{\text{large}} = N_{\text{small}} \left(\frac{2.15}{1.30} \right)^{\frac{1}{0.67}} = 191 rpm$$

5.8 The heat balances for the fluid in the chemostat and the fluid in the jacket are:

$$w_{\text{feed}} C_{P\text{feed}} (T_{IN} - T_R) + (-r_S)(-\Delta H_R) V = UA(T_R - T_J)$$

$$UA(T_R - T_J) = W_J \cdot C_{PJ} (T_J - T_{JIN})$$

which can be combined into:

$$w_{\text{feed}} C_{P\text{feed}} (T_{IN} - T_R) + (-r_S)(-\Delta H_R) V = W_J \cdot C_{PJ} (T_J - T_{JIN})$$

From this equation, we can calculate the enthalpy of reaction:

$$(-\Delta H_R) = \frac{W_J \cdot C_{PJ} (T_J - T_{JIN}) - w_{\text{feed}} C_{P\text{feed}} (T_{IN} - T_R)}{(-r_S) V}$$

$$= \frac{100{,}000 \frac{kg}{d} \cdot 4{,}186 \frac{J}{kg} \cdot (17 - 10)\,^oC - 80{,}000 \frac{kg}{d} \cdot 4{,}186 \frac{J}{kg} \cdot (15 - 28)\,^oC}{20 \frac{m^3}{d} \cdot 80 \frac{kg}{m^3}}$$

$$= 4.55 \cdot 10^6 \frac{J}{kg}$$

5.9 From the total mass balance for this reaction, we calculate the amount of produced water per kg of reacted substrate. The water produced is 0.15 kg/kg. The enthalpy of reaction per unit mass of reacted substrate is:

$$\Delta H_R (35^oC) \left(\frac{J}{kg} \right) = 0.12 H_{\text{biom}} (35^oC) + 0.40 H_{\text{product}} (35^oC) + 0.35 H_{CO2} (35^oC)$$
$$+ 0.15 H_{\text{water}} (35^oC) - H_{\text{subs}} (35^oC) - H_{NH3} (35^oC)$$

where all the enthalpies are in J/kg. The enthalpies are calculated from the general equation:

$$H_i(35^oC) = \Delta H_{fi}(25^oC) + C_{Pi}(35-25)$$

Using this equation, the enthalpies of the reaction components in their reference states at 35°C are (in J/kg):

$$H_{\text{subs}} = -7.485 \cdot 10^6$$
$$H_{\text{NH3}} = -4.463 \cdot 10^6$$
$$H_{\text{biom}} = -7.97 \cdot 10^6$$
$$H_{\text{product}} = -1.0788 \cdot 10^7$$
$$H_{\text{CO2}} = -8.99 \cdot 10^6$$
$$H_{\text{water}} = -1.596 \cdot 10^7$$

The enthalpy of the reaction is, therefore:

$$\Delta H_R = -3.238 \cdot 10^6 \frac{J}{kg}$$

For an adiabatic system, the heat balance for the fluid in the reactor becomes:

$$w_{\text{feed}} C_{P\text{feed}}(T_{IN} - T_R) + (-r_S)(-\Delta H_R)V = 0$$

The term $(-r_S)V$ is equal to QS_0, under the assumption that all the substrate is removed.

Therefore, the heat balance becomes:

$$Q\rho C_{P\text{feed}}(T_{IN} - T_R) = -(-\Delta H_R)QS_0 \quad \Rightarrow \quad S_0 = \frac{\rho C_{P\text{feed}}(T_R - T_{IN})}{(-\Delta H_R)} = 16.8 \frac{kg}{m^3}$$

This is the maximum substrate concentration we can have in the feed with this fermentation reaction and these temperatures for the fluid in the reactor and for the feed.

5.10 The first step is the calculation of the required U when the reactor temperature is at its maximum value, 30°C. The heat balances for the fluid in the vessel and the jacket are:

$$w_{\text{feed}} C_{P\text{feed}}(T_{IN} - T_R) + (-r_S)(-\Delta H_R)V = UA(T_R - T_J)$$

$$UA(T_R - T_J) = W_J \cdot C_{PJ}(T_J - T_{JIN})$$

With the information given in this problem, this is a system of two equations with the two unknowns: T_J and U. With the given numerical values, we have:

$$A = \pi DH = 3.14 \cdot 3.2 \cdot 3.1 = 31.15 m^2$$

$$(-r_S)V = 75 \cdot 70 = 5,250 \frac{kg}{d}$$

The two equations can be written as:

$$75,000 \frac{kg}{d} 4,190 \frac{J}{kg\,^oC}(20-30)\,^oC + 5,250 \frac{kg}{d} \cdot 9.50 \cdot 10^5 \frac{J}{kg}$$

$$= U \cdot 31.25 m^2 (30 - T_J)\,^oC$$

$$U \cdot 31.25 m^2 (30 - T_J)\,^oC = 50,000 \frac{kg}{d} \cdot 4,190 \frac{J}{kg\,^oC}(T_J - 5)\,^oC$$

Solving these two equations, we get: T_J = 13.8°C and U = 89.1 W/m²·°C.

We need now to calculate the actual U that can be achieved in this vessel. U can be obtained from the combination of the heat transfer coefficients for the reactor and for the jacket:

$$\frac{1}{U} = \frac{1}{h_{jacket}} + \frac{D_{ext}}{D_{int} h_{reactor}} \cong \frac{1}{h_{jacket}} + \frac{1}{h_{reactor}}$$

h_{reactor} is obtained from the correlation:

$$\frac{h_{\text{reactor}} D_{\text{vessel}}}{k_L} = 0.74 Re_{ag}^{0.67} Pr_{ag}^{0.33} \quad \Rightarrow \quad h_{reactor} = \frac{k_L 0.74 \left(\frac{N D_{ag}^2 \rho_L}{\mu_L}\right)^{0.67} \left(\frac{c_{PL} \mu_L}{k_L}\right)^{0.33}}{D_{\text{vessel}}}$$

The rotational speed of the agitator can be calculated from the available power (15 kW) and from the power number $P_0 = \frac{P}{\rho N^3 D_{ag}^5}$ which gives:

$$N = \left(\frac{P}{\rho P_0 D_{ag}^5}\right)^{0.33} = 1.62 s^{-1} = 97 rpm$$

Substituting this value for N and the other geometrical parameters and physical properties in the equation for h_{reactor}, we obtain:

$$H_{react} = 697 \text{ W/m}^2\,^\circ\text{C}$$

The heat transfer coefficient for the fluid in the jacket is given by:

$$Nu_{\text{jacket}} = \frac{h_{\text{jacket}} D_e}{k_J} = 0.023 Re_J^{0.8} Pr_J^{0.33} \Rightarrow h_{\text{jacket}}$$

$$= \frac{k_J 0.023\left(\frac{\rho_J v D_e}{\mu_J}\right)^{0.8}\left(\frac{c_{PJ}\mu_J}{k_J}\right)^{0.33}}{D_e}$$

Each spiral has a rectangular section with sides of 0.05 and 0.04 m, therefore, the cross-sectional area is $0.05 \cdot 0.04 = 0.002\,m^2$ and the wetted perimeter is $2 \cdot (0.05 + 0.04) = 0.18\,m$. The equivalent diameter is:

$$De = \frac{4 \times 0.002}{0.18} = 0.044m$$

The fluid velocity in the jacket is:

$$v = \frac{\frac{50}{86{,}400}}{0.002} = 0.29\frac{m}{s}$$

Replacing the numerical values, we obtain for the fluid in the jacket:

$$h_{\text{jacket}} = 1{,}217\ \text{W/m}^2.^{o}\text{C}.$$

Combining the values obtained for h_{jacket} and h_{reactor}, we obtain:

$$U = 443\ \text{W/m}^2.^{o}C$$

This value of U is considerably higher than the value of U required for this process, so the fermentation can be carried out in the proposed vessel without exceeding the maximum allowable temperature for the reaction. The process can be carried out with an agitation rate lower than the maximum value or with a lower value for the flow rate of the jacket fluid.

5.11 The heat balances for the batch fermentation with a jacketed vessel are:

$$M_R C_{PR}\frac{dT_R}{dt} = r_H \cdot V - UA(T_R - T_J) \Rightarrow 0$$

$$= \frac{\mu_{\max} S}{K_S + S}\frac{X}{Y_{X/S}}(-\Delta H_R)V - UA(T_R - T_J)$$

$$M_J C_{PJ}\frac{dT_J}{dt} = W_J \cdot C_{PJ}\left(T_{JIN} - T_J\right) + UA\left(T_R - T_J\right)$$

These must be solved together with the mass balances for the substrate and microorganisms in the reactor:

$$\frac{dS}{dt} = -\frac{\mu_{max}}{K_S + S}\frac{X}{Y_{X/S}}$$

$$\frac{dX}{dt} = \frac{\mu_{max}}{K_S + S}X$$

The volume of the fluid in the reactor is:

$$V = \frac{\pi D^2 H}{4} = 3.51 m^3$$

The mass of fluid in the jacket is:

$$M_J = \pi D H w \rho = 471 kg$$

The area for heat transfer is:

$$A = \pi D H = 9.42 m^2$$

The heat balance for the fluid in the vessel can be re-arranged to calculate the jacket temperature T_J:

$$T_J = T_R - \frac{\mu_{max} S}{K_S + S}\frac{X}{Y_{X/S}}(-\Delta H_R)\frac{V}{UA} = T_R - \left(-\frac{dS}{dt}\right)(-\Delta H_R)\frac{V}{UA}$$

This equation allows for the calculation of T_J from the substrate profile. Once the time profile of T_J is calculated, $\frac{dT_J}{dt}$ can be calculated numerically and the required time profile for w_J can be calculated from the heat balance of the fluid in the jacket:

$$W_J = \frac{M_J C_{PJ} \frac{dT_J}{dt} - UA(T_R - T_J)}{C_{PJ}(T_{JIN} - T_J)}$$

In the numerical solution, the first step is to calculate the time profiles of the substrate and biomass from the following equations:

$$\frac{dS}{dt} = -\frac{3.8S}{0.06 + S}\frac{X}{0.14}$$

$$\frac{dX}{dt} = -\frac{3.8S}{0.06 + S}X$$

Once the time profiles of S and X are calculated, the time profile of T_J is obtained from:

$$T_J = 30 - \frac{3.8S}{0.06+S}\frac{X}{0.14} 9.67\cdot 10^5 \frac{3.51}{100\cdot 86{,}400\cdot 9.42}$$

From the time profile of T_J vs t, $\frac{dT_J}{dt}$ can be calculated and finally the profile of w_J vs time is obtained from:

$$W_J = \frac{471\cdot 4{,}190\frac{dT_J}{dt} - 100\cdot 86{,}400\cdot 9.42(30-T_J)}{4{,}190(15-T_J)}$$

The plots of the calculated variables are shown in Figure C.11.

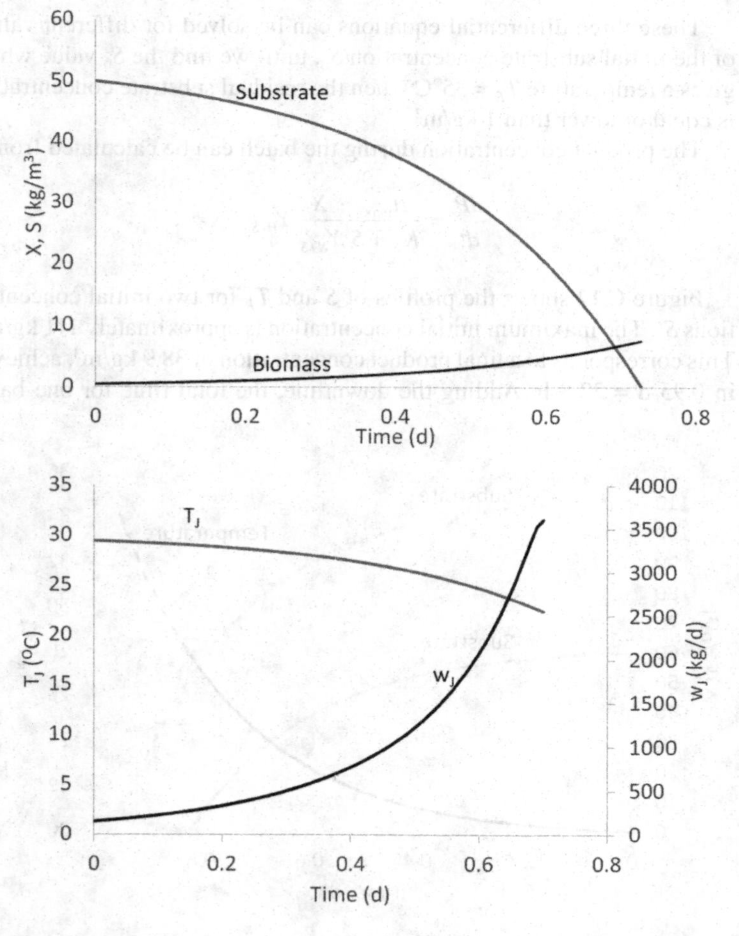

FIGURE C.11 Plots for Problem 5.11.

5.12 For an adiabatic batch fermentation, the heat balance is:

$$M_R C_{PR} \frac{dT_R}{dt} = r_H \cdot V \quad \Rightarrow \quad \frac{dT_R}{dt} = \frac{\frac{\mu_{max}}{K_S+S}\frac{X}{Y_{X/S}}(-\Delta H_R)}{M_R C_{PR}}$$

This equation can be solved simultaneously to the mass balances for X and S:

$$\frac{dS}{dt} = -\frac{\mu_{max}}{K_S+S}\frac{X}{Y_{X/S}}$$

$$\frac{dX}{dt} = \frac{\mu_{max}}{K_S+S}X$$

These three differential equations can be solved for different values of the initial substrate concentration S_0, until we find the S_0 value which gives a temperature $T_R = 35°C$ when the residual substrate concentration is equal or lower than 1 kg/m^3.

The product concentration during the batch can be calculated from:

$$\frac{dP}{dt} = \frac{\mu_{max}}{K_S+S}\frac{X}{Y_{X/S}}Y_{P/S}$$

Figure C.12 shows the profiles of S and T_R for two initial concentrations S_0. The maximum initial concentration is approximately 111 kg/m^3. This corresponds to a final product concentration of 38.9 kg/m^3, achieved in 0.95 d = 22.8 h. Adding the downtime, the total time for one batch

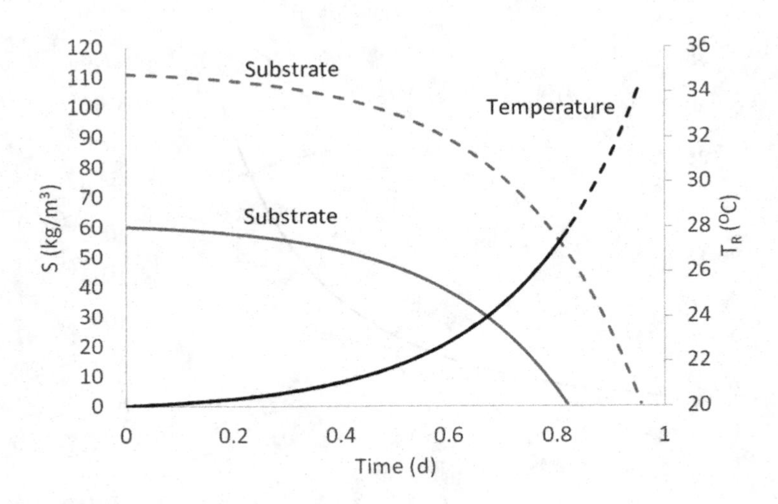

FIGURE C.12 Plot for Problem 5.12.

is 24.8 h. The volume of one batch is 10 m³, so one batch produces 389 kg of product. In 8,000 h per year, we can do 322 batches, with a product productivity of 125,258 kg/year or 343.2 kg/d.

5.13 The heat balances for a jacketed chemostat are:

$$w_{\text{feed}} C_{P\text{feed}} \left(T_{IN} - T_R\right) + \left(-r_S\right)\left(-\Delta H_R\right) V = UA\left(T_R - T_J\right)$$

$$UA\left(T_R - T_J\right) = W_J \cdot C_{PJ}\left(T_J - T_{JIN}\right)$$

From these two equations under the initial conditions, we can calculate the terms UA and $(-\Delta H_R)$:

$$UA = \frac{W_J \cdot C_{PJ}\left(T_J - T_{JIN}\right)}{\left(T_R - T_J\right)} = 9{,}699 \frac{W}{{}^{o}C}$$

$$\left(-\Delta H_R\right) = \frac{UA\left(T_R - T_J\right) - w_{\text{feed}} C_{P\text{feed}}\left(T_{IN} - T_R\right)}{\left(-r_S\right)V} = 6.36 \cdot 10^{6} \frac{J}{kg}$$

Under the new conditions of increased substrate concentration, U will have to increase because the substrate removal rate $(-r_S)$ increases. We can use the two heat balances under the new conditions to calculate the new values of UA and T_J.

T_J can be obtained from:

$$T_J = T_{JIN} + \frac{w_{\text{feed}} C_{P\text{feed}}\left(T_{IN} - T_R\right) + \left(-r_S\right)\left(-\Delta H_R\right)V}{W_J C_{PJ}} = 21.9{}^{o}C$$

And the new value of UA from:

$$UA = \frac{W_J \cdot C_{PJ}\left(T_J - T_{JIN}\right)}{\left(T_R - T_J\right)} = 14{,}249 \frac{W}{m^2 \, {}^{o}C}$$

Since A doesn't change under the new conditions, U will have to increase:

$$\frac{U_{\text{new}}}{U_{\text{initial}}} = 1.47 = \frac{h_{\text{reactor,new}}}{h_{\text{reactor,initial}}}$$

Using the given correlation for h_{reactor}, assuming that the only change is for the agitation speed:

$$\frac{h_{\text{reactor,new}}}{h_{\text{reactor,initial}}} = \left(\frac{N_{\text{new}}}{N_{\text{initial}}}\right)^{0.67} \Rightarrow N_{\text{new}} = N_{\text{old}}\left(\frac{h_{\text{reactor,new}}}{h_{\text{reactor,initial}}}\right)^{\frac{1}{0.67}} = 142\,rpm$$

To calculate the product productivity, we need the parameter $Y_{P/S}$ which can be calculated under the initial conditions:

$$Y_{P/S} = \frac{P}{(S_0 - S)} = 0.43\frac{kg}{kg}$$

The product productivity under the new conditions (70 kg/m^3 of substrate in the feed) is:

$$QP = Q(S_0 - S)Y_{P/S} = 731\frac{kg}{d}$$

(increased from 624 kg/d under the initial conditions).
The increase in the power can be calculated from the power number:

$$P = P_0 \rho N^3 D_{ag}^5 \quad \Rightarrow \quad \frac{P_{\text{new}}}{P_{\text{initial}}} = \left(\frac{N_{\text{new}}}{N_{\text{initial}}}\right)^3 = 5.59$$

The agitator power required under the new conditions is 5.59 times higher than the initial one. This is a very considerable increase in power requirements which is likely to negatively affect the economics of the process.

CHAPTER 6

6.1 Bioethanol is a primary metabolite, produced during the main metabolism of the microorganisms. A chemostat is suitable for its production and would generally give higher productivities than a batch process. Penicillin is a secondary metabolite, mainly produced when growth has stopped or has significantly slowed down. Production of penicillin is a two-phase process and, therefore, the use of a chemostat, characterised by steady-state conditions, is not recommended. It is better to use batch processes, where microorganisms can grow when substrate is in excess and can produce penicillin when the substrate is (almost) completely removed.

6.2 The most typical factors that trigger L-glutamic acid production with the microorganism *Corynebacterium glutamicum* are biotin limitation, presence of surfactants or of penicillin. A fed-batch process for the production of L-glutamic acid can be carried out starting with a balanced feed (with enough biotin, without surfactants or penicillin) in order to increase the number and mass of microorganisms, followed by an unbalanced feed with biotin limitation (or with penicillin or surfactants) to trigger the extraction of the amino acid.

6.3 Citric acid production is favoured by low concentrations of certain metals (Fe, Mn, Zn) and by relatively high oxygen concentrations in the medium (at least 25% of saturation).

6.4 We know from the data that 200 kg/h of carbon dioxide are produced. According to the stoichiometry, 0.44 kg carbon dioxide are produced per kg of glucose that reacts. Therefore, the glucose reaction rate is 455 kg/h. We have 50 kg/h of glucose leaving the reactor, therefore, the mass flow rate of glucose in the feed is 505 kg/h.

With the reaction of 455 kg/h of glucose we produce, according to the stoichiometry, 209 kg of ethanol. Of these, 25 are lost with stream 3 and 10 with stream 4, therefore, the mass flow rate of ethanol in stream 4 is 174 kg/h.

6.5 The penicillin production rate is in total 3.4 kg/h. From the reaction stoichiometry 0.087 kg penicillin are produced per kg of oxygen consumed. So, the oxygen consumption is 3.4/0.087 = 39.1 kgO_2/h.

From the reaction stoichiometry 0.166 kg penicillin/kg microorganisms are produced. So, for the production of 3.4 kg/h of penicillin, 20.5 kg/h microorganisms are produced. Of these, 1.5 leave the system with stream 4, so 19.0 kg/h of microorganisms leave with stream 2.

6.6 From the stream table, we see that 9,300 kg of starch per day are converted. From the stoichiometry of starch hydrolysis this corresponds to a production of glucose of 10,333 kg/d. 1,000 kg/d of glucose leave the system unreacted, so 9,333 kg/d of glucose are converted. We produce 0.20 kg of microorganisms per kg of glucose converted.

We use this information to calculate the stoichiometry of glucose fermentation to ethanol:

$$C_6H_{12}O_6 + aNH_3 \rightarrow bC_5H_7O_2N + cC_2H_6O + dCO_2 + eH_2O$$

$b = a = 0.32$ (from the N balance)
combining the elemental balances for O, C and H, we obtain:
$c = 1.47$, $d = 1.46$ and $e = 0.97$ and the balanced stoichiometry is:

$$C_6H_{12}O_6 + 0.32NH_3 \rightarrow 0.32C_5H_7O_2N + 1.47C_2H_6O + 1.46CO_2 + 0.97H_2O$$

From this stoichiometry 0.38 kg of ethanol are produced per kg of glucose converted, so in total we produce 9,333 × 0.38 = 3,547 kg of ethanol per day. The total loss of ethanol with streams 3 and 4 is 313 kg/d, so the ethanol in stream 5 is 3,234 kg/d.

The production of carbon dioxide is 0.36 kg/kg glucose, so the total production of carbon dioxide (all in stream 2) is 3,360 kg/d.

b. We know from [illegible] that 200 kg/h of carbon dioxide are produced. According to the stoichiometry, 144 kg [of] carbon dioxide are produced per [illegible] of glucose that reacts. Therefore the glucose consumption is 455 kg/h. We have 70 kg/h of glucose [illegible] ; therefore, the mass flow rate of glucose in the [illegible] is [illegible] kg/h.

With the consumption of 455 kg/h of glucose we produce, according to the stoichiometry, [illegible] kg of ethanol. [illegible] 15 are [illegible] streams 1 and 10 [illegible]; therefore, the mass flow of ethanol in stream [illegible] g/h.

c. The [illegible] production [illegible] kg/h [illegible] production [illegible] 0.087 kg [illegible] are consumed, so the oxygen consumption is [illegible] kg O_2/h.

From the [illegible] production of [illegible] kg/h of penicillin [illegible] kg/h [illegible] leave the system with stream [illegible] g/h [illegible] with stream [illegible] of [illegible] per day [illegible] which [illegible] corresponds to a production of glucose of [illegible] kg [illegible] glucose leave the system [illegible] of [illegible]. We produce [illegible] converted.

We [illegible] this information [illegible] stoichiometry of glucose fermentation [illegible] ethanol:

$$C_6H_{12}O_6 + [illegible] \rightarrow [illegible]$$

[illegible]

[illegible] the elemental balances [illegible] we obtain [illegible] and [illegible] the balanced [illegible] is:

$$[illegible]\ H_{12}O_6 + 0.[illegible]NH_3 \rightarrow [illegible] + 1.46CO_2 + 0.9[illegible]H_2O$$

From this stoichiometry [illegible] are produced per [illegible] of [illegible] we produce [illegible] = [illegible] kg ethanol per day [illegible] with streams [illegible] and 4 is [illegible] kg/h [illegible] ethanol [illegible].

The production [illegible] production of [illegible] kg.

Bibliography

JOURNALS AND BOOKS

D.C. Armstrong, M.R. Johns. Culture conditions affect the molecular weight properties of hyaluronic acid produced by *Streptococcus zooepidemicus*. Appl. Env. Microbiol., 63, 2759–2764, 1997.

S.H.M. Azhar, R. Abdulla, S.A. Jambo, H. Marbawi, J.A. Gansau, A.A.M. Faik, K.F. Rodrigues. Yeasts in sustainable bioethanol production: A review. Biochem. Biophys. Rep., 10, 52–61, 2017.

J.E. Bailey, D.F. Ollis. Biochemical Engineering Fundamentals. Second Edition. McGraw-Hill, New York 1986.

G.T. Benz. Piloting bioreactors for agitation scale-up. *CEP* 104, 32–34, 2008.

C.A. Cardona, O.J. Sanchez; L.F. Gutierrez. Process Synthesis for Fuel Ethanol Production. CRC Press, Boca Raton, 2010.

D. Dionisi. Biological Wastewater Treatment Processes-Mass and Heat Balances. CRC Press, Boca Raton, 2017.

D. Dionisi, I.M.O. Silva, Production of ethanol, organic acids and hydrogen: an opportunity for mixed culture biotechnology?. Rev. Env. Sci. Biotechnol., 15, 213–242, 2016.

E.M.T. El-Mansi, J. Nielsen, D. Mousdale, R.P. Carlson (Eds.). Fermentation Microbiology and Biotechnology. CRC Press, Boca Raton, 2018.

F. Garcia-Ochoa, E. Gomez. Bioreactor scale-up and oxygen transfer rate in microbial processes: an overview. Biotechnol. Adv. 27.2, 153–176, 2009.

D. Humbird, R. Davis, J.D. McMillan. Aeration costs in stirred-tank and bubble column bioreactors. Biochem. Eng. J., 127, 161–166, 2017.

J.R. Kwiatkowski, A.J. McAloon, F. Taylor, D. Johnston. Modeling the process and costs of fuel ethanol production by the corn dry-grind process. Ind. Crops Prod., 23, 288–296, 2006.

B. Kalbfuss, M. Wolff, R. Morenweiser, U. Reichl. Purification of cell culture-derived human influenza A virus by size-exclusion and anion-exchange chromatography. Biotechnol. Bioeng., 96(5), 932–944, 2007a.

B. Kalbfuss, Y. Genzel, M. Wolff, A. Zimmermann, R. Morenweiser, U. Reichl. Harvesting and concentration of human influenza A virus produced in serum-free mammalian cell culture for the production of vaccines. Biotechnol. Bioeng., 97(1), 73–85, 2007b.

H.S. Oberoi, P.V. Vadlani, R.L. Madl, L. Saida, J.P. Abeykoon. Ethanol production from orange peels: two-stage hydrolysis and fermentation studies using optimized parameters through experimental design. J. Agric. Food Chem., 58, 3422–3429, 2010.

G.C. Okpokwasili, C.O. Nweke. Microbial growth and substrate utilization kinetics. African J. Biotechnol., 5, 305–317, 2005.

A. Pérez Rubio, J.M. Eiros. Cell culture-derived flu vaccine: present and future. Hum. Vaccines Immunother. 14, 8, 2018.

D.J. Pollard, G. Hunt, T. Kirschner, P. Salmon. Rheological characterization of a fungal fermentation for the production of pneumocandins. Bioproc. Biosyst. Eng., 24(6), 373–383, 2002.

C. Ratledge, B. Kristiansen (Eds). Basic Biotechnology. Second Edition. Cambridge University Press, Cambridge, UK, 2001.

M.L. Shuler, F. Kargi, M. DeLisa. Bioprocess Engineering: Basic Concepts. Third Edition. Prentice Hall, Hoboken, 2017.

K.L. Schulze, R.S. Lipe. Relationship between substrate concentration, growth rate, and respiration rate of *Escherichia coli* in continuous culture. *Archiv für Mikrobiologie* 48, 1–20, 1964.

P. Stanbury, A. Whitaker, S. Hall. Principles of Fermentation Technology, 3rd Edition. Butterworth-Heinemann, Oxford, 2016.

H. Taguchi, A.E. Humphrey. Dynamic measurement of the volumetric oxygen transfer coefficient in fermentation systems. J. Ferment. Technol., 44, 881–889, 1966.

C.C. Todaro, H.C. Vogel (Eds). Fermentation and Biochemical Engineering Handbook. Elsevier Science & Technology Books, the Netherlands, 2014.

M.R. Van Leeuwen, P. Krijgsheld, T.T. Wyatt, E.A. Golovina, H. Menke, A. Dekker, J. Stark, H. Stam, R. Bleichrodt, H.A.B. Wösten, J. Dijksterhuis. The effect of natamycin on the transcriptome of conidia of *Aspergillus niger*. Stud. Mycol. 74, 71–85, 2013.

M. Villano, F. Valentino, A. Barbetta, L. Martino, M. Scandola, M. Majone. Polyhydroxyalkanoates production with mixed microbial cultures: from culture selection to polymer recovery in a high-rate continuous process. New Biotechnol. 31, 289–296, 2014.

M.J. Whytes, N.L. Morgan, J.S. Rockey, G. Higton. Industrial Microbiology: An Introduction. Blackwell Science, Oxford, UK, 2001.

INTERNET REFERENCES

- https://commons.wikimedia.org/wiki/File:20101212_200110_Lactobacillus Acidophilus.jpg
- https://bacdive.dsmz.de/strain/7096
- https://commons.wikimedia.org/w/index.php?curid=52254246
- https://www.technologynetworks.com/cell-science/articles/prokaryotes-vs-eukaryotes-what-are-the-key-differences-336095
- https://commons.wikimedia.org/wiki/File:Cell_membrane_detailed_diagram_4.svg
- Molview.org
- https://commons.wikimedia.org/wiki/File:Hexokinase_induced_fit.svg
- https://en.wikipedia.org/wiki/Nucleotide#/media/File:0322_DNA_Nucleotides.jpg
- https://commons.wikimedia.org/wiki/File:RNA_chemical_structure.GIF
- http://2010.igem.org/Team:UCL_London/Fermenter_Mechanics
- https://sites.google.com/site/fermentersin/types

COMMERCIAL PROVIDERS OF LAB, PILOT AND FULL-SCALE FERMENTERS AND FERMENTATION EQUIPMENT

Many companies provide fermenters and fermentation equipment at lab, pilot and/or full scale. The list below, not exhaustive, mentions some providers active worldwide.

- Applikon Biotechnology, applikon-biotechnology.com
- De Dietrich Process Systems, dedietrich.com
- Gea, gea.com

- GPC Bio, gpc-bio.com
- INFORS HT, infors-ht.com
- Pall, pall.com
- Paul Mueller Company, paulmueller.com
- Pierre Guerin Technologies, pierreguerin.com
- Sartorius, sartorius.com
- Solida Biotech, solidabiotech.com

Index

Note: Page numbers in *italics* and **bold** refer to figures and tables, respectively.

A

Acetic acid, 14, 189
ADP, ATP, 8–14, 22
Anabolism, 11–13
Andrews model, *see* Haldane model
Antibiotics, 3, 4, 53, 181
Archaea, 4, 5

B

Bacillus subtilis, **5**, 94, **179**
Bacteria, 4, 5, 27
Bioethanol, 1, 53, 76, 129, 176–181
Butyric acid, 14

C

Candida antarctica, **5**
Candida utilis, **31**, **44**
Catabolism, 11, 13, 14
Cell membrane, *6*
Citric acid, 15, 53, 129, 184–185
Cultivation, 22
Cytoplasm, 6

D

DNA, 7, 8

E

Endogenous metabolism, 15, 43–47, 67–72
Enterobacter aerogenes, **31**
Enthalpy of reaction, 129–139
Enzymes, 7, 14, 15, 22, 35, 178, 179, 182, 188
Escherichia coli, 27, **44**, 53, 83
Eukarya, 4, 5
Extraction, 23
  solvent extraction, 23, 182

G

Geometric similarity, 93, 99, 116–119
Glucose metabolism, 11–14
Glutamic acid, 4, 15, 185–186
Growth, 1, 10, 14–17, 27–52

H

Haldane model, 34
Heat balances, 139–159
Heat generation, 129–138
Heat transfer coefficients, 160–164

I

Inhibitors, 34–35
Inoculum, 16, 54, 75

K

Kinetics, 27–52, 57–63, 72–76, 103, 160
Klebsiella, **31**, **44**

L

Lactic acid, 3, **5**
Lactobacillus delbruekii, **5**

M

Maintenance metabolism, 15, 43–47, 67–72
Mass balances, 53–72, 103–114
Mass transfer coefficients, 84–95
Metabolism, 1, 2, 10–15
Metabolites, 11, 14–16, 23, 43, 67, 76
Monod model, 28, 29, 58, 103

N

NAD^+, NADH, 10–13

O

Organelles, 7
Oxygen transfer, 83–120

P

Penicillin, **5**, 15, 53, 181–182
Penicillium, **5**, **44**, 181
pH, 34–35
Power draw, 92–93, 95–102, 115–120
Product, 1, 3–5, 10, 14–19, 27, 30–33, 41–43
Propionic acid, 14
Pseudomonas, **30**, **31**
Pyruvate, 11–14

R

Reactors
airlift reactor, 20
batch reactor, 16, 36, 54–57, 75–76, 103–110, 155–159
bubble column reactor, 20
continuous reactor, 16–17, 36, 53, 57–63, 68–76, 110–115, 139–155
fed-batch reactor, 18, 63–67, 75–76
fluidised bed reactor, 19
packed bed reactor, 18
stirred tank reactor, 18
RNA, 7, 8

S

Saccharomyces cerevisiae, 2, 4, **5**, **32**, 53, 129, 179, 182
Scale-down, scale-up, 115–120, 164–167
Scheffersomyces stipitis, **32**
Separation, 15, 179–182
Sterilisation, 24–25
Stoichiometry, 1, 10–12, 27, 39
Streptococcus zooepidemicus, **30**, **32**

T

Temperature, 34–36, 92, 95, 129, 139–168

U

Undefined mixed cultures, 187–189

V

Vaccines, 3, 4, 186–187
Virus, 186–187

W

Whisky, 182–183

Y

Yeasts, 1, 4, **5**, 21, 184
Yield, 30–42

Printed in the United States
by Baker & Taylor Publisher Services